Enhancing Water and Food Security Through Improved Agricultural Water Productivity

Tafadzwanashe Mabhaudhi
Vimbayi G. P. Chimonyo
Aidan Senzanje • Pauline P. Chivenge
Editors

Enhancing Water and Food Security Through Improved Agricultural Water Productivity

New Knowledge, Innovations and Applications

Springer

Editors
Tafadzwanashe Mabhaudhi
Centre on Climate Change and Planetary Health
London School of Hygiene and Tropical Medicine
London, UK

Centre for Transformative Agricultural and Food Systems
University of KwaZulu-Natal, Carbis Road, Scottsville
Pietermaritzburg, KwaZulu-Natal, South Africa

Aidan Senzanje
School of Engineering
University of KwaZulu-Natal
Pietermaritzburg, South Africa

Vimbayi G. P. Chimonyo
International Maize and Wheat Improvement
Harare, Zimbabwe

Pauline P. Chivenge
International Rice Research Institute
Dar Salam, Tanzania

ISBN 978-981-96-1850-7 ISBN 978-981-96-1848-4 (eBook)
https://doi.org/10.1007/978-981-96-1848-4

This work was supported by Tafadzwanashe Mabhaudhi.

This Springer imprint is published by the registered company Springer Nature Singapore Pte Ltd.
The registered company address is: 152 Beach Road, #21-01/04 Gateway East, Singapore 189721, Singapore

Foreword

Optimising agricultural water management has become paramount in addressing global challenges such as water and food insecurity, climate change and sustainable development. As we strive to address these grand challenges, there is a growing recognition that enhancing water productivity (WP) in agriculture holds immense potential for achieving greater efficiency, resilience and sustainability in agri-food systems in the global South. This book, "Enhancing Water and Food Security through Improved Agricultural Water Productivity: New Knowledge, Innovations, and Applications," provides an evidence-based knowledge synthesis on WP and agricultural water management in the global South.

The global South, encompassing regions with diverse agroecological settings and complex socio-economic dynamics, is particularly vulnerable to water and food insecurity. Rapid population growth, migration of people from rural to urban areas or urbanisation and changing consumption patterns or dietary requirements exert tremendous pressure on limited water resources, while climate change exacerbates these challenges. To overcome these obstacles, we must adopt a holistic approach that integrates scientific research, innovation and practical applications in transforming agricultural water management.

This book provides multidisciplinary perspectives, practical experiences and case studies from scholars, practitioners and policymakers from the global South. The book showcases how enhancing WP is a viable solution to address water and food insecurity in water-scarce environments by showcasing evidence-based knowledge. The book delves into various dimensions of agricultural water productivity, exploring cutting-edge research, practical interventions and innovative technologies. We examine the interplay between WP and water management practices, agricultural production systems and socio-economic contexts. Doing so aims to present a nuanced understanding of the complex relationships and trade-offs involved, empowering readers to make informed decisions and implement effective strategies. Furthermore, this approach aligns with several of the Sustainable Development Goals (SDGs), including SDG 2 (Zero Hunger), SDG 6 (Clean Water and Sanitation) and SDG 13 (Climate Action), among others.

While our challenges are daunting, the book also showcases inspiring success stories and transformative interventions that have enhanced WP in diverse settings. This serves to inspire further innovation and encourage adopting best practices. By promoting sustainable intensification, resource-efficient technologies and inclusive governance, we can foster agricultural systems that produce more with less water, protect ecosystems and improve livelihoods. As such, the book significantly contributes to the discourse on agricultural water management in the global South. By synthesising evidence-based research, sharing practical experiences and promoting innovation, this book provides a comprehensive guide for policymakers, researchers, practitioners and stakeholders involved in water and food security. Together, we can harness the potential of enhanced WP to address the grand challenges of our time, paving the way for a more sustainable, resilient and equitable future.

Collectively, the chapters in this book have illuminated the significance of agricultural water productivity as a powerful tool in addressing water and food security challenges in the global South. It is evident from the diverse range of topics covered that enhancing agricultural water productivity requires a multifaceted approach that considers technological advancements, policy frameworks, socio-economic dynamics and local contexts.

The book underscores the need for systematic and transdisciplinary approaches to irrigation development and agricultural water management, innovative technologies, knowledge exchange platforms and inclusive governance to foster sustainable agricultural water management.

Executive Manager: Water Use in Agriculture
Water Research Commission, Vice President:
International Commission on Irrigation and Drainage
Board Member: South African Weather Service
Cape Town, South Africa

Sylvester Mpandeli

Preface

This book aims to share knowledge on agricultural water productivity (WP) and showcase how knowledge on increasing WP can be part of addressing grand global challenges related to increasing water scarcity and food insecurity under climate change while contributing to several Sustainable Development Goals in the global South. This book can be used as an evidence-based guide on agricultural water management in the global South and, hopefully, inform regional and sectoral policies in Africa on sustainable development solutions through better management of resources.

The chapters in this book highlight the significance of WP as a powerful tool in addressing water and food security challenges in the global South. By showcasing evidence-based research, case studies and innovative approaches, the book provides a synthesis of knowledge to guide policymakers, researchers, practitioners and stakeholders in making informed decisions and implementing effective strategies. It is evident from the diverse range of topics covered that enhancing agricultural water productivity requires a multifaceted approach that considers technological advancements, policy frameworks, socio-economic dynamics and local contexts. The book emphasises the need for integrated agricultural water management practices, innovative technologies, knowledge exchange platforms and inclusive governance to foster sustainable agricultural systems.

The knowledge, innovations and applications of WP are broken down into (i) case studies, (ii) critical analyses and (iii) metanalyses. As such, the book explores various topics, case studies and perspectives to address the grand challenges of water and food insecurity in the global South. As we conclude this book, we reflect on the collective insights gained and highlight the key takeaways from each chapter.

The insights from this book resonate strongly with several Sustainable Development Goals, including Zero Hunger (SDG 2), Clean Water and Sanitation (SDG 6), Climate Action (SDG 13) and others. We can make significant strides towards achieving these global targets by promoting water-smart agricultural practices, resource optimisation and ecosystem protection.

London, UK — Tafadzwanashe Mabhaudhi
Harare, Zimbabwe — Vimbayi G. P. Chimonyo
Pietermaritzburg, South Africa — Aidan Senzanje
Dar Salam, Tanzania — Pauline P. Chivenge

Acknowledgements

The Water Research Commission of South Africa is acknowledged for supporting the book's development, together with the Sustainable and Health Food Systems (SHEFS) Programme, supported through the Wellcome Trust's Our Planet, Our Health Programme [Grant No 205200/Z/16/Z].

We want to acknowledge the anonymous reviewers and the Reference Group of Experts assembled by the Water Research Commission to review the book:

Name	Organisation
Prof NS Mpandeli	Water Research Commission (Chairperson)
Dr L Nhamo	Water Research Commission
Dr SN Hlophe Ginindza	Water Research Commission
Prof S Midgley	Department of Agriculture, Western Cape
Prof M van der Laan	Agricultural Research Council
Dr NJ Taylor	University of Pretoria
Dr KA Tshikolomo	Department of Agriculture, Limpopo
Dr TP Chibarabada	Zimbabwe Sugar Association

Lastly, we thank the Nexus Gains, Excellence in Agronomy and Mixed Farming Systems Initiatives of the CGIAR for significant contributions to the book.

Contents

Editors and Contributors

About the Editors

Tafadzwanashe Mabhaudhi is a professor of Climate Change, Food Systems and Health at the London School of Hygiene and Tropical Medicine and Director of the Institute for Natural Resources (NPC) in South Africa. In addition, he is also the lead—Water, Energy, Food and Environment (WEFE) Nexus at the United Nations University Institute for Water, Environment and Health (UNU-INWEH). He is a globally recognised researcher, boundary spanner and science advocate, working on complex water, energy, food and environmental systems at the interface of science, policy and society. He has a commendable track record of scientific leadership, capacity building and partnership development crucial for realising the benefits of research, development and innovation. He aims to spearhead research and development initiatives that influence policy, promote equality and drive transformation in Africa and globally.

Vimbayi G. P. Chimonyo is a sustainable agriculture scientist at CIMMYT. She specialises in the intersections of water, nutrition and climate resilience. With a solid foundation in crop science, she has dedicated her career to tackling the multifaceted challenges of climate change, with a particular focus on semi-arid regions. Her research is centred on developing and implementing adaptive strategies that enhance the resilience of agricultural systems in the face of water scarcity and extreme weather events while simultaneously promoting nutrition and human well-being.

Aidan Senzanje is a senior research fellow in the School of Engineering, at the University of KwaZulu-Natal (South Africa). His research interests are in irrigation technology, irrigation and agricultural water management and the water-energy-food (WEF) nexus. He has published over 60 journal articles, authored/co-authored 23 book chapters and co-edited 4 books. He was previously a visiting scientist at the International Water Management Institute (IWMI) in Pretoria, South Africa. He has practical experience in undertaking technical consultancies in irrigation engineering, agricultural water management, project evaluation and environmental management systems. He was involved in the planning, developing and launching of new academic programmes in Land and Water Management for the SADC region and the BSc Agricultural Engineering degree at the University of Zimbabwe.

Pauline P. Chivenge is a senior scientist in Cropping Systems Agronomy/Climate Change for the International Rice Research Institute (IRRI) based in Tanzania. Pauline earned a PhD in Soils and Biogeochemistry at the University of California, Davis. Most of Pauline's research has focused on soil and nutrient management in smallholder farming systems in sub-Saharan Africa and Asia. Soil fertility improvement research is core to her research, including the development and utilisation of digital decision support tools for dissemination of nutrient and crop management recommendations based on knowledge-intensive scientific principles to increase agricultural productivity. She has led a couple of international projects, working in diverse farming systems on topics that include biogeochemical nutrient cycling, natural resource management, watershed management, carbon sequestration, greenhouse gas emissions and sustainable ecosystem functioning.

Contributors

Rasheed A. Adeleke North-West University, Potchefstroom, South Africa

National Centre for Food Manufacturing, University of Lincoln, Holbeach, United Kingdom

Muiz O. Akinyemi North-West University, Potchefstroom, South Africa

National Centre for Food Manufacturing, University of Lincoln, Holbeach, United Kingdom

Niranga Alahakoon International Water Management Institute, Colombo, Sri Lanka

Kazeem A. Alayande North-West University, Potchefstroom, South Africa

National Centre for Food Manufacturing, University of Lincoln, Holbeach, United Kingdom

Upali Amarasinghe International Water Management Institute, Colombo, Sri Lanka

L. Chikwiramakomo Department of Construction and Civil Engineering, University of Zimbabwe, Harare, Zimbabwe

Vimbayi G. P. Chimonyo International Maize and Wheat Improvement Centre (CIMMYT), Harare, Zimbabwe

Centre for Transformative Agricultural and Food Systems School of Agricultural, Earth and Environmental Sciences, University of KwaZulu-Natal, Pietermaritzburg, South Africa

O. Cofie International Water Management Institute (IWMI), West Africa Regional Office, Accra, Ghana

George Danso Trackpact, Edmonton, AB, Canada

T. Desalegn International Water Management Institute (IWMI), Addis Ababa, Ethiopia

T. L. Dirwai International Water Management Institute, Pretoria, South Africa

Centre for Transformative Agricultural and Food Systems, University of KwaZulu-Natal, Pietermaritzburg, South Africa

Pay Drechsel International Water Management Institute, Accra, Ghana

W. Gumindoga Department of Construction and Civil Engineering, University of Zimbabwe, Harare, Zimbabwe

S. T. Hadebe Department of Plant Production, Soil Science and Agricultural Engineering, University of Limpopo, Sovenga, South Africa

H. E. Igbadun Department of Agricultural and Bioresources Engineering, Ahmadu Bello University, Zaria, Nigeria

E. K. Kanda Department of Civil and Structural Engineering, Masinde Muliro University of Science and Technology, Kakamega, Kenya

M. Kapari Water Research Commission of South Africa, Pretoria, South Africa

University of the Western Cape, Bellville, Cape Town, South Africa

J. Lautze International Water Management Institute (IWMI), Pretoria, South Africa

Francis B. Lewu Department of Agriculture, Faculty of Applied Sciences, Cape Peninsula University of Technology, Cape Town, South Africa

Tafadzwanashe Mabhaudhi Centre on Climate Change and Planetary Health, London School of Hygiene and Tropical Medicine, London, UK

Centre for Transformative Agricultural and Food Systems, University of KwaZulu-Natal, Carbis Road, Scottsville, Pietermaritzburg, KwaZulu-Natal, South Africa

J. Magidi Tshwane University of Technology, Pretoria, South Africa

S. Mpandeli Department of Environmental, Water and Earth Sciences, Tshwane University of Technology (TUT), Pretoria, South Africa

Faculty of Science, Engineering and Agriculture, University of Venda, Thohoyandou, South Africa

B. Mukandiwa Department of Construction and Civil Engineering, University of Zimbabwe, Harare, Zimbabwe

L. Nhamo Water Research Commission of South Africa, Pretoria, South Africa

Centre for Transformative Agricultural and Food Systems, University of KwaZulu-Natal (UKZN), Pietermaritzburg, South Africa

T. N. M. Nzimande Centre for Transformative Agricultural and Food Systems, School of Agricultural, Earth & Environmental Sciences, Pietermaritzburg, South Africa

Linda U. Obi North-West University, Potchefstroom, South Africa

O. A. Ojeleye Department of Agricultural Economics, Ahmadu Bello University, Zaria, Nigeria

Andrew Emmanuel Okem International Water Management Institute, Accra, Ghana

Chika E. Oyeagu Department of Agriculture, Faculty of Applied Sciences, Cape Peninsula University of Technology, Cape Town, South Africa

Madhusha Perera International Water Management Institute, Colombo, Sri Lanka

D. T. Rwasoka Department of Construction and Civil Engineering, University of Zimbabwe, Harare, Zimbabwe

B. Sharma International Water Management Institute (IWMI), New Delhi, India

M. Sibanda University of the Western Cape, Bellville, Cape Town, South Africa

Alok Sikka International Water Management Institute, New Delhi, India

C. Taguta Centre for Transformative Agricultural and Food Systems, University of KwaZulu-Natal, Pietermaritzburg, South Africa

E. M. Wimalasiri Crops For the Future Research Centre, Semenyih, Malaysia

Abbreviations

4IR	4th industrial revolution
ACPC	African Climate Policy Centre
AgNP	Silver nanoparticles
AI	Artificial intelligence
Al_2O_3NP	Aluminium oxide nanoparticles
ANN	Artificial neural network
AU	African Union
AUC	African Union Commission
AWM	Agricultural water management
BDF	Business Development Fund
BEWAP	BEsproeiingsWAterBestuursprogram
BMP	Best management practices
BREBS	Bowen ratio energy balance system
BRT	Boosted regression tree
BTEX	Benzene, toluene, ethylbenzene and xylenes
CA	Conservation agriculture
CAADP	Comprehensive African Agriculture Development Programme
CBZ	Carbamazephine
CC	Climate change
Cd	Cadmium
CDM	Clean development mechanism
CERES	Crop environment resource synthesis
CFS	Cubic feet per second
CGMs	Crop growth models
CGWB	Central Ground Water Board (of India)
CH_4	Methane
CIAT	International Center for Tropical Agriculture
CIP	Crop Intensification Programme
CLS	Integrated crop-livestock system
CLWP	Crop-livestock-water productivity
CLWS	Crop-livestock-water-soil nexus

CO_2	Carbon dioxide
CommonKADS	Common Knowledge Acquisition and Documentation Structure
Cr	Chromium
CS	Chameleon sensor
CSA	Climate-smart agriculture
CSAIP	Climate-Smart Agriculture Investment Plan
CSA-PF	Climate-smart agriculture prioritization framework
CV	Climate variability
CWUI	Crop water use index
DACT	Degrees above canopy threshold
DOI	Digital object identifier
DS	Divisional secretariat
DSS/IDSS	Decision support system/intelligent DSS
DSSAT	Decision support system agrotechnology transfer
DWW	Domestic wastewater
E/Ex -	Relative soil evaporation (100 E/Ex)
EC	Eddy covariance
ECOSAN-EU	Ecological sanitation
EF	Endophytic fungi
EIRRs	Economic internal rate of returns
ELM/KELM	Extreme learning machine/kernel ELM
EPS	Extracellular polymeric substances
ES	Expert systems
ET	Evapo-transpiration
ET_a	Actual evapotranspiration
ETM	Enhanced thematic mapper
ETo	Reference evapotranspiration
EWP	Economic water productivity
EWUI	Economic water use index
FAO	Food and Agriculture Organization of the United Nations
FAO	Food and agriculture organizations
FLID	Farmer-led irrigation development
FMNR	Farmer-managed (assisted) natural regeneration
FTOPSIS	Fuzzy technique for order performance by similarity to ideal solution
GCMs	General circulation models
GD	Growing degrees
GDP	Gross domestic product
GDP	Growth domestic productivity
GHG	Greenhouse gas
GIS	Geographical Information Systems
GIZ	Deutsche Gesellschaft für Internationale Zusammenarbeit
GNDVI	Green normalised difference vegetation index
GOR	Government of Rwanda
GWP	Global water partnership

ha	Hectare
HI	Harvest index
HYV	High yielding
IBRD	International Bank for Reconstruction and Development
IDAWM	Irrigation development and agricultural water management
ILWIS	Integrated land and water information system
IMF	International Monetary Fund (IMF)
IQQM	Integrated water quantity and quality simulation model
IRB	Indus river basin
IRR	Internal rate of return
IRS	Indian remote sensing (satellite)
ITCZ	Inter-tropical convergence zone
IWMI	International Water Management Institute
IWRM	Integrated Water Resources Management
IWUA	Irrigation water users' associations
kg	Kilo gram
kPa	Kilo Pascal
LANDSAT	Low altitude satellite
LBDC	Lower Bari Doab Canal
LCI	Leaf chlorophyll index
LOA	Agricultural Orientation Law
LUC	Land use consolidation
LWC	Leaf water content
LWH	Land husbandry, water harvesting and hillside
LWP	Livestock water productivity
MAE	Mean absolute error
Mali Meteo	National Meteorological Agency
MAP	Mean annual precipitation
MAPE	Mean absolute percentage error
METRIC	Mapping evapotranspiration at high resolution using internalized calibration
MFS	Mixed farming system
MG	Malachite green
MINAGRI	Ministry of Agriculture and Animal Resources of Rwanda
mm	Millimetres
MnO	Manganese oxide
MODFLOW	Modular finite-difference flow model (USGS)
MODIS	Moderate resolution imaging spectroradiometer
MSc	Master of Science
MSD	Meteorological service department
MVMD	Multivariate variational mode decomposition
MWCNT	Multi-walled carbon nanotubes
N_2O	Nitrous oxide (dinitrogen monoxide)
NAEB	National Agricultural Export Board
NAPA	National Adaptation Programme of Action

NBI	Nile Basin Initiative
NDCs	Nationally determined contributions
NDP	National Development Plan
NDRE	Normalized difference red-edge index
NEMA	National Environmental Management Act
NEPAD	New Partnership for Africa's Development
NEPAD	New Partnership for African Development's
NGOs	Non-governmental organizations
NRP	National Rice Policy
NUS	Neglected and underutilized species
nZVI	Zero-valent iron nanoparticles
OECD	Organisation for Economic Co-operation and Development
OLI	Operational land imager
ON	Office du Niger
OPTRAM	Optical trapezoid model
PGP	Plant growth-promoting
PhD	Doctor of Philosophy
PICSA	Participatory Integrated Climate Services for Agriculture
PNISA	National Investment Plan in the Agricultural Sector
PPPs	Private-public partnerships
PRI	Photochemical reflectance index
PRISMA–P	Preferred Reporting Items for Systematic Review and Meta-Analysis Protocols
PSTA	Strategic Plan for Agricultural Transformation
RAB	Rwanda Agriculture and Animal Resources Development Board
RR	Ridge regression
RRR	Resource recovery and reuse
SA	South Africa
SA	South Asia
SAPWAT	South African Program Water
SAR	Synthetic aperture radar
SAVI	Soil adjusted vegetation index
SCOPE	Soil-canopy-observation of photosynthesis and energy fluxes
SE	Soil evaporation
SEBAL	Surface energy balance algorithm for land
SEBS	Surface energy balance systems algorithms
SH	Smallholder
SPAM	Soil-plant-atmosphere module
SPIDER	Sample, phenomenon of interest, design, evaluation and research
SPOT	Satellite Pour l'Observation de la Terre
SRI	System of rice intensification
SSA	Sub-Saharan Africa
SSA	Sub-Saharan African
SSIT	Small-scale irrigation technology
SWC	Soil and water conservation

SWOT	Strengths, weaknesses, opportunities and threats
SWP	Sustainable water partnership
TNA	Technology needs assessments
UAV	Unmanned aerial vehicles
UNEP	United Nation Environmental Problem
UNEP-CCC	United Nations Environment Programme Copenhagen Climate Centre
USAID	United States Agency for International Development
USGS	United States Geological Survey
VTCI	Vegetation temperature condition index
VWC	Vegetation water content
WaPOR	Water productivity open-access portal
WFD	Wetting front detectors
WHC	Water harvesting and control
WOFOST	World Food Studies
WP	Water productivity
WP/EWP/PWP	Water productivity/Economic WP/Physical WP
WPET	Water productivity concerning evapotranspiration
WRC	Water Research Commission
WU	Crop-water use
WUA	Water User Association
WUE	Water-use efficiency
ZIM ASSET	Zimbabwe Agenda for Sustainable Socio-Economic Transformation
ZIMSTAT	Zimbabwe National Statistics Agency
ZIMVAC	Zimbabwe Vulnerability Assessment Committee
Zn	Zinc
ZnONP	Zinc oxide nanoparticles

Chapter 1
A Review of Existing Knowledge on Water Use and Nutritional Water Productivity in South Africa

Vimbayi G. P. Chimonyo and Tafadzwanashe Mabhaudhi

Abstract In South Africa, while the nation is food secure at the national level, a substantial portion of the population lives in poverty and faces food insecurity, particularly in rural areas. The country's water scarcity, variable rainfall patterns, and unequal distribution of irrigation resources further complicate the issue. Agriculture, which relies heavily on irrigation, is a major water consumer and contributor to food production. Efforts to address these challenges include exploring strategies like rainwater harvesting and improving water productivity in agriculture. The Water Research Commission (WRC) has played a pivotal role in developing innovative solutions to enhance water productivity, focussing on environmentally sensitive approaches and the water–energy–food nexus. Achieving food security in water-scarce regions like South Africa necessitates dynamic institutions, improved water productivity, and sustainable agricultural practices. This chapter emphasises the urgency of addressing these issues to ensure a healthier and more productive life for rural communities while promoting sustainable development and water-based agriculture.

Keywords Resource use efficiency · Water scarcity · Sustainable development · Resource security · Irrigated agriculture

V. G. P. Chimonyo (✉)
International Maize and Wheat Improvement Centre (CIMMYT), Harare, Zimbabwe

Centre for Transformative Agricultural and Food Systems School of Agricultural, Earth and Environmental Sciences, University of KwaZulu-Natal, Pietermaritzburg, South Africa
e-mail: v.chimonyo@cgiar.org

T. Mabhaudhi
Centre on Climate Change and Planetary Health, London School of Hygiene and Tropical Medicine, London, UK

Centre for Transformative Agricultural and Food Systems, University of KwaZulu-Natal, Carbis Road, Scottsville, Pietermaritzburg, KwaZulu-Natal, South Africa

T. Mabhaudhi et al. (eds.), *Enhancing Water and Food Security Through Improved Agricultural Water Productivity*,
https://doi.org/10.1007/978-981-96-1848-4_1

1.1 Introduction

Although water is a renewable resource, its availability for human use is limited. Almost 70% of the globe is covered in water, 97.5% is seawater, and only 2.5% is available as freshwater (Shiklomanov 1991). Not all of the 2.5% fresh water is available for humans. Some of it is locked as frozen water or groundwater. Although water is a renewable resource, its accessibility is limited due to geographical distribution and human activities polluting it. Almost 60% of fresh water is available in only nine countries, representing 44% of the earth's surface, namely, Brazil, Canada, China, Columbia, the Democratic Republic of Congo, India, Indonesia, Russia, and the United States (Oki and Kanae 2006). The rest of the world's population faces physical or economic water stress/scarcity, with two-thirds experiencing water shortages for at least a month in a 12-month cycle (Mekonnen and Hoekstra 2016). Many developing countries face poverty, high unemployment rates, and, more importantly, food and nutrition insecurity within this two-thirds.

Water and food security is attained when there is an acceptable quantity and quality for health, livelihoods, ecosystems and production. However, today's grand challenges brought about climate change, technological development, rapid urbanisation and increasing population and incomes threaten the future availability of water for food. The demand for more water in agriculture has intensified competition for the resource within the agriculture value chain and across domestic and industrial sectors. Furthermore, existing challenges related to water insecurity are impacting the energy sector as it is used in its production (hydropower generation and biofuel production or for cooling in nuclear and geothermal power plants). There is, therefore, a need to produce more food to meet the growing demands of an increasing population using the same or even less water. Achieving optimal water productivity results in availing water resources to other equally important sectors including (i) enhancing the health of the environment, (ii) reducing pressure on other sectors of industry, and (iii) ensuring sustained economic development and growth.

Addressing the water and food insecurity challenges requires the strengthening of institutions and the promotion of transdisciplinary actions aimed at balancing water demand and supply. This is particularly relevant in the agriculture sector as over 60% of the available freshwater resources are used in the sector. There is a need for both the public and private sectors to work together in providing the relevant information that allows decision-makers to assess and respond effectively to the worsening water and flood risks (Masipa 2017). Agricultural production and livelihoods in water-scarce regions can be sustained only if priority improves water productivity and enhances water research and development (Cai et al. 2011). This book provides an overview of the role of agricultural water productivity in improving water-use efficiency in the whole food value chain under increasing water scarcity. The term water productivity is used in rainfed and irrigated agricultural sub-sectors. The book explores the implications of improved water productivity and then provides a systematic review to understand the limits and opportunities that are derived from improved water productivity. The discussion will focus on South Africa, a country whose status is water-stressed and has high food and nutrition insecurity at the household level.

1.1.1 An Overview of Water and Food Security in South Africa

South Africa is food secure at the national level, but over 50% of households are food insecure as evidenced by extreme poverty and economic inequality, particularly in densely populated, rural areas and informal settlements (Stats-Sa 2019). According to Statistics South Africa, almost 25.2% of the population in the country live below the poverty line (Stats-Sa 2019). Most households are faced with food insecurity (Chakona and Shackleton 2019), about 28.3% of the population are at risk of hunger, and 26% are food insecure (SADHS 2017). This has seen the government give monthly grants to many due to the prevalence of poverty. Besides these alarming statistics, rainfed agriculture is the third most important means of livelihood, yet it contributes only 10% to household income (Stats-Sa 2019). These challenges of resource insecurity are being exacerbated by the emergence of novel infectious diseases such as the COVID-19 pandemic. This requires an urgent need to provide smart technologies and practices that are cost-effective and support smallholder agriculture. The National Development Plan (NDP) acknowledges agriculture and rural development as key to job creation, economic growth, poverty reduction, and, more importantly, addressing household food and nutrition security (National Planning Commission 2012).

However, water security is one of many challenges affecting the implementation of many agricultural reform programmes. South Africa is water-scarce, the 30th driest country in the world. Rainfall in the country is highly variable and unevenly distributed in space and time across the country. The average annual rainfall received by over 60% of the country is less than 500 mm, which is the minimum needed for dryland farming, whilst 21% of the country's land area receives less than 200 mm (De Villiers et al. 2004). Seventy percent of crop production is rainfed, yet only 35% of the land area receives enough rainfall that is needed for rainfed crop production (CSIR 2010). The water deficit resulting from the low rainfall, which is compounded by climate change and exacerbated by the high evaporative rate, limits crop production under the rainfed agricultural system in the country (Van Averbeke et al. 2011). These challenges of water deficits are posing the greatest risk to national food production and security. The agriculture sector in South Africa is faced with complex challenges that need transitional pathways if the country is to continue to meet the growing food demands of an increasing population. These transformations include producing more nutritious food with less water per unit of output. This transition requires the adoption of novel and climate-smart technologies and practices. Provide. As the water challenge for agricultural production increases, the importance of irrigated agriculture in South Africa is expected to rise.

As already alluded to, over 60% of the available freshwater resources in South Africa are allocated to agriculture to irrigate about 1.3 million ha, which represents just under 10% of the arable land (Cousins 2007; Nhamo et al. 2016). Irrigated agriculture supports between 25% and 30% of South Africa's agricultural production and contributes about 90% of the high-value crops like potatoes, vegetables, and fruits, and between 25% and 40% of industrial crops that include sugarcane and cotton. The distribution of water for irrigation is determined along racial lines as most of the water is allocated to the white commercial farmers who own vast agricultural land (Lahiff and Cousins 2005). About 90% of the water supply is used in this commercial agriculture sector which is supported by massive state investment in infrastructure (Lahiff and Cousins 2005). Smallholder farmers, who are in the majority, are mostly found in resource-insecure former homelands or Bantustans (Van Averbeke and Khosa 2007), where they are only able to use about 7.7% of irrigated land, which is about 100,000 hectares (Hardy et al. 2011). The strategy of the National Development Plan (NDP) to increase the irrigated area under smallholder farming to about 200,000 ha has the risk of further straining the already scarce water and energy resources (Cai et al. 2017). This requires interventions through transformative and circular approaches that address synergies and trade-offs.

Some of the available options to increase water availability for irrigation include the construction of new dams to capture runoff and to increase the use of groundwater resources (Mabhaudhi et al. 2018). However, the country has limited options to construct more dams as it already has a high number of dam density. Therefore, the only option is to use groundwater resources; however, there is a lack of knowledge on the amount of groundwater stored in aquifers. Besides, the smallholder farmers lack water, energy, infrastructure and technical skills to irrigate, resulting in some of the land equipped for irrigation not being irrigated (Nhamo et al. 2024). Another alternative is for farmers to explore opportunities from rainwater harvesting and soil water conservation techniques. These challenges are being compounded by decreasing rainfall totals and the frequency of droughts and heatwaves that increase evapotranspiration (Fig. 1.1). Hence, the practicality of this approach under the increasing frequency of drought is quite limiting. Thus, smart technologies that enhance water productivity are envisaged to contribute immensely to water-use efficiency (Mabhaudhi et al. 2018). Therefore, system efficiencies can significantly improve water productivity and sustainable food production under water-limited conditions.

For over 50 years, the Water Research Commission (WRC) has developed effective, holistic, water-smart solutions and innovations to ensure water and food security. Interventions spearheaded by the WRC have included, but are not limited to, innovations in water quality and environmentally sensitive water development, a fundamental diversification of water-supply options, innovative water-sensitive designs, embracing the water–energy–food nexus and fully implementing a fourth industrial revolution (4IR) approach to agricultural water management. Enhanced WP has been central to these initiatives, spanning different temporal, spatial, and socio-economic scales. The mapping of research funded by the WRC can showcase the existing research networks on agricultural WP, emerging thematic areas and possible research and knowledge gaps in the global South. Such an exercise can aid

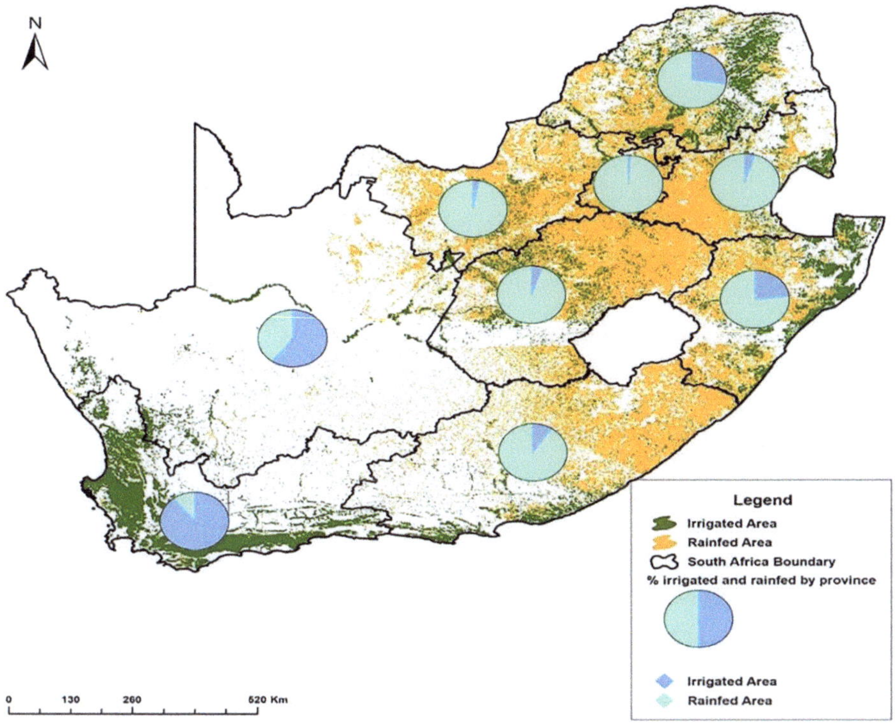

Fig. 1.1 The distribution of rainfed and irrigated areas in South Africa. The pie charts represent the percentage of irrigated and rainfed land by province (Mabhaudhi et al. 2019)

in aligning key global, national, and provincial plans, priorities and policy documents for ensuring food and water security, sustainable development and empowering rural communities, and other WRC's programmes for sustainable water-based agriculture in rural areas.

1.1.2 Aims and Objectives

This chapter aims to:

(i) To showcase the research (grey literature, journal publications, reports, popular articles, etc.) on water productivity emanating from South Africa
(ii) Identify gaps in the existing knowledge, clarify WP, and report on the types of evidence needed to address and inform the advancement of the WEF nexus

1.1.3 Methodology

The study conducted a state-of-the-art literature review on the current status of water productivity research within South Africa. The methodology is structured into four phases, namely, (i) a literature review of key terms and definitions used in the book, (ii) a mixed-methods approach to establish the status of water productivity research and the gaps, and (iii) a determination of the pathways to operationalise smart-technologies and practices that improve water productivity. These phases are detailed below.

Phase 1—Definition of Terms

Various definitions have been used to describe water productivity. Each of these has its meaning and context, which has caused incoherency at times. This book has identified the most relevant and contextualised definitions of key water productivity terms. The definitions of the terms provide a short formal definition, some additional characteristics and the available references. The terms to be defined include 'water productivity', 'water use efficiency', and 'water footprint'. Official guidelines, position papers, research and journal articles, and statements and reports from various sources were consulted.

Phase 2—Status of WP Research in South Africa

In this phase, an assessment of the thematic research areas emerged from the articles and scientific reports on WP within the WRC and globally. Firstly, scientific reports were downloaded from the WRC knowledge hub. The keywords used to search for these articles included 'water productivity', 'water use efficiency', and 'water footprint'. Three hundred forty-five reports and popular articles were retrieved and published between 1974 and 2021. To retrieve the global research trends, the same keywords were then inputted into SCOPUS. A total of 3540 articles were retrieved, and we extracted the top 200 cited articles to include in the study. It should be noted that the global research database on WP was not subjected to the same assessment since it only served to identify themes that show the advancements made in the subject area.

Phase 3

Identified literature and extracted data were subjected to qualitative analysis. To analyse the data obtained from the WRC knowledge hub, a word cloud was initially generated from the titles of research reports, articles and popular articles. A word cloud was used as a special visualisation of text whereby more frequently used words are boldly highlighted. The frequency at which a word appears within a document under review, the larger and bolder it appears in the image generated. This provides an overview of the main themes being addressed in the document. Word clouds are increasingly becoming popular in understanding the main theme of written material. However, word clouds have several limitations including their failure to group words with the same or similar meaning, such as 'water use' and 'water utilisation'. Further, the words were retrieved out of context as the technique omits the semantics and phrases they comprise. To overcome these challenges, retrieved titles were subjected to bibliometric analysis.

Two separate bibliometric analyses were also conducted to reveal key terms in water productivity research within the global context and South Africa. Bibliometric analysis quantitatively assesses published articles and facilitates the evaluation of peer-reviewed studies in a specific subject of research (Rey-Martí et al. 2016; Small 1973). It examines secondary data obtained from digital databases from a quantitative and objective perspective (Albort-Morant and Ribeiro-Soriano 2016). Such an analysis facilitates the structuring of the evolution of a focal research area (Cobo et al. 2011; Klavans and Boyack 2006). The VOSviewer software was used for the analysis in this study and performed the key term analysis and network visualisation of relevant documents related to water productivity.

1.1.4 Results and Discussion

1.1.5 Terminology

1.1.5.1 Water Productivity Versus Water Use Efficiency

While the terms *water use efficiency* (WUE) and *water productivity* (WP) are aimed at addressing the term 'more crop per drop', they are being used interchangeably and, as a result, seemingly lack a clear definition. Experts from irrigation engineering, crop physiologists, and water managers tend to have opposing points of view on the correct terminology. Molden et al. (2003) came up with a detailed conceptual framework to communicate water productivity. Although water use efficiency and water productivity are used interchangeably, they are two distinct words.

Water use efficiency is the ratio of biomass or yield to water applied (Eq. 1.1), while WP is the ratio of biomass or yield to actual water used (Eq. 1.2).

$$\mathrm{WUE} = \frac{Y_a \text{ or } B_a}{\text{water applied}} \mathrm{kg\,m^{-3}} \tag{1.1}$$

$$\mathrm{WP} = \frac{Y_a \text{ or } B_a}{\mathrm{ET_a}} \mathrm{kg\,m^{-3}} \text{ or } \mathrm{kg\,ha^{-1}mm^{-1}} \tag{1.2}$$

Y_a and B_a are the actual yield and biomass (kg), respectively, and $\mathrm{ET_a}$ is the actual evapotranspiration (mm $\mathrm{ha^{-1}}$ or $\mathrm{m^{-3}\,ha^{-1}}$) or water consumed. For the calculation of WUE, water applied suggests water entering the systems. It is silent on the unproductive water loss such as runoff, deep percolation, and capillary rise, and changes in subsurface flow, since it is challenging to quantify these.

1.1.5.2 Irrigation Efficiency

Irrigation efficiency is a water use efficiency that measures an irrigation system's effectiveness within specific limits. It determines the ratio of the amount of water consumed by the crop to the amount supplied through Irrigation (Irmak et al. 2011). It encompasses all types of irrigation, including surface, sprinkler, or drip irrigation. The term is important for characterising irrigation performance and evaluating irrigation water use (Howell 2003). Irrigation efficiency is measured according to irrigation system performance, homogeneity of water application, and crops' response to Irrigation (Irmak et al. 2011). The measurements are interconnected but differ with space or time. On a temporal scale, the measurements can be considered for a single application to a season or a year, while that on the spatial scale vary from a single field to as large as an irrigation district.

1.1.5.3 Water Footprint

According to Waterfootprint.org, *water footprint* (WF) measures the amount of water used to produce each of our goods and services. It can be measured for a single process, such as growing maize, or an entire value chain related to food production. The water footprint indicates the amount of water consumed by a country, or globally at any spatial scale (river basin or an aquifer). Water footprint, therefore, represents 'blue' water (sourced from surface or groundwater), 'green' water (precipitation and runoff used directly by plants), and 'grey' (water needed to assimilate pollutants) to provide information on volumetric water use and pollution from the perspectives of consumers, supply chains and products, or even specific geographies (Hoekstra 2017). Researchers agree that reducing the water footprint of crop production, that is, increasing water productivity can aid in increasing future food demands in the context of global water scarcity (Nyathi et al. 2019).

$$\mathrm{WF} = \frac{\mathrm{ET_a}}{Y_a \text{ or } B_a}\left(\mathrm{m^3 kg^{-1}}\right) \quad (1.3)$$

1.1.6 *Status of Research on Water Productivity in South Africa*

The current section highlights the work funded by the WRC to address water productivity challenges. Water and food are among some of the most important resources that sustain human life; therefore, their security is vital for achieving Sustainable Development Goals (SDGs) (Mensah and Ricart Casadevall 2019). This is highlighted in the collection of words identified by the word cloud and is consistent with one of the main goals of WRC. However, the security of water and food is threaded by increasing population and climate change which contribute to their depletion and degradation. Both water and agriculture are culprits and victims of climate change and changing how they are managed has the potential to alleviate the current challenges and contribute towards achieving the SDGs. Continued agricultural developments, intensification, and extensification have the risk of aggravating existing challenges related to land and water insecurity (Allen and Prosperi 2016). Furthermore, existing and future socio-ecological changes that are being driven by climate change require novel adaptation strategies that do not compromise livelihoods (UNGA 2015). The WRC has supported various frameworks, methods, practices and indicators aimed to support water use efficiency. Most of the financial resources from the WRC has achieved in establishing a criterion for water security based on four themes: the state of the water environment; human health and well-being; the sustainability of livelihoods; and the stability, functions, and responsibility of society (Figs. 1.2 and 1.3). These novel climate-smart technologies are enhancing the adaptation and resilience strategies and reducing the impacts of the prevailing socio-ecological changes including in the water and agriculture sectors. These interventions have seen an improvement in understanding the linkages and connections between agronomic practices and agricultural water management. All this is emanating from the knowledge that water and food continue to deteriorate as several countries face the double burden of hunger and undernutrition with overweight and obesity (IFPRI 2016). Research within the WRC has been proactive in addressing these challenges.

The analysis produced three thematic clusters of keywords that co-occur in literature, namely, the response of crops to water stresses (green), interactions between the environment and water use (blue), and water management in the production of agronomic crops (red).

(a) *Green Cluster*: Response of crops to water stress, particularly droughts (Green). Countries that contributed to research were mostly conducted in Southern

Fig. 1.2 Key terms identified in scientific report titles on Water Productivity submitted to the WRC

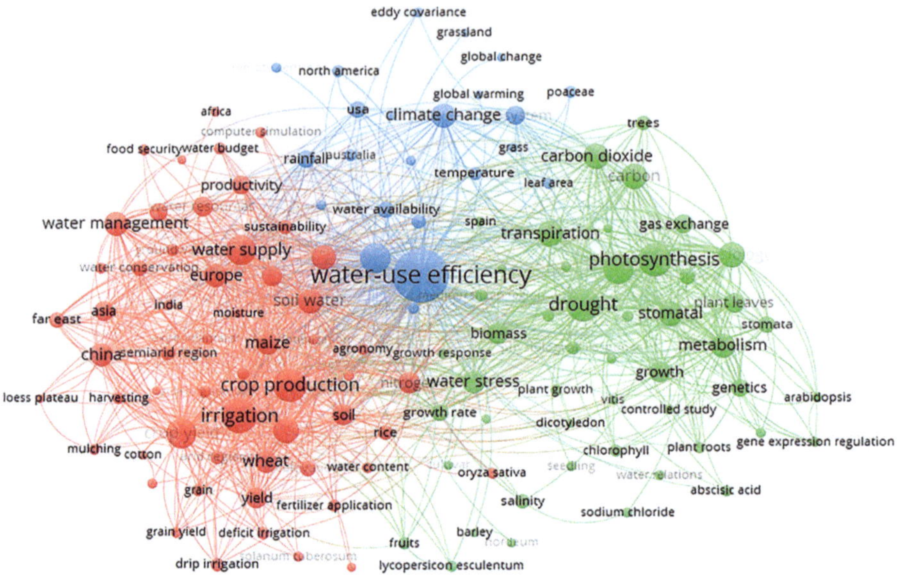

Fig. 1.3 Visualisation of thematic areas assessed across 200 relevant scientific reports on water productivity. The network map identified 30 terms to be the most relevant across identified scientific report titles. These key terms were designated into three thematic areas, namely, crop water use (green), water and climate change (blue), and agricultural water management (red)

Europe and Mediterranean regions. Central to the cluster are droughts and water, which together were closely linked. The keywords drought and water were linked to physiological responses such as photosynthesis, transpiration, and gaseous exchange. Most physiological response studies were conducted on roots, seedlings, and leaves, particularly stomata. This finding is not surprising since the leaf is the site for photosynthesis, and the stoma is the site for transpiration and gaseous exchange.

The strength of the connection between droughts and trees is weak, as denoted by the distance between both nodes. This may show that most researchers are concerned about crops' responses since they directly impact food security.

Although it is positioned in the upper extremity of the cluster, the keyword 'carbon dioxide' has been cited in many papers, as illustrated by the large size of the node. This may indicate that researchers acknowledge the significance of carbon dioxide concentration in water use and drought mitigation studies.

In research on response to water stress, plants included dicotyledons, Arabidopsis, *Vitis*, barley *(Hordeum)*, fruits, and *Lycopersicon* (tomato). All the plants in the network map were situated in the lower proximity of the green cluster. Studies that cited fruits as a keyword were connected to keywords in other clusters, except for the key terms 'water stress' and 'lycopersicon', which belong to the green cluster. The importance of fruits in food security cannot be underestimated; hence, this map features research on the effects of water stress on fruits.

Located on the outskirts of the map is the key term 'gene expression regulation'. Keywords cited together with 'gene expression regulation' are drought, water, physiology, metabolism, and genetics. The key term and its linkages suggest regulation of gene expression as another angle of tackling the plant's response to drought. The approach was associated with Arabidopsis, an important model plant in genetic experiments for its small genome that has already been sequenced. The plant is small and does not occupy much space in growth facilities. Arabidopsis also has high proliferation, which makes it an ideal plant for genetic studies.

(b) *Blue Cluster*: The blue cluster in the middle of the map consisted of key terms from research that focussed on the interactions between environmental elements and water use. The cluster contains key terms such as water use efficiency, climate change, ecosystem, temperature, rainfall, global warming, and evapotranspiration. The water use efficiency node represents the nucleus of the map, showing its central relationship with all terms from the three clusters in the map. The node's size also denotes the large number of articles that have cited it (1017 occurrences).

The studies in the blue cluster were conducted in the United States and Australia. North America, in particular, was linked to studies on water use efficiency and evapotranspiration.

Climate change, the second largest node (205 occurrences), was linked to terms inside and outside the cluster. Inside the blue cluster, climate change had strong linkages with terms such as water use efficiency, ecosystems, evapotranspiration, and rainfall. Externally, climate change studies were associated with studies on physiological responses such as carbon dioxide enrichment, transpiration and photosynthesis (green cluster) and water use, agricultural management, productivity and food security in the red cluster.

Grasses (Poaceae) and trees are particularly dominant in this cluster. The only linkage of grasses that existed was with the term 'water use efficiency'. Trees (forestry) were used collaboratively with water use efficiency and evapotranspiration and externally with transpiration, biomass, and drought in the green cluster, while soil water and water supply in the red cluster.

The eddy covariance is the only measurement technique associated with water use efficiency and evapotranspiration in the blue cluster. Although there is a linkage among the eddy covariance, water use efficiency and evapotranspiration, the distance between these nodes shows that the linkage's strength is weak; thus, the long distance indicates fewer co-occurrences between the terms.

(c) *Red Cluster*: The common thematic area created by keywords in the red cluster was water management in producing crops of agronomic value. The most cited keywords (denoted by the size of the nodes) were irrigation (436 occurrences), crop production (391 occurrences), crop yield (310 occurrences), water supply (305 occurrences), wheat (232 occurrences), water management (211 occurrences), and soil water (204 occurrences), among others. The regions that contributed to the studies were Europe, Asia (China and India), the Far East, and the semi-arid and arid regions. The crops identified in the cluster included *Gossypium hirsutum* (cotton), *Solanum tuberosum* (potato), *Triticum* (wheat), rice, and maize. From the most cited keywords, it is clear that the cluster's objective was to investigate water availability and management in areas prone to physical water scarcity to ensure high yields of food crops. Irrigation was essential to sustain agronomy in semi-arid and arid regions (denoted by linkages between these regions and the irrigation node). Deficit and drip irrigation were researched to determine their applicability in arid and semi-arid regions. Drip irrigation is located at the cluster's periphery, possibly for experimental purposes, indicating its low use in agronomy. Most of the nodes were linked with those in other clusters. For example, irrigation co-occurred with water stress and growth rate in the green cluster and water use efficiency, rainfall and evapotranspiration in the blue cluster. This indicates the interdependency of the themes to ensure global food security.

Within the context of global research on WP, the results of the analysed keywords showed three thematic research clusters that co-occur in literature, namely, crop water use (green), water and climate change (blue), and agricultural water management (red) (Supplementary information 1). On the contrary, keyword analysis from WRC-funded research on WP produced five clusters, which were water governance (red), water management (yellow), water and climate change (green), women and water (purple), and smallholder irrigation (blue cluster) (Fig. 1.4). While these thematic areas may be regarded as different from those depicted by results from the global search, they represent sub-themes of broader research areas under climate, water, and agriculture. The clearer articulation of more defined research themes within the WRC research portfolio shows that research has been

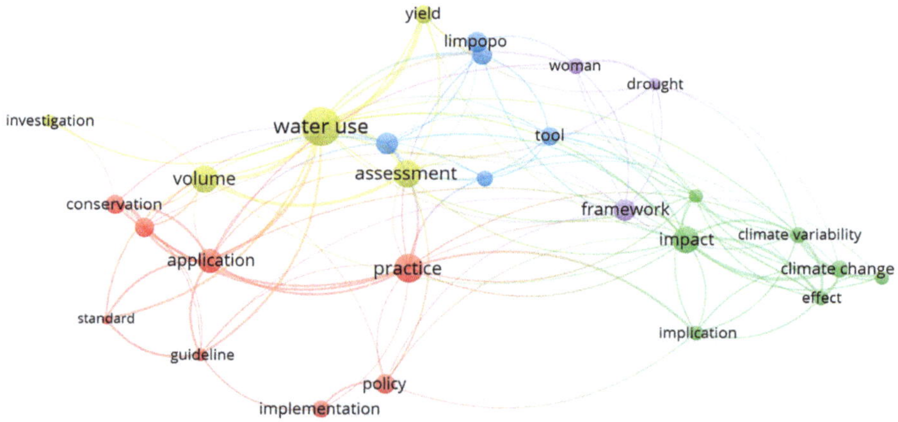

Fig. 1.4 Visualisation of the WRC-funded thematic areas assessed across 345 scientific reports on water productivity. The network map identified 30 terms to be the most relevant across identified scientific report titles. These key terms were designated into five thematic areas, namely, water governance (red), water management (yellow), water and climate change (green), women and water (purple), and smallholder irrigation (blue cluster)

contextualised around specific research areas. However, the research clusters are not all-inclusive, suggesting gaps in research focus.

1.1.6.1 Crop Water Use

Under the thematic area of crop water use (WU), the main keywords were in line with the definition of the term, which is the amount of water lost by the crop through evapotranspiration (transpiration, gas exchange, evaporation) in exchange for biomass accumulation (carbon dioxide, photosynthesis, gas exchange) (Dong et al. 2018; Kassam and Smith 2001). Crop water use (WU) is linked to the interconnectedness between plant roots and their ability to search for water in the soil, including the ability of the corresponding canopy to transpire (Morris and Garrity 1993). Plant water uptake is a function of the root density, soil–root system conductivities, and soil-available water, as determined by crop management and genotype (Ogindo and Walker 2005). An assessment of water use by crops is linked to biomass production and crop yield. Many of the keywords identified in this cluster suggested that research has been done from a physiological perspective by examining the interconnectedness between carbon uptake, growth, and water loss (Fig. 1.4). Agronomic and physiological understanding improvements have led to recent crop water productivity increases. The appearance of words relating to plants (genetics, gene expression regulation) shows that there is room for further improvements due to advances in understanding the physiological responses of plants to water supply, as well as the promise in the latest molecular genetic approaches.

Contrary to the observed global trends, only 23% (79 out of 345) of the WRC-funded projects focussed on crop water use of several fields, fodder, pasture, tree, and horticultural crops. Most of the showcased research used conventional methods of quantifying WU. On the contrary, global research has moved into the use of GIS and remote sensing (Haseeb et al. 2023).

1.1.6.2 Climate Change, Water, and Agriculture

Climate change has seen an increase in temperatures, deviations in rainfall patterns and variability, and increased intensity and frequency of droughts and floods. Furthermore, changing climate is compounding water stress and hydrologic variability, especially in semi-arid and arid regions. Under this cluster, several keywords related to climate change's impacts on water and water use efficiency (water availability, climate change, ecosystem, temperature, rainfall, global warming, and evapotranspiration). Climate change is risking water and food security, threatening to reverse the progress achieved in the past and the attainment of developmental outcomes like poverty reduction and sustainable development. It is generally understood that the water profiles of many water-stressed countries will worsen due to increasing temperature and temperature extremes. As such, most of the research articles under this cluster aimed at understanding the mechanisms of climate variability and climate change on water resources and crop yield. These articles showed the importance of multidisciplinary studies that involve agronomy, climatology, and hydrology.

Similarly, South Africa's research has focussed on climate change's impacts on water (green), emphasising climate-smart technologies to enhance WP. For instance, reports by Mabhaudhi and Modi (2016, 2019, 2020) suggest using underutilised crop species to increase crop water use in areas exposed to high climate risks. Crops that have been considered in water use efficiency projects included indigenous tree and crop species (Dye 2008; Oelofse 2008; Everson et al. 2015; Modi 2013, 2017; Modi and Mabhaudhi 2020), fruit trees such as apple, pomegranate, avocado, and macadamia (Dzikiti 2018; Taylor 2021), crops and trees for biofuels (Jewitt 2009), vegetables (Korsten 2015), and pastures (Truter et al. 2016). Many of these projects went on further to investigate the interaction between water use and biomass (Dye 2008; Gush 2014) and value chains (Grove 2012). Overall, water availability and subsequent use have highlighted the potential to alleviate poverty among disadvantaged groups through realising gender equity, capacity building, and entrepreneurial development.

Food and nutrient security have been at the forefront of most Agricultural studies funded by the WRC has been focussing on food, water, and nutrition security. A case study approach was applied on research focussing on improving food security (Chitja 2015; Denison 2015) and a scoping review approach (Wenhold et al. 2012). The topics aimed to empower vulnerable groups including women and children (Chitja 2015; Mudhara 2020) and smallholder farmers (Chitja 2015; Denison 2015). However the scoping study focussed on the whole country (Wenhold et al. 2012). Research on past and present 'food value chains' has also been key to improving water, food and nutritional security among the most vulnerable groups (de Lange 2014; Letty 2014). Studies on food security proposed future studies on developing pathways that promote the bottom-up approach starting from the grassroots level (Chitja 2020). Smallholder farming, particularly at the homestead food gardening level, has been critical for subsistence food security with no option of generating income. Titles with the phrase 'homestead food gardening' indicated the attempt of researchers to grow homestead food gardening to contribute to household food security.

1.1.6.3 Agricultural Water Management

The last cluster consisted of articles that focussed on agricultural water management. Overall, global research is stratified into research looking into (i) methods to increase crop yield for enhanced WP (agronomy, fertiliser), (ii) a decrease in water losses through soil evaporation (mulching, drip and deficit irrigation), and (iii) soil water storage (soil water, moisture). These thematic areas align with the research focus across many WRC-funded projects.

Irrigation Water Management

In South Africa, irrigation plays an important role in ensuring food security as it stabilises food production by covering up the water deficit due to unreliable rainfall. Irrigation supports 25–30% of South Africa's agricultural production as the subsector sustains about 90% of high-value crops (including potatoes, vegetables and fruit) and 25–40% of industrial crops (including sugarcane and cotton). Since its inception in 1971, the WRC has been supporting research that improves overall irrigation efficiency for sustained water, food, and nutrition security. Within the context of irrigation and irrigation efficiency, WRC-funded research has extensively covered the management, appropriateness, economics, and sustainability of irrigated agriculture (Figs. 1.3 and 1.4).

For instance, project TT465/10 suggested assessing irrigation efficiency from the source to the root zone (Reinders 2010). Reinders et al. (2012a, b) assessed the cost-effectiveness of large- and small-scale irrigation systems. Several irrigation technologies have also been evaluated (Reinders 2004). Other studies have applied models and measurement tools for evaluating water flow (Hlela-Mwanyama 2004) and the performance of filters (Volschenk et al. 2003). Decision support tools, such as OPERA (de Clercq 2019), SAPWAT (van Heerden 2008, 2020), PLANWAT (van Heerden 2008), and FARMS (Volschenk 2005) have been proposed, while (Singels 2008) offered a tool which offers real-time advice to farmers in issues on irrigation water management. The projects further developed and tested models that determined the economic effectiveness of water rationing in irrigation systems (Pott 2012).

Smallholder Irrigation Schemes

In South Africa, smallholder irrigation schemes cover about 3.3% of the total irrigated area (1.5 million ha). It is an important sector for job creation and poverty reduction. However, most smallholder irrigation schemes have collapsed while the rest suffer reduced efficiency for various reasons including incorrect water allocation, poor leadership among elected representatives, lack of understanding of governance issues, etc. Due to these schemes' importance, effective revitalisation is extremely important. According to van Verbeke (2012), the WRC first enquired into smallholder irrigation schemes in 1985, more than 10 years after their establishment. Legoupil (1985) concluded that smallholder Irrigation is struggling to produce high yields due to technical, management, training, agricultural policy, and financing issues. It was only in the late 1990s and early 2000s when research around smallholder irrigation schemes and the need to revitalise them took centre stage (see De Lange 1994; IPTRID 2000; Du Plessis et al. 2002; Shah et al. 2002; Backeberg 2006).

For more than 20 years, smallholder irrigation schemes have been the subject of research from the WRC. Most projects focussed on revitalising irrigation schemes indicate that the WRC is at the forefront of these efforts as shown by the theme's prominence (Fig. 1.3) and related key terms (Fig. 1.4). These studies have been done to address challenges related to rural poverty and unemployment and increased food and nutrition insecurity in former homelands (Zegeye and Chipfupa 2018). Denison (2007) studied 317 schemes showing the limitations of prior efforts to revitalise them including limited consultation, engagement with the intended beneficiary, and human and social capital. Some of the projects mapped the empowerment and development pathways associated with water productivity with the smallholder schemes to move towards commercial-based crop production (Jiyane 2019).

Aquaculture

Freshwater aquaculture is also key to economic development and food security in rural areas. It is important in sustainable agriculture as water is recirculated within the systems, partly addressing the challenge of water scarcity. According to Rouhani and Britz (2004), during the early 1980s, several fish hatcheries and production units were set up in the former homelands (Gazankulu, Ciskei, Transkei, and Venda) to contribute to food security. However, all these initiatives collapsed in the 1990s due to little or no backup in extension services, financial support, and a lack of political will. The WRC came in to support research that identified key research priority areas in the sector. An important area that was identified emphasised water quality in aquaculture (Salie 2008, 2013, 2017). These studies also singled out livestock production, and a risk-based approach for assessing livestock watering and aquaculture water quality guidelines was adopted (Moodley 2021). A smartphone application was also recently developed (Rouhani 2021) to convey a WRC manual on aquaculture to small-scale farmers.

Rainwater Harvesting

Several studies have focussed on rainwater harvesting (RWH) for agriculture and domestic water use in South Africa (Botha 2014; Mzezewa et al. 2011). However, its uptake in rural areas is very slow. Kahinda et al. (2011) highlighted the high labour requirements in constructing and maintaining RWH structures as a major constraint on the adoption of the technique. Some challenges with RWH include low human capacity and lack of economic aspects during the implementation of RWH technologies (Kahinda et al. 2007).

Water Security and Empowerment

Social protection, water insecurity, and gender are topical issues in South Africa, especially within social justice and equity. This is based on that water insecurity restricts women's participation in social protection (and related education and employment opportunities) and undermines efforts to promote health, nutrition, and food security (Mudhara 2020). The WRC has supported gender-sensitive improvements in water security by enhancing and empowering women's access to water (Denison 2015; Chitja et al. 2016; Oladele 2015, Figs. 1.2 and 1.3). WRC-funded research has focussed on increasing the capacity of marginalised communities and supporting initiatives to access markets (van Schalkwyk 2007; Korsten 2016). Overall, water security and empowerment studies focus on challenges related to water use, entrepreneurial development, and sustainable water use (Asiwe 2020; Zegeye and Chipfupa 2018).

1.1.7 Considerations for Improving Water Productivity

The key principles that have been identified to improve water productivity at any spatial scale include (i) improving the marketable yield of crops for each unit of water transpired by the same crop, (ii) reducing runoff, drainage, seepage, percolation, and evaporative), and (iii) increasing the effective use of rainfall, stored water, and water of marginal quality (Kijne 2003). These principles are applicable in both rainfed and irrigated sectors. However, options and practices associated with these principles require different approaches and technologies at different spatial scales.

1.1.7.1 Enhancing Water Productivity at the Plant Level

Plant-level options of crop-water productivity have been studied extensively and these studies relied mainly on germplasm improvements (Condon 2004; Morison et al. 2008; Richards et al. 2002, 2010; Richards 2006; Zoebl 2006). Improvements in crop-water productivity depend on early seedling establishment, improving seedling vigour, increasing rooting depth, increasing the harvest index, and enhancing photosynthetic efficiency (Table 1.1). For example, sorghum breeding has resulted in about a threefold increase in water productivity compared with traditional varieties (Xin et al. 2009). A range of crop varieties that match growth cycles with the expected water supply have been developed without any harm to both humans and the crop (Richards et al. 2010). Short- to medium-duration varieties are increasing water productivity as they escape late-season water stress (Kijne 2003). Breeding for and selecting crop species through enhanced architecture is also improving water productivity as deep root systems increase water uptake in deeper horizons while increasing adventitious roots will allow more water to be captured in the top layer before evaporative losses (Kell 2011; Richards et al. 2002). These advances are envisaged to be applied to many crop types with the aim of not only improving water productivity but also improving productivity and water and food security (Bennett 2003). This is based on the fact that drought tolerance and escape have been strategic in increasing water productivity and crop yield (Araus et al. 2002; Kulathunga 2013).

Table 1.1 Principles, strategies, options, and practices for enhancing crop water productivity at the plant, field, and basin scales

Principle	Strategy	Scale: Plant	Scale: Field	Scale: Basin
Enhancing the marketable yield of crops for each unit of crop transpiration	Increasing the yield or value of the product	Increasing harvest index, increasing photosynthesis, increasing sink strength	Crop and resource management for enhancing yield, synchronising water application with crop water demand, changing to high-value crops	Improving water management to synchronise system water supply and field-level water demand, reallocating water from value to higher-value uses, spatial analyses for maximum production and minimum transpiration
	Reducing transpiration	Reducing non-stomatal transpiration, reducing stomatal transpiration, shortening crop growth duration	Crop scheduling to match season with low evaporative demand; deficit irrigation	
Reducing non-beneficial atmospheric depletions and the outflows from the domain of interest	Reducing evaporation from soil and water	Early shading, seedling vigour	Crop scheduling to reduce evaporation during the fallow period, plant spacing and row orientation, tillage and soil management (e.g. minimum tillage, mulching) to reduce evaporation, Irrigation techniques (e.g. drip, subsurface irrigation), saturated culture with rice on a bed	Land use planning over the whole domain of interest reduces evaporation from fallow land and decreases free water surface
	Reducing transpiration from weeds	Increasing weed competitiveness	Weed management, levelling and precision irrigation, water-saving irrigation in rice	Land use planning to reduce weeds and other non-beneficial vegetation
	Reducing percolation	Seedling vigour, deep roots, aerobic rice	Levelling and precision irrigation, water-saving irrigation in rice	Reducing percolation and runoff into sinks, water reuse
	Reducing runoff		Water harvesting, tillage to increase infiltration	

Enhancing the effective use of rainfall, water with marginal quality and water stored in the domain of interest	Effective use of rainfall	Drought escape, drought tolerance, submergence tolerance	Risk management in rain-fed agriculture, synchronising crop demand and rainfall, nutrient management to reduce drought effects, drainage	Irrigation scheduling to account for rainfall variability, utilisation of medium and long-term weather forecasts for reducing risk
	Effective use of water storage	Deep rooting for drought avoidance	Water harvesting and supplementary irrigation	Conjunctive use of surface water and groundwater, increasing water storage within the domain to capture runoff
	Effective use of water with marginal water quality	Salinity stress tolerance	Mixing marginal water with water of good quality, crop management to reduce salinity effects	Land management to reduce salinisation hazard

Source: Kijne et al. (2003)

1.1.7.2 Improving Water Productivity at the Field Level

Improvements in crop, soil, and water management at the field level have been shown to enhance water productivity under irrigated and rainfed agriculture. These improvements include the selection of the right crops and cultivars, planting methods that reduce tillage, irrigation scheduling, nutrient management, and improved drainage for water table control (Table 1.1). In rainfed agriculture where farmers face the risk of intra-seasonal drought or rainfall variability, crop water productivity is improved by selecting adapted water-efficient crops that reduce unproductive water losses and ensuring optimum agronomic conditions for crop production, (Chimonyo et al. 2016a, b; Kijne 2003; Mabhaudhi et al. 2018; Rockström et al. 2003) Table 1.1.

Rainwater harvesting plays an important role in enhancing crop-water productivity (Kahinda et al. 2011). Coupled with soil conservation practices, RWH increases land productivity in terms of crop-water productivity (Mupangwa et al. 2006). The main RWH techniques are micro- and macro-catchment, which are indicated in Table 1.1. The promotion of RWH is based on the fact that irrigation currently uses more water than all other users, and agriculture faces competing demands for water from other sectors. About 1.3 million hectares, or under 10% of all arable land, are under Irrigation (Cousins 2007; Nhamo et al. 2016).

While the National Development Plan (NDP) projects to increase the area under irrigation to about 200,000 ha, there is a risk of straining other sectors like energy which is already failing to meet the national requirements (Cai et al. 2017). The construction of new dams may not be viable as the country is already over-dammed (Mabhaudhi et al. 2018). *Deficit irrigation* is an alternative but smallholder farmers lack the required knowledge yet it enhances water productivity (WHO 2003). Research has also shown that *drip irrigation* increases water use efficiency more than any other type of irrigation; its yield gains can be 100% and water savings about 40–80% (Rao et al. 2016; Ali and Talukder 2008; Zwart and Bastiaanssen 2004). *Supplementary Irrigation* is an option to reduce water stress in rainfed agriculture as it provides sufficient moisture for normal plant growth to improve and stabilise yields (ICID 2012). However, it depends on the precipitation of a basic source of water for the crop (Grafton et al. 2018; Guendouz et al. 2016; Li and Sun 2016; Zhang et al. 2017a, b; Steduto et al. 2012). Supplementary irrigation significantly increases crop-water productivity if water is applied at the moisture-sensitive stages of plant growth (Oweis and Hachum 2006; Ali and Talukder 2008).

1.1.7.3 System and Basin Level Water Productivity

Agricultural water management needs to advance and support ecological and human services under climate change (Rockström et al. 2010; Menéndez et al. 2016). Water users need to look beyond their farms and consider other users (Faurès et al. 2003; Kijne 2003). Upstream users need to also consider users downstream. The challenges require cross-sectoral interventions and move away from the current linear approaches (Blignaut et al. 2007; Griggs and Golet 2002; Molden et al. 2010).

Further, water management practitioners need to appreciate the scales at which water productivity interventions can be implemented.

There are four key strategies for improving water productivity at the basin level, including (i) increasing water productivity at plant and field levels, (ii) reducing non-productive loses of water flows by minimising water runoffs to sinks, (iii) improving the management of existing irrigation facilities and reusing return flows by controlling, diverting and storing drainage flows, and (iv) identifying water users and, relocate and allocate water among uses based on the value of use, e.g. relocating lower value to higher-value uses within and between sectors (Mabhaudhi et al. 2018; Mpandeli et al. 2018).

1.1.7.4 Policy and Water Productivity

Nhamo et al. (2018) explored opportunities for the WEF nexus to promote cross-sectoral policy linkages among the water, energy, and food sectors to achieve regional integration and sustainable development. The main recommendations were on transboundary water management for improved resource use efficiency. Mabhaudhi et al. (2018) assessed the status of irrigated agriculture in southern Africa from a water–energy–food (WEF) nexus perspective. They emphasised the need to increase water storage and human capacity and broaden the energy base to increase the area under irrigation. The WEF nexus approach addresses the multifaceted and interrelated nature of resource systems for any intended outcome or impact (Bizikova et al. 2013; Entholzner and Reeve 2016; Mpandeli et al. 2018).

The prevalent governance systems in an area determine access to water. Studies have focussed on climate change highlighting drought and its response in southern Africa (Davies 2000; Vogel et al. 2010; Vogel and Olivier 2019). The emphasis has been on assessing the administrative role of institutions and governance systems (Baudoin et al. 2017), interrogating past response mechanisms to reduce risk at different spatial governance scales. The most highlighted subject has been moving from reactive water management to a more proactive approach guided by working and strong water governance systems (Vogel et al. 2010). One example of a more proactive approach to water management is the National Water Act (NWA; Act 36 of 1998) and Water Services Act (WSA; Act 108 of 1997), administered by the Department of Water and Sanitation (DWS) at its core (RSA 1997, 1998; Vogel and van Zyl 2016). A significant gap, already noted by Hornby et al. (2016), appears to be the coordination of the widespread but localised efforts by the government, civil society, and private sector, particularly in identifying and responding to the areas and people most in need.

1.1.8 Water Management and the SDGs

The 2030 Agenda and Sustainable Development Goals (SDGs) are threaded together by a common denominator: water. Although SDG 6, the Water Goal, specifically addresses promoting sustainable water management, water is embedded in the rest of the SDGs, particularly those that focus on food, energy and the environment (Ait-Kadi 2016). Sadoff et al. (2015) established the symbiotic association between water and development. As such, achieving SDG 6 can be realised by fulfilling the other SDGs and vice versa. The WRC has also taken the same notion. The WRC research over the years speaks to the role of water in development. Improving water productivity for food security addresses SDG 2, which calls for zero hunger. The theme 'water security and empowerment' addresses gender disparity through women empowerment (SDG 5) by highlighting women's roles in water management. The thematic area' climate change water and agriculture' links with SDG 13, which mandates nations to work actively against climate change. SDG 14, which calls for sustainable marine resources, has been acknowledged in the WRC research theme, aquaculture. The agricultural water management research area has fulfilled SDGs 1 to 6 directly or indirectly.

1.1.9 Recommendations

Underwater scarcity, the success of agriculture lies in increasing water productivity. While food security is a priority, an integrated approach to addressing water productivity will ensure trade-offs with the environment and socio-economic constructs. The following recommendations are suggested:

- Policies and strategies must create an enabling environment for supporting investments in irrigation development and agricultural water management.
- Approaches should be transdisciplinary and involve stakeholders such as scientists, farmers, and public, private and civil partners.
- Capacity building at multiple levels is needed to support implementation and transformation at different scales.
- Promoting good agronomic practices, including climate-smart agriculture techniques that increase crop productivity and enhance efficient resource use under rainfed and irrigated conditions.

References

Ait-Kadi M (2016) Water for development and development for water: realizing the sustainable development goals (SDGs) vision. Aquat Procedia 6:106–110

Albort-Morant G, Ribeiro-Soriano D (2016) A bibliometric analysis of international impact of business incubators. J Bus Res 69(5):1775–1779

Ali MH, Talukder MSU (2008) Increasing water productivity in crop production—a synthesis. Agric Water Manag 95:1201–1213. https://doi.org/10.1016/j.agwat.2008.06.008

Allen T, Prosperi P (2016) Modeling sustainable food systems. Environ Manag 57:956–975

Araus JL, Slafer GA, Reynolds MP, Royo C. (2002) Plant breeding and drought in C3 cereals: what should we breed for?. Annals of botany. 89(7):925-40. https://doi.org/10.1093/aob/mcf049

Asiwe J (2020) Enhancing food security and nutrition of selected rural communities in Limpopo Province using high yielding and water use efficient grain legume varieties. WRC project no: TT 829/1/20

Backeberg GR (2006) Water institutional reforms in South Africa. Water Policy 7:107–123

Baudoin MA, Vogel C, Nortje K, Naik M (2017) Living with drought in South Africa: lessons learnt from the recent El Niño drought period. Int J Disaster Risk Reduct 23:128–137

Bennett J (2003) Opportunities for increasing water productivity of CGIAR crops through plant breeding and molecular biology. In: Water productivity in agriculture: limits and opportunities for improvement. CABI publishing, Wallingford, pp 103–126

Bizikova L, Roy D, Swanson D, Venema HD, McCandless M (2013) The water-energy-food security nexus: Towards a practical planning and decision-support framework for landscape investment and risk management. International Institute for Sustainable Development. https://www.researchgate.net/publication/356879079_The_Water-Energy-Food_Security_Nexus_Towards_a_practical_planning_and_decision-support_framework_for_landscape_investment_and_risk_management_for_Sustainable_Development

Blignaut JN, Marais C, Turpie JK (2007) Determining a charge for the clearing of invasive alien plant species (IAPs) to augment water supply in South Africa. Water SA 33. https://doi.org/10.4314/wsa.v33i1.47868

Botha J (2014) Rainwater harvesting and conservation for rangeland and cropland productivity in communal areas. WRC report no: 1775/1/14. Pretoria

Cai X, Molden D, Mainuddin M, Sharma B, Ahmad M ud D, & Karimi P. (2011). Producing more food with less water in a changing world: assessment of water productivity in 10 major river basins. Water International, 36(1), 42–62. https://doi.org/10.1080/02508060.2011.542403

Cai X, Magidi J, Nhamo L, van Koppen B (2017) Mapping irrigated areas in the Limpopo Province, South Africa. Colombo, Sri Lanka: International Water Management Institute (IWMI).. 37p. (IWMI Working Paper 172) https://doi.org/10.5337/2017.205

Chakona G, Shackleton CM (2019) Food insecurity in South Africa: to what extent can social grants and consumption of wild foods eradicate hunger? World Dev Perspect 13:87–94

Chimonyo VGP, Modi AT, Mabhaudhi T (2016a) Assessment of sorghum–cowpea intercrop system under water limited conditions using a decision support tool. Water SA 42:316. https://doi.org/10.4314/wsa.v42i2.15

Chimonyo VGP, Modi AT, Mabhaudhi T (2016b) Simulating yield and water use of a sorghum–cowpea intercrop using APSIM. Agric Water Manag 177:317–328

Chitja JM (2015) Empowerment of women through water use security, land use security and knowledge generation for improved household food security and sustainable rural livelihoods in selected areas of Limpopo Province. WRC project no: 2082/1/15

Chitja JM (2020) Water use for food and nutrition security at the start-up of food value chains. WRC project no: 255/1/20

Chitja J, Mthiyane CCN, Mariga IK, Shimelis H, Murugani VG, Morojele PJ, Aphane OD (2016) Empowerment of women through water use security, land use security and knowledge generation for improved household food security and sustainable rural livelihoods in selected areas of Limpopo: report to the Water Research Commission. Water Research Commission

Chivenge P, Mabhaudhi T, Modi AT, Mafongoya P (2015) The potential role of neglected and underutilised crop species as future crops under water scarce conditions in Sub-Saharan Africa. International Journal of Environmental Research and Public Health, 12(6):5685–5711. https://doi.org/10.3390/ijerph120605685

Cobo MJ et al (2011) An approach for detecting, quantifying, and visualizing the evolution of a research field: a practical application to the fuzzy sets theory field. J Informet 5(1):146–166

Condon AG (2004) Breeding for high water-use efficiency. J Exp Bot 55:2447–2460. https://doi.org/10.1093/jxb/erh277

Council for Scientific and Industrial Research (CSIR) (2010) A CSIR Perspective on Water in South Africa – 2010. CSIR Report No. CSIR/NRE/PW/IR/2011/0012/A.

Cousins B (2007) More than socially embedded: the distinctive character of communal tenure regimes in South Africa and its implications for land policy. J Agrar Change, vol 7, p 281

Davies A (2000) Waste management: The duty of care—a code of practice. The Stationery Office. https://www.daera-ni.gov.uk/sites/default/files/publications/doe/duty-of-care-code-of-practice-june2016.pdf

De Clercq W (2019) Operationalizing the increase of water use efficiency and resilience in irrigation (OPERA). WRC report no: 2788/1/20

De Lange M (1994) Small scale irrigation in South Africa. WRC report no. 578/1/94

De Lange W (2014) Investigation of water conservation in food value chains by beneficiaries of water allocation reform and land reform programmes in South Africa. WRC project no: 1958/1/14

Denison JA (2007) Principles, approaches and guidelines for participatory revitalisation of smallholder irrigation schemes. WRC report no: TT308/07

Denison JA (2015) Empowerment of women through water use security, land use security and knowledge generation for improved household food security and sustainable rural livelihoods in selected areas of the Eastern Cape Province. WRC project no: 2083/1/15

Dong J, Steele-Dunne SC, Ochsner TE, van der Schalie R, Wen J, van de Giesen N (2018) Use of satellite soil moisture to diagnose climate model representations of European soil moisture–air temperature coupling strength. Geophysical Research Letters 45(22)12:875–12,885. https://doi.org/10.1029/2018GL080547

Du Plessis FJ, Van Averbeke W, Van Der Stoep I (2002) Micro-irrigation for smallholders: guidelines for funders, planners, designers and support staff in South Africa. WRC report no: TT 164/01

Dye P (2008) Water use in relation to biomass of indigenous tree species in woodland, forest and/or plantation conditions. WRC report no: TT361/08

Dzikiti S (2018) Quantifying water use of high performing commercial apple orchards in the winter rainfall area of South Africa WRC project no: TT 751/18

Entholzner A, Reeve C (2016) Building climate resilience through virtual water and nexus thinking in the Southern African Development Community

Faurès JM, Hoogeveen J, Bruinsma J (2003) The FAO irrigated area forecast for 2030. FAO, Citeseer, Rome

Grafton RQ et al (2018) The paradox of irrigation efficiency. Science 80(361):748–750. https://doi.org/10.1126/science.aat9314

Griggs FT, Golet GH (2002) Riparian valley oak (Quercus lobata) forest restoration on the Middle Sacramento River, California, In: Standiford, Richard B, et al. Proceedings of the fifth symposium on Oak Woodlands: Oaks in California's challenging landscape. Gen. Tech. Rep. PSW-GTR-184, Albany: Pacific Southwest Research Station, Forest Service, US Department

Grove B (2012) Assessment of the contribution of water use to value chains in agriculture. WRC report no: 1779/1/12

Guendouz A et al (2016) The effect of supplementary irrigation on leaf area, specific leaf weight, grain yield and water use efficiency in durum wheat (Triticum durum Desf.) cultivars. Ekin J Crop Breed Genetic 2:82–89

Gush MB (2014) Water use of fruit tree orchards. WRC report no: 1770/1/14

Hardy M et al (2011) Rainfed farming systems in South Africa. In: Tow P, Cooper I, Partridge I, Birch C (eds) Rain fed farming systems. Springer Netherlands, Dordrecht. https://doi.org/10.1007/978-1-4020-9132-2

Haseeb M et al (2023) Quantifying irrigation water demand and supply gap using remote sensing and GIS in Multan. Pak Environ Monit Assess 195(8):990

Hlela-Mwanyama O (2004) Investigation into the potential of sustainable irrigation in black developing communities of two sub-catchments of the Pongola and Thukela Rivers. Water Research Commission, Pretoria. WRC report, (1138/1), 04

Hoekstra AY (2017) Water footprint assessment: evolvement of a new research field. Water Resour Manag 31(10):3061–3081

Howell TA (2003) Irrigation efficiency. In: Encyclopedia of water science, vol 467. Marcel Dekker, New York, p 500

ICID (2012) Supplementary irrigation

IFPRI (2016) From promise to impact: ending malnutrition by 2030. In: International Food Policy Research Institute (IFPRI), Washington
IPTRID (2000) Affordable irrigation technologies for smallholders: opportunities for technology adaptation and capacity building. FAO, Rome, p 35
Irmak S, Odhiambo LO, Kranz WL, Eisenhauer DE (2011) Irrigation efficiency and uniformity, and crop water use efficiency. Circular EC732. University of Nebraska Extension, Lincoln
Jewitt GPW (2009) Scoping study on water use of crops trees for biofuels. WRC report no: 1772/1/09
Jiyane J (2019) Investigating the factors influencing under-utilisation of existing smallholder irrigation schemes and opportunities for improved future use in Limpopo Province WRC report no: TT 787/19
Kahinda JMM, Taigbenu AE, Boroto JR (2007) Domestic rainwater harvesting to improve water supply in rural South Africa. Phys Chem Earth, Parts A/B/C 32(15–18):1050–1057
Kahinda J et al (2011) Rainwater harvesting in South Africa: challenges and opportunities. Phys Chem Earth, Parts A/B/C 36:968–976. https://doi.org/10.1016/j.pce.2011.08.011
Kassam A, Smith M (2001) FAO methodologies on crop water use and crop water productivity. In: Expert meeting on crop water productivity, Rome, p 18
Kell DB (2011) Breeding crop plants with deep roots: their role in sustainable carbon, nutrient and water sequestration. Ann Bot 108:407–418. https://doi.org/10.1093/aob/mcr175
Kijne JW (2003) Unlocking the water potential of agriculture. Food & Agriculture Org
Kijne JW et al (2003) Ensuring food security via improvement in crop water productivity. Chall Progr Water Food Backgr Pap 1:20–26
Klavans R, Boyack KW (2006) Identifying a better measure of relatedness for mapping science. J Am Soc Inf Sci Technol 57:251–263
Korsten L (2015) An investigation into the link between water quality and microbiological safety of fruit and vegetables from the farming to the processing stages of production and marketing. WRC project no: 1875/1/15
Korsten I (2016) Evaluation of the risks associated with the use of rain-water, harvested from roof tops, for domestic use and homestead food gardens; and groundwater for domestic use and livestock watering. WRC project no: 2175/1/16
Kulathunga MRDL (2013) Traits associated for adaptation to water limited environment of cereal crops a review of literature. Int J Sci Technol Res 2:73–81
Lahiff E, Cousins B (2005) Smallholder agriculture and land reform in South Africa. IDS Bull 36(2):127–131
Legoupil JC (1985) Some comments and recommendations about irrigation schemes in South Africa: report of mission, 11 February–3 March 1985. Water Research Commission, Pretoria, South Africa, p 35
Letty BA (2014) Analysis of food value chains in rain-fed and irrigated agriculture to include emerging farmers in the mainstream of the economy. WRC project no: 1879/1/14
Li Z, Sun Z (2016) Optimized single irrigation can achieve high corn yield, and water use efficiency in the Corn Belt of Northeast China. Eur J Agron 75:12–24. https://doi.org/10.1016/j.eja.2015.12.015
Mabhaudhi T, Chibarabada TP, Modi AT (2016) Water–food–nutrition–health nexus: Linking water to improving food, nutrition and health in Sub-Saharan Africa. International Journal of Environmental Research and Public Health, 13(1):107. https://doi.org/10.3390/ijerph13010107
Mabhaudhi T, Chimonyo VGP, Modi AT (2017) Status of underutilised crops in South Africa: Opportunities for developing research capacity. Sustainability, 9(9), 1569. https://doi.org/10.3390/su9091569
Mabhaudhi T et al (2018) Prospects for improving irrigated agriculture in southern Africa: linking water energy and food. Water 10:1881. https://doi.org/10.3390/w10121881
Mabhaudhi T, Chibarabada TP, Chimonyo VGP, Murugani VG, Pereira LM, Sobratee N, Govender L, Slotow R, Modi AT (2019) Mainstreaming underutilized indigenous and traditional

crops into food systems: A South African perspective. Sustainability, 11(1):172. https://doi.org/10.3390/su11010172

Mabhaudhi T, Mpandeli S, Nhamo L, Senzanje A, Chimonyo VGP, Modi AT (2019, September) Options for improving agricultural water productivity under increasing water scarcity in South Africa. In 3rd World Irrigation Forum (WIF3) on development for water, food and nutrition security in a competitive environment, Bali, pp 1–7

Masipa TS (2017) The impact of climate change on food security in South Africa: current realities and challenges ahead. Jàmbá J Disaster Risk Stud 9:411. https://doi.org/10.4102/jamba.v9i1.411

Mekonnen MM, Hoekstra AY (2016) Four billion people facing severe water scarcity. Sci Adv 2:e1500323

Menéndez C et al (2016) Hydrological cycle, temperature, and land surface-atmosphere interaction in the La Plata Basin during summer: response to climate change. Clim Res 68:231–241. https://doi.org/10.3354/cr01373

Mensah J, Ricart Casadevall S (2019) Sustainable development: meaning, history, principles, pillars, and implications for human action: literature review. Cogent Soc Sci 5:1653531

Modi AT (2013) Water use of drought tolerant food crops. WRC report no: 1771/1/13

Modi AT (2017) Determining water use of indigenous grain and legume food crops WRC report no: TT 710/17

Modi AT, Mabhaudhi T (2020) Water use of crops and nutritional water productivity for food production, nutrition and health in poor rural communities. WRC project no: 2493/1/20

Molden D et al (2003) A water-productivity framework for understanding and action. In: Water productivity in agriculture: limits and opportunities for improvement. Cabi Publishing, Wallingford, pp 1–18. https://doi.org/10.1079/9780851996691.0001

Molden D et al (2010) Improving agricultural water productivity: between optimism and caution. Agric Water Manag 97:528–535. https://doi.org/10.1016/j.agwat.2009.03.023

Moodley P (2021) Development of a risk based approach for assessing livestock watering and aquaculture water quality guidelines

Morris RA, Garrity DP (1993) Resource capture and utilization in intercropping: Non-nitrogen nutrients. Field Crops Research, 34(3–4):319–334. https://doi.org/10.1016/0378-4290(93)90119-8

Morison JI et al (2008) Improving water use in crop production. Philos Trans R Soc B Biol Sci 363:639–658. https://doi.org/10.1098/rstb.2007.2175

Mpandeli S et al (2018) Climate change adaptation through the water-energy-food nexus in Southern Africa. Int J Environ Res Public Health 15. https://doi.org/10.3390/ijerph15102306

Mudhara M (2020) Assessment of the effectiveness of policies and strategies for governance of smallholder irrigation farming in KwaZulu-Natal Province, South Africa. WRC project no: 2556/1/20

Mupangwa W, Love D, Twomlow S (2006) Soil–water conservation and rainwater harvesting strategies in the semi-arid Mzingwane Catchment, Limpopo Basin, Zimbabwe. Phys Chem 31:893

Mzezewa J, Gwata ET, Van Rensburg LD (2011) Yield and seasonal water productivity of sunflower as affected by tillage and cropping systems under dryland conditions in the Limpopo Province of South Africa. Agric Water Manag 98:1641–1648. https://doi.org/10.1016/j.agwat.2011.06.003

National Planning Commission (2012) National development plan 2030: our future–make it work

Nhamo L et al (2016) The impact of investment in smallholder irrigation schemes on irrigation expansion and crop productivity in Malawi. Afr J Agric Resour Econ 11:141–153

Nhamo L et al (2018) The water-energy-food nexus: climate risks and opportunities in southern Africa. Water 10(5):567. https://doi.org/10.3390/w10050567

Nhamo L et al (2024) Why do farmers not irrigate all the areas equipped for irrigation? Lessons from Southern Africa. Agriculture 14:1218. https://doi.org/10.3390/agriculture14081218

Nyathi S, Olowoyo JO, Oludare A (2019) Perception of scavengers and occupational health hazards associated with scavenging from a waste dumpsite in Pretoria, South Africa. Int J Environ Res Public Health 16:4277

Oelofse A (2008) Nutritional value and water use of indigenous crops for improved rural livelihoods. WRC report no: TT362/P/08

Ogindo H, Walker S (2005) Comparison of measured changes in seasonal soil water content by rainfed maize-bean intercrop and component cropping systems in a semi-arid region of southern Africa. Phys Chem Earth, Parts A/B/C 30(11–16):799–808

Oki T, Kanae S (2006) Global hydrological cycles and world water resources. Science 313:1068–1072

Oladele O (2015) Empowerment of women in rural areas through water use security and agricultural skills training for gender equity and poverty reduction in KwaZulu-Natal and North West Province. WRC project no: 2176/1/16

Oweis T, Hachum A (2006) Water harvesting and supplemental irrigation for improved water productivity of dry farming systems in West Asia and North Africa. Agric Water Manag 80:57–73. https://doi.org/10.1016/j.agwat.2005.07.004

Pott A (2012) The development and testing of an integrated set of models to evaluate the financial/economic impact of irrigation water curtailment decisions on participant farm case studies in the crocodile catchment. WRC report no: 1805/1/12

Rao KVRB et al (2016) Maximising water productivity of wheat crop by adopting drip irrigation. Res Crop 17:163. https://doi.org/10.5958/2348-7542.2016.00029.2

Reinders FB (2004) Surface drip irrigation. WRC report no: 1189/1/04

Reinders FB (2010) Standards and guidelines for improved efficiency of irrigation water use from dam wall release to root zone application. WRC report no: TT465/10

Reinders F et al (2012a) Technical aspects and cost estimating procedures of surface and subsurface drip irrigation systems. Report no. TT 526/12. Water Research Commission, Pretoria

Reinders FB et al (2012b) Technology transfer on the technical aspects and cost estimating procedures of surface and sub-surface drip irrigation systems. WRC report no: TT524/12

Rey-Martí A, Ribeiro-Soriano D, Palacios-Marqués D (2016) A bibliometric analysis of social entrepreneurship. J Bus Res 69(5):1651–1655

Richards RA (2006) Physiological traits used in the breeding of new cultivars for water-scarce environments. Agric Water Manag 80:197–211. https://doi.org/10.1016/j.agwat.2005.07.013

Richards R, Rebetzke G, Condon A (2002) Breeding opportunities for increasing the efficiency of water use and crop yield in temperate cereals. Crop Sci 42(1):111–121. https://doi.org/10.2135/cropsci2002.0111

Richards R et al (2010) Breeding for improved water productivity in temperate cereals: phenotyping, quantitative trait loci, markers and the selection environment. Funct Plant Biol 37:85. https://doi.org/10.1071/FP09219

Rockström J, Barron J, Fox P (2003) Water productivity in rain-fed agriculture: challenges and opportunities for smallholder farmers in drought-prone tropical agroecosystems. In: Kijne JW, Barker R, Molden D (eds) Water productivity in agriculture: limits and opportunities for improvement. CAB International, pp 145–162

Rockström J et al (2010) Managing water in rainfed agriculture—the need for a paradigm shift. Agric Water Manag 97:543–550. https://doi.org/10.1016/j.agwat.2009.09.009

Rouhani Q (2021) Developing a smart phone app for small scale fish farmers and government aquaculture extension officers to deliver an existing WRC manual for small scale farmers

Rouhani QA, Britz PJ (2004) Contribution of aquaculture to rural livelihoods in South Africa: a baseline study. WRC report no: TT 235/04

Sadoff S, Samek A, Sprenger C (2015) Dynamic inconsistency in food choice: experimental evidence from a food desert. Becker Friedman Institute for Research in Economics Working Paper (2572821)

Salie K (2008) Assessment of the interaction between aquaculture and water quality in on-farm irrigation dams. WRC report no: TT369/08

Salie K (2013) Interaction between aquaculture and water quality in on-farm irrigation dams: extended monitoring and mitigating procedures to manage environmental impact. WRC report no: 1802/1/13

Salie K (2017) Knowledge transfer on water quality management for improved integrated aquaculture and agriculture systems. WRC report no: TT718/17
Shah T et al (2002) Institutional alternatives in African smallholder irrigation: lessons from international experience with irrigation management transfer. IWMI Research Report 60. International Irrigation Management Institute, Colombo
Shiklomanov IA, (1991) "World fresh water resources." In Water in Crisis: A Guide to the World's Fresh Water Resources, edited by Peter H. Gleick, 13–24. Oxford University Press.
Singels A (2008) Real-time irrigation advice for small scale sugarcane production using a crop model, weather data and cellular communication. WRC report no: 1576/1/08
Small H (1973) Co-citation in the scientific literature: a new measure of the relationship between two documents. J Am Soc Inf Sci 24(4):265–269
South Africa Demographic and Health Survey (2017) Key indicators report. Pretoria
Stats-Sa (2019) Mid-year population estimates 2019. Statistics South Africa (StatsSA), Pretoria, p 26
Steduto P et al (2012) Crop yield response to water, FAO irrigation and drainage paper no. 66
Taylor NJ (2021) Water use of avocado and macadamia orchard
Truter et al (2016) irrigation guidelines for mixed Pastures and Lucerne, vol 20. Water Research Commission, Pretoria, pp 697–616
Unga (2015) Transforming our world: the 2030 agenda for sustainable development, resolution adopted by the General Assembly (UNGA). United Nations General Assembly, New York, p 35
Van Averbeke W, Khosa B (2007) The contribution of smallholder agriculture to the nutrition of rural households in a semi-arid environment in South Africa. Water SA 33:413–418. https://doi.org/10.4314/wsa.v33i3.49158
Van Averbeke W, Denison J, Mnkeni PNS (2011) Smallholder irrigation schemes in South Africa: a review of knowledge generated by the Water Research Commission. Water SA 37(5):797–808
Van Heerden PS (2008) Integrating and upgrading of SAPWAT and PLANWAT to create a powerful and user-friendly irrigation planning tool. WRC report no: TT391/08
Van Heerden PS (2020) Assessment of the basis of authorised water use on selected water stressed irrigation schemes in humid, semi-humid, semi-arid and arid areas through the application of SAPWAT4. WRC report no: 2822/1/20
Van Schalkwyk HD, Groenewald JA, Fraser GCG, Obi A, Van Tilburg A (Eds.) (2012) Unlocking markets to smallholders: Lessons from South Africa. Springer Science & Business Media. http://library.oapen.org/handle/20.500.12657/34544
de Villiers EM, Fauquet C, Broker TR, Bernard HU, zur Hausen H (2004) Classification of papillomaviruses. Virology. 324(1):17–27. https://doi.org/10.1016/j.virol.2004.03.033. PMID: 15183049.
Van Verbeke W (2012) Agriculture and the food industry in the information age. European Review of Agricultural Economics, 39(3):385–407. https://doi.org/10.1093/erae/jbr056
Vogel C, Koch I, Van Zyl K (2010) "A persistent truth"—reflections on drought risk management in Southern Africa. Weather Clim Soc 2(1):9–22
Vogel C, Olivier D (2019) Re-imagining the potential of effective drought responses in South Africa. Regional Environmental Change, 19(6):1561–1570. https://doi.org/10.1007/s10113-019-01499-0
Volschenk ES (2005) A new technique for examining surface morphosculpture of scorpions. J Arachnol 33(3):820–825
Vogel C, van Zyl K (2016) Drought: In search of sustainable solutions to a persistent,'wicked'problem in South Africa. Climate change adaptation strategies–An upstream-downstream perspective 195–211.
Volschenk T, De Villiers JF, Beukes O (2003) The selection and calibration of a model for irrigation scheduling of deciduous fruit orchards. WRC report no. 892/1/03. Water Research Commission, Pretoria, p 126
Wenhold F, Annandale J, Faber M, Hart T (2012) Water use and nutrient content of crop and animal food products for improved household food security: a scoping study. Water Research Commission, Pretoria

World Health Organization (2003) Diet, nutrition, and the prevention of chronic diseases: report of a joint WHO/FAO expert consultation, vol 916. World Health Organization
Xin Z, Aiken R, Burke J (2009) Genetic diversity of transpiration efficiency in sorghum. Field Crops Res 111:74–80. https://doi.org/10.1016/j.fcr.2008.10.010
Zegeye EW, Chipfupa U (2018) Appropriate entrepreneurial development paths for homestead food gardening and smallholder irrigation crop farming in KwaZulu-Natal Province: report to the Water Research Commission. Water Research Commission
Zhang H et al (2017a) Rational water and nitrogen management improves root growth, increases yield and maintains water use efficiency of cotton under mulch drip irrigation. Front Plant Sci 8:912. https://doi.org/10.3389/fpls.2017.00912
Zhang Y et al (2017b) Reduced irrigation increases the water use efficiency and productivity of winter wheat-summer maize rotation on the North China Plain. Sci Total Environ 618:112–120. https://doi.org/10.1016/j.scitotenv.2017.10.284
Zoebl D (2006) Is water productivity a useful concept in agricultural water management? Agric Water Manag 84(3):265–273. https://doi.org/10.1016/j.agwat.2006.03.002
Zwart SJ, Bastiaanssen WGMM (2004) Review of measured crop water productivity values for irrigated wheat, rice, cotton and maize. Agric Water Manag 69:115–133. https://doi.org/10.1016/J.AGWAT.2004.04.007

Chapter 2
Why Agricultural Water Productivity Matters: A Review

Chika E. Oyeagu and Francis B. Lewu

Abstract Due to the increasing population, economic development, and climate change, providing food and water security requires significant changes in the technologies that propel present-day water management. Irrigation is the largest and most inefficient water user, and there is an expectation that even small improvements in agricultural water productivity will enhance water security. Increasing water productivity is a novel practice needed to improved water management for sustainable agriculture, food security and healthy ecosystem. The explanation of water productivity involves the quantity of agricultural output per unit of water loss, which can be applied to crops, livestock and aquaculture. This chapter reviews the challenges and opportunities for enhancing water productivity in socially equitable and sustainable ways. Water productivity can be improved in irrigated and rainfed cropping systems by choosing a well-adopted crop type, reducing unproductive water losses, and maintaining healthy, vigorous growing crops through optimized water, nutrient, and agronomic management. Livestock water productivity can be increased through improved feed management and animal husbandry, reduced animal mortality, appropriate livestock watering, and sustainable grazing management. In aquaculture systems, most water is lost indirectly for feed production via seepage and evaporation from water bodies, and through polluted water discharge, and efforts to improve water productivity should be directed at minimizing those losses. The review also identifies different strategies for improving agricultural water productivity and biophysical and climate change influence on agricultural water productivity. Effort has been made to address why agricultural water productivity matters.

Keywords Agriculture · Animal husbandry · Food security · Irrigation · Livestock · Water productivity

---

C. E. Oyeagu (✉) · F. B. Lewu
Department of Agriculture, Faculty of Applied Sciences, Cape Peninsula University of Technology, Cape Town, South Africa

T. Mabhaudhi et al. (eds.), *Enhancing Water and Food Security Through Improved Agricultural Water Productivity*,
https://doi.org/10.1007/978-981-96-1848-4_2

2.1 Introduction

It is important to make efficient use of scarce water resources globally to ensure food and nutrient security with a stable ecosystem. The increase in the world's population and expansion of economic prosperity are putting more of a burden on the limited land and water resources worldwide. About 9.3 billion people are expected to be on earth by 2050 (PRB 2013). There has been a global surge in food demand in the last five decades because of the increased population growth (1.69% per annum) and low per capita consumption (Bouman 2007). Even now, agricultural activities consume more water globally than other water-consuming sectors. Among different sub-sectors of agriculture, more water is needed in the food production sub-sector. Apart from the global population increase, there is a dietary change that includes more meat (Gardner et al. 2019). The increased consumption of calories, protein, and more complex foods is due to income growth in developing countries. As a result of this pressure, it is estimated that agricultural production will need to expand by approximately 70% by 2050 to satisfy the growing demands (Akhtar et al. 2021). The most important input in the production of food and feed is water. In other words, water promotes food and nutrition security, which helps solve the global food demand. Yet many parts of the world experience a higher shortage of this water, and its availability differs widely over temporal and spatial scales (Huang et al. 2021). Again, other challenges, such as climate change and increasing competition with other users (e.g. urbanization, industrial growth, and environmental water flow requirement), increase water scarcity and its effect on food production and the health of freshwater ecosystems (Mancosu et al. 2015).

One urgent challenge in sub-Saharan Africa is increasing crop yields and, simultaneously, reducing the water used in crop production, that is, more crops per drop of water (Fan et al. 2012). This is urgent, particularly in smallholder farming, which experiences marginal production due to biophysical and management-related factors (Mungai et al. 2016). In most arid and semi-arid regions, water is a scarce resource which improves crop and livestock water productivity (Du et al. 2015). In addition, changes in the pattern of rainfall in different regions and the water level in local water bodies threaten agricultural sustainability, leading to reductions in food production (Nhemachena et al. 2020). However, the reduced yield of the most important crops may also be linked to a deviation in funding for agricultural research and development since 1990, which has complicated the situation (Gurdeep et al. 2021).

To develop farming systems strategies to cope with water scarcity and climate change, it is important to understand better, quantify and assess water use and its productivity in different food production systems (He and Rosa 2023). It was estimated that the production of animal products takes up to one-third of the water used in agriculture globally (Gerbens-Leenes et al. 2013). The relative share of water used by animal products is expected to grow because of changes in economic prosperity and consumption patterns, particularly in developing countries. Most of the water used in livestock production is consumed to produce feed. Water use for cleaning, cooling, and drinking should not be overlooked because of the relevance of water and its intensified competition with other sectors, such as industrial and municipal uses (Akhtar et al. 2021).

Indiscriminate water use in all sectors may result in severe water scarcity for useful production, such as crops and animals. Therefore, improving our ability to produce food with less water is crucial. Improving agricultural productivity on existing lands using the same amount of water will be essential. This means using less water to complete a particular task or using the same amount but producing more. In addition, there is improved food security and livelihoods due to increased water productivity (Nhamo et al. 2016). In addition, agricultural water productivity is about increasing the production of rainfed or irrigated crops and maximizing the products and services from livestock, trees, and fish per unit of water use. Again, crop water productivity has been the subject of many years of research, and its assessment and improvement means are well documented (Bouman 2007; Molden et al., 2007a). However, the study of improving water productivity is still in its infancy for other agricultural outputs and systems, such as livestock, agroforestry, fisheries, and aquaculture. Besides discussing the concept of agricultural water productivity, this chapter also emphasizes different ways of improving agricultural water productivity and the challenges surrounding agricultural water productivity. We also discussed the rationale for agricultural water productivity, which confirms why agricultural water productivity matters.

2.2 Water in Agriculture

Water is essential in agriculture as it supports the growth of vegetables and fruits and raises livestock. It is an irreplaceable resource that sustains life, the environment and the functioning of societies. Humankind uses water resources for drinking, municipal needs, and other economic activities. Water demand in agriculture is as high as 85% more than human water consumption (Falkenmark and Rockström 2004), and some of the sub-sectors of agriculture that use water include irrigation, pesticides, and fertilizer. Generally, water use refers to water utilization in households, industry, and agriculture. The term 'water use' relates to human activity involvement in water use. Water use also includes but is not limited to, water withdrawal, water release, or other activities in the drainage basin that influence water flows and/or quality, including in-stream uses such as fishing, recreation, and transportation (Driver et al. 2020). Removing water from a particular source without replacing it explains water consumption. It occurs during evaporation, transpiration, integration into a product, or release into a different drainage basin (e.g. inter-basin transfer) or the sea (Zheng et al. 2020).

The temporary or permanent removal of water by humans from a particular source explains water withdrawal (Rooks 2020). Even with the importance of water in agricultural sub-sectors such as crop production, food security, and rural livelihoods, its economic valuation always remains latent, considering its effect within and outside the agricultural sector. Unfortunately, water is not treated as a commodity, unlike oil, considering its huge effect (Selby 2005). Instead, it remains underpriced since users do not pay for its real value (Hoekstra 2013; Anisfeld 2010). However, as crops use large quantities of water, the price of agricultural products seldom accounts for the cost of water consumption. The recognition of water as an asset in agriculture and its scarcity or limited supply in most of the world should trigger the need for farmers and researchers to maximize the limited amount of water to achieve water productivity.

2.2.1 Water Productivity

Water productivity represents the quantity of agricultural output per unit of water loss. In a broad sense, water productivity measures the value or benefit (product or services) derived per unit of water used. It also reflects the objectives of producing more food and the associated income, livelihood, and ecological benefits at a lower social and environmental cost per water unit (Molden et al. 2007b). Water productivity involves a mass (kg) and monetary value of production per unit of evapotranspiration water (Molden et al. 2009) and, as such, it is a measure of the ability of agricultural systems to convert into food. Water productivity is usually estimated as the agricultural output produced per unit of water consumed. In a broader sense, water productivity is the ratio of the net benefits from the crop, forestry, fishery, livestock, and mixed agricultural systems to the amount of water required to produce those benefits (Descheemaeker et al. 2013). In summary, water productivity increases food, income, livelihood, and ecological benefits by using less social and environmental costs per water unit.

Higher water productivity shows that more products and services are produced using the same amount of water, or the same quantity of products and services can be created with less water (Nhamo et al. 2016). Traditionally, water use efficiency has been used to assess water use in agricultural production systems. Hence, water use efficiency and water productivity are two diverse concepts; water use efficiency is the ratio or percentage of water that a plant productively consumes. For instance, if the water use efficiency is 80%, 8 of the 10 mm of irrigation water applied to a crop is used through plant root uptake. At the same time, the remaining 2 mm is lost to drainage below the root zone or due to fruitless soil evaporation (Liu et al. 2023). Water productivity refers to the ratio of output produced to water consumed. For instance, when water productivity is 50 kg/m^3, 50 kg of grain is produced per 1 m^3 of water consumed.

Meanwhile, water productivity offers a conceptual framework that can be seen using different terms for the numerator (e.g. biomass, harvestable yield, economic value). Also, it is the denominator (e.g. transpiration, evapotranspiration, irrigation, water inflows) (Bouman 2007). Crop physiologists express water use efficiency as carbon assimilated and crop yield per unit of transpiration (Hatfield and Dold 2019). They later describe it as the amount of produce per unit of evapotranspiration. On the other hand, irrigation specialists have used water use efficiency to express how effectively water is delivered to crops and to show the amount of water wasted.

2.2.2 *Water Productivity, Irrigation Efficiency, and Water Use Efficiency*

The terms irrigation efficiency, water use efficiency, and water productivity cause some confusion among researchers. However, it is vital to note that all these terminologies have different definitions and applications. For example, some researchers considered water productivity and water use efficiency the same, but there is a substantial difference between irrigation efficiency, water use efficiency, and water productivity. Irrigation efficiency aims at evaluating the performance of the irrigation system. According to Sielska and Nojszewska (2022), productivity takes different forms with different units, but efficiency has only one form (dimensionless). Borowiecki (2013) showed that water productivity differs from water use efficiency since water productivity refers to crop production about total water consumed.

On the other hand, water use efficiency is a dimensionless ratio of the total amount of water used to the total amount of water applied. Water productivity is not dimensionless; this shows that it cannot be considered or characterized in terms of efficiency. They are just some ratios with different units in the numerator and denominator.

Table 2.1 shows a clear difference regarding water productivity, use, and irrigation efficiency. Interestingly, the application of these terms is completely different. Irrigation is more important to irrigation engineers when assessing the performance of irrigation systems (de Jong et al. 2020). Water use efficiency is more beneficial to crop physiologists to assess the efficiency with which crops utilize the applied water and its applicability at individual plant scale. However, water productivity is a broader term that includes all the system's benefits in one representative index (Borowiecki 2013).

2.2.3 *Agricultural Water Productivity*

Agricultural water productivity has to do with the quantity of water consumed (evapotranspiration) in its denominator and different quantities in its numerator (Kumar and van Dam 2013). The different quantities in the numerator can include crop yield, net income (profit), produced amount of calorific energy, value-added, etc. From a broader perspective, physical and economic water productivity are more common for water management analyses and decision-making (de Jong et al. 2020). The physical water productivity is the produced crop (biomass) per unit volume of water consumed, mostly expressed in kg m^{-3}. Economic water productivity, on the

Table 2.1 Difference between water productivity, irrigation efficiency, and water use efficiency

Water productivity	Irrigation efficiency	Water use efficiency	References
Aim—getting the best return from applied water	Aim—water saving	Only assessing the amount of water taken up by the plant	Khoshnavaz et al. (2016)
Loss accounting depends on the context—supply or depletion of water productivity	It considers losses—seepage, soil evaporation	It does not consider losses	de Jong et al. (2020)
It is concerned with crop production/benefits in relation to the total water consumed	It is concerned with water consumed by crops to water diverted from the source	The directionless ratio of the total amount of water used to the total amount of water applied	Zhang et al. (2019)
It is related to the benefits of a system	It applies to an irrigation system	It applies to crop	Ware (2023)
It has a dimension of $[M^1L^{-3} T^0]$	Non-dimensional	Non-dimensional	de Jong et al. (2020)
Performance of the production system as a whole	Used to evaluate the performance of the water system	Performance of crop	Garrido et al. (2005)

Source: Hatfield and Dold (2019)

other hand, considers the economic value of the benefits produced per unit of water used. It expresses the economic values generated for water consumed (expressed in monetary terms, US\$ m^{-3} or ZAR m^{-3}). However, finding the parts of the crop that yield the final product is vital to show a complete view of agricultural water productivity. Although all plant parts need water before growing, only specific parts generate profit (has or carry economic value) for the farmer (Zwart and Bastiaanssen 2004). Therefore, depending on the type of the cultivated plant, various farming operations, cropping patterns, and merchantability of the produced crop all contribute greatly to the concept of agricultural water productivity.

2.2.4 *Different Approaches to Improving Agricultural Water Productivity*

Since water resources worldwide are threatened by scarcity, degradation, and overuse, food demands are projected to increase (Descheemaeker et al. 2013). It is important to improve food production with less water. A huge crop yield per unit of water supply is very important as it helps farmers make decisions, promote economic growth, and improve the environment (Glenn Schaible and Aillery 2012). Increasing the value per unit of water used can be achieved through evolving strategies for increasing yield per unit of water supply. To maximize the value of water, there is a need to cultivate high-value crops as an alternative to low-value crops, which offer less economic returns (Glenn Schaible and Aillery 2012). Integration of fish and livestock in production systems can guarantee multiple water uses, improving the value derived per unit of water used (Molden et al. 2009). Therefore, improving agricultural productivity on existing lands using the same amount of water will be vital. Improving agricultural water productivity has been linked with improved food security and livelihoods (Nhamo et al. 2016).

Different strategies to improve agricultural water productivity at a global scale include the following:

(a) Improving the efficiency of water use at the level of irrigation schemes through improvement in the efficiency of water utilisation from reservoirs (by stepping down reservoir evaporation) and reduction in no recoverable seepage losses in the conveyance (Henchion et al. 2017).
(b) Improving the utilization of available water in the basins, which otherwise go uncaptured and flow into natural sinks for crop production by enlargement of rainfed crops or diverting the developed water resources for irrigated crop production (Molden et al. 2007b).
(c) Growing crops in regions with advantages in low aridity, high solar energy (Zwart and Bastiaanssen 2004) and good soils.
(d) The plants' water use efficiency by improving transpiration efficiency and raising the harvest index (Hatfield and Dold 2019).
(e) Improving water use efficiency at the farm level by diverting more water for economically more water-efficient crops (Gichuki et al. 2006).
(f) Improving water use efficiency at the field/plot level by reducing the evaporation/evapotranspiration ratio and reducing the nonrecoverable deep percolation (Kumar and Van Dam 2013).

2.2.4.1 Improving Agricultural Water Productivity

An understanding of basic biological and hydrological crop–water relations is needed for profit in water productivity. How much more water will be needed for agriculture in the future is influenced largely by links between water, food, and diet changes (Molden et al. 2009). The demand for livestock products is growing and expected to continue in the coming decades, particularly in developing countries, due to a wide spectrum of drivers such as population increase, urbanization, and income (Henchion et al. 2017). As income grows, people tend to change their eating habits, as they spend more on livestock products (Herrero et al. 2023). To satisfy these demands, the livestock sector must achieve increased productivity and efficiently utilise feed and water.

For a given crop variety and climate, there is a well-established linear relationship between plant biomass and transpiration (Steduto et al. 2007). Different crop types are more water-efficient regarding the ratio between biomass and transpiration. More biomass production needs more transpiration because when stomata open, carbon dioxide flows into the leaves for photosynthesis, and water flows out (Molden et al. 2009). Water outflow is important for cooling and creating liquid movement in the plant to transport nutrients. Stomata close during drought, limiting transpiration, photosynthesis, and production (Saradadevi et al. 2017). The most common crops, C3 crops such as wheat and barley, are less water efficient than C4 crops such as maize and sugarcane. Meanwhile, the most water-efficient crops are CAM (crassulacean acid metabolism) crops such as cactus and pineapple (Matsuno et al. 2006). These different types of crops (C3, C4, and CAM) have evolved based on their different environments and are grouped primarily based on how they fix carbon dioxide in the photosynthetic process (Steduto 1996).

2.2.4.2 Increasing Crop Water Productivity

Improving crop water productivity mainly lies in choosing adapted, water-efficient plants, stepping down unproductive water losses and ensuring ideal agronomic conditions for crop production (Bouman 2007). Notably, agronomic measures directed at healthy, vigorous growing crops favour transpiration and productive water losses over unproductive losses. A vital principle for crop water productivity is eliminating water stress. There will be further improvement if other stresses (nutrient deficiencies, weeds and diseases) are also taken away or removed (Bouman 2007). The idea is that water management should go together with nutrient management, soil management, and pest management (Scholz 2022).

Genetic enhancement of crops for higher yield and natural resource management has increased agricultural water productivity (Karavolias et al. 2021). Efforts are needed to decrease losses such as evaporation and deep percolation to reduce the denominator term in the water productivity equation. Researchers suggest different practices to enhance crop water productivity, including improving nutrient status, minimum tillage, zero tillage, and soil moisture conservation (Meena et al. 2023). Supplemental irrigation, deficit irrigation and irrigation during critical stages of crop growth can significantly influence crop yield, which translates to improved water productivity (Ibba et al. 2023). Crop water productivity has everything to do with crop breeding programs geared towards improving the proportion of economic biomass, reducing the growing period, tolerance to higher and lower temperatures, and resistance to various diseases and pests (Driedonks et al. 2019). The gains in economic water productivity can be achieved through improvements in soil fertility and management of rainwater to step down evaporation and divert more flows to transpiration (Jägermeyr 2020).

2.2.4.3 Water Productivity of Livestock

Livestock products provide one-third of humans' protein; hence, they use up to one-third of the water used in agriculture worldwide (Herrero et al. 2023). Most global animal production comes from the rainfed mixed crop–livestock systems in the developing world and the intensive production in the developed world (Amede et al. 2009a). The increasing demand for animal products, higher global water scarcity and competition for water have triggered the need to improve livestock water productivity (Descheemaeker et al. 2013). Livestock is a vital part of agricultural production systems. The feed animals consume and maintaining hygienic conditions at the pens require water (Peden et al. 2007). The livestock water intake is negligible compared with the water needed for animal feed production. Hence, improving crop management practices, carefully selecting the kind of feed, efficiently reusing crop residues, and improving the feed processing system will increase water productivity (Amede et al. 2009b). Livestock improves the productivity of the whole system by complementing other production systems. Livestock has impacted the production system in many ways and plays a vital role in livelihood strategies, contributing to overall productivity and welfare gains (de Fraiture et al. 2007). Other benefits include supporting smallholder farmers in irrigated and rainfed systems. The processing of animal products and value addition can improve livestock water productivity (Molden et al. 2009).

Livestock is one of the aspects of smallholder farming systems, and they (livestock) often provide farm power for cultivation, transportation, and manure for soil fertility management (Thornton et al. 2018). Acknowledging manure as one of the beneficial outputs of livestock systems will produce a much higher livestock water productivity than when only meat, milk, and eggs are considered (Bekele et al. 2013). The primary water depletion regarding livestock production is the evapotranspiration of water for feed production (Peden et al. 2007). The large world differences in feed water productivity (Table 2.2) are not only a sign of different methodologies; however, it shows that livestock water productivity depends on the type, the growing conditions and the management of forage production (Descheemaeker et al. 2013). There is some literature on the global range of feed water productivity for different types of feed, as presented in Table 2.2.

Some researchers (Amole et al. 2021) developed some scientific strategies to enhance livestock water productivity, including:

(i) Feed-related strategy helps to enhance livestock water productivity, and it involves carefully selecting crop residues and other waste products; enhancing feed quality; improving feed water productivity by selecting appropriate crops, cultivars and improved agronomic management; and observing the sustainability and management of grazing practices.

Table 2.2 Feed water productivity world range of different feed types

Type of feed	FWP (kg/m^3)	References
Irrigated Lucerne	0.81–2.40	Amole et al. (2021)
Food-feed crops (total biomass)	1.21–4.03	Ayantunde et al. (2018)
Cereal grains	0.36–1.20	Ayele (2012)
Pasture	0.35–2.26	Bayala et al. (2014)
Cereal forages	0.34–2.17	Bekele et al. (2017)
(Semi)-arid rangeland	0.15–0.61	Zampaligré et al. (2013)

Source: Mekonnen and Hoekstra (2012) and Amole et al. (2021)

(ii) A water-related strategy helps promote higher livestock water productivity through water conservation and harvesting, strategically placing and monitoring watering points, and incorporating livestock production into irrigation schemes.
(iii) The animal management strategy to enhance livestock water productivity includes the use of improved breeds, disease prevention and control, appropriate animal husbandry, and increased awareness among livestock keepers that the same benefit can be obtained from smaller and fewer but more productive herds.

2.2.5 Increasing Water Productivity in Aquaculture

The benefits of aquaculture include food production, improved livelihoods, nutrition, and health (Gephart et al. 2021; Hasimuna et al. 2023). Water productivity of aquaculture is expressed as the mass or value of the aquaculture produced divided by the quantity of water (Verdegem et al. 2006). Water use includes the water needed in feed production and fish consumption. Evaporation from the pond is also included in water use as this can be the primary component in arid and semi-arid regions (Abudu et al. 2019). Irrigation reservoirs can be used for culturing fish, offering a higher profit to the farmer (Pueppke et al. 2020). Approving cage aquaculture in irrigation canals is also a potential option, considering the operational issues. The value and livelihood benefits of fisheries are high and sometimes ignored or underestimated; however, considering only the value of fish produced would grossly underestimate the value of water in the aquatic ecosystems. Fishery water productivity needs to be considered in terms of ecosystem services and livelihoods supported per unit of water. However, maintaining wetlands and biodiversity should be considered as potential benefits of having water in aquatic ecosystems (Molden et al. 2009). It is necessary to note that agricultural water management practices can provide multiple ecosystem services beyond food production. Any practices that step down environmental costs and enhance ecosystem services increase the value generated from agricultural water management (Matsuno et al. 2006).

There is a challenge in determining the water productivity in cage or pen aquaculture. Aquaculture in cages allows natural water exchange and, like capture fisheries, does not induce significant water losses to the system. Cage aquaculture has the disadvantage of releasing large amount of nutrients and metabolites that empty into its aquatic environment (Descheemaeker et al. 2013). The relative influence on the environment per ton of cage and pen aquaculture product in inland waters is much higher than that of any other aquatic production system (Hasimuna et al. 2023). Water use efficiency differs markedly between aquaculture production systems (Table 2.3), although fish and crustaceans are more efficient than terrestrial animals in feed-associated water use. However, on-farm utilization of non-feed associated water in aquaculture can be very high, attaining up to 45 m^3 per kg produced in ponds.

Table 2.3 Water use efficiency (m^3 water/kg fresh weight) in aquaculture systems

AS	WUE	Water management traits	References
Traditional extensive fish pond culture	45	Rainwater and drainage water are routinely channelled into the fish ponds to compensate for drainage and evaporation losses; excessive water exchange is detrimental as it is desirable to retain nutrients within the pond	Verdegem et al. (2006) and Wang et al. (2018)
Flow-through ponds	30.1	Water exchange of 20% of the pond volume/day eliminates waste and replenishes oxygen levels; annual production of 30 t/ha is achievable, but seepage and evaporation contribute to water loss in the system	Verdegem et al. (2006) and Wang et al. (2018)
Semi-intensive fish ponds	11.5	Fish ponds fed with formulated pellet feed can yield 6 t/ha while producing two crops annually and with complete drainage to facilitate harvest; one-fifth of water consumption is linked with feed inputs	Verdegem et al. (2006) and Wang et al. (2018)
Wastewater-fed aquaculture	11.4	Wastewater is routinely fed into fish ponds to make up the water to a needed level; estimate shows that 550,000 m^3/day of wastewater is used to produce 18,000 t/year of fish in 3900 ha of ponds	Béné et al. (2016) and Wang et al. (2018)
Intensively managed ponds	2.7	Lined ponds offer an annual production of 100 t/ha, while intensive mixing results in the evaporation of 2000 mm/year	Verdegem et al. (2006) and Wang et al. (2018)
Super-intensive recirculation system	0.5–1.4	Process water is recirculated with pumps and treated with mechanical filters, biofilters and disinfection technology; stocked animals are entirely dependent on high-protein formulated feed inputs	Verdegem et al. (2006) and Wang et al. (2018)

Source: Sewilam et al. (2023)
AS aquaculture system, *WUE* water use efficiency

2.2.6 Why Agricultural Water Productivity Matters

Engaging in increased water productivity is greatly vital because of the scarcity of water in some areas compared with other resources involved in production. Therefore, investing in agriculture and agricultural water is the best way to free up water for other purposes, which explains why agricultural water productivity matters. Reasons to improve agricultural water productivity include: (i) to meet the increasing food demands in a growing, wealthier and increasingly urbanized population in light of water scarcity (Mancosu et al. 2015), (ii) to respond to pressures to re-allocate water from agriculture to cities and ensure that water is available for environmental uses (de Fraiture et al. 2007), (iii) to contribute to poverty reduction and economic growth (Mupaso et al. 2023), (iv) to compensate for the reduction of agricultural production areas as a result of urban encroachment and soil degradation (Molden et al. 2009), (v) to increase access of water for the rural poor and vulnerable groups, (vi) to generate wealthier farming systems, and (vii) to improve water management and ecosystems for sustainable agriculture and food security (Nhamo et al. 2016).

More productive water use can mean better family nutrition, income, and productive employment for the rural poor. The focus on increased water productivity can reduce the investment cost by reducing the amount of water that must be withdrawn. Higher water productivity decreases the need for extra water and land resources in irrigated and rainfed systems (Breman et al. 2001). Enhancing water productivity is thus a critical response to the growing water scarcity, including the need to leave enough water in rivers to sustain ecosystems to meet the growing demands of cities and industries (Tzanakakis et al. 2020).

2.2.7 The Influence of Climate Change and Other Emerging Drivers on Agricultural Water Productivity

Climate change brings further uncertainty to elevating agricultural water productivity because the climate is central to physical agricultural water productivity. Higher biomass production per unit of transpiration is achievable at a lower vapour pressure deficit (Vadez et al. 2021), which is common at higher latitudes (Zwart and Bastiaanssen 2004). It is scientifically accepted that increased carbon dioxide levels are linked with climate change. It is also linked with increasing water productivity per unit of evapotranspiration because more carbon can enter the plant for more photosynthesis (Droogers and Aerts 2005). According to Long et al. (2006) higher temperature will substantially offset gains in water productivity. The Intergovernmental Panel on Climate Change (IPCC) reported a decrease in potential yields in sub-Saharan Africa, implying a decrease in potential rates of water productivity.

2.2.8 Future Perspectives

The use of water and competition over water is expected to increase further. Innovation technology and changes in the policy environment will need to play an increasingly vital role in agricultural water management (Irmak 2012). Advances in the use of remote-sensing technologies are now making it possible to cost-effectively estimate crop evapotranspiration (the sum of evaporation and plant transpiration to the atmosphere) from the farmers' fields and to improve water accounting and management at the regional and basin-wide levels (Diani et al. 2004; Sharma 2012). The technological advances also enable impact assessments of agricultural water productivity. There is a need to fortify institutions, including associations of water users and councils and agencies for river basin management, and institutional and policy reforms need to be pursued and scaled to support the improved capacity.

The effects of different options on future water demands from agriculture can be analysed through scenario analysis (Amin et al. 2018). The inclusion of other sectors, such as livestock, fisheries/aquaculture as well as non-provisioning ecosystem services, makes it possible for such scenario analyses to contribute to a better understanding of the trade-offs between food, environment and the equitable distribution of gains (Cohen et al. 2019). Different strategies for improving agricultural water productivity have been highlighted in this text. Increased water productivity is very important in the face of serious water scarcity. We have already highlighted the reasons for improved water productivity. Some of them are improved food security and livelihoods, which saves fresh water and makes it available for healthy ecosystem functioning and other uses. These benefits accrued from improved water productivity are why agricultural water productivity matters. However, further study is required on the implications of various (integrated) interventions and improved agricultural water productivity on poverty, food security, economic growth, and landscape functioning.

References

Abudu S, Sheng Z, King JP, Ahn SR (2019) A Karez System's dilemma: a cultural heritage on a shelf or still a viable technique for water resiliency in arid regions. In: Yang L, Bork HR, Fang X, Mischke S (eds) Socio-environmental dynamics along the historical silk road. Springer, Cham. https://doi.org/10.1007/978-3-030-00728-7_22

Akhtar N, Syakir Ishak MI, Bhawani SA, Umar K (2021) Various natural and anthropogenic factors responsible for water quality degradation: a review. Water 13(19):2660. https://doi.org/10.3390/w13192660

Amede T, Descheemaeker K, Peden D, van Rooyen A (2009a) Harnessing benefits from improved livestock water productivity in crop–livestock systems of sub-Saharan Africa: a synthesis. Rangel J 31:169–178. https://doi.org/10.1071/RJ09023

Amede T, Geheb K, Douthwaite B (2009b) Enabling the uptake of livestock–water productivity interventions in the crop–livestock systems of sub-Saharan Africa. Rangel J 31:223–230. Dynamics in secondary reservoir for intensive fish production. Abstract 176. Paper

Amin A, Iqbal J, Asghar A, Ribbe L (2018) Analysis of current and future water demands in the upper Indus Basin under IPCC climate and socio-economic scenarios using a hydro-economic WEAP model. Water 10(5):537. https://doi.org/10.3390/w10050537

Amole TA, Adekeye A, Ayantunde AA (2021) Assessment of livestock water productivity in Seno and Yatenga provinces of Burkina Faso. Front Sustain Food Syst 5:632624. https://doi.org/10.3389/fsufs.2021.632624

Anisfeld SC (2010) Water resources. Island Press, Washington

Ayantunde AA, Cofie O, Barron J (2018) Multiple uses of small reservoirs in crop-livestock agro-ecosystems of Volta basin: implications for livestock management. Agric Water Manag 204:81–90. https://doi.org/10.1016/j.agwat.2018.04.010

Ayele A (2012) Study of small-holder farms livestock management practices and their implications on livestock water productivity in mixed crop-livestock systems in the highlands of Blue Nile basin: a case study from Fogera, Diga and Jeldu districts (Ethiopia). MSc. Thesis, Hawassa University Ethiopia, Hawassa

Bayala J, Ky-Dembele C, Kalinganire A, Olivier A, Nantoume H (2014) A review of pasture and fodder production and productivity for small ruminants in the Sahel, ICRAF occasional paper no. 21. World Agroforestry Centre, Nairobi

Bekele W, Melaku S, Mekasha Y (2013) Effect of substitution of concentrate mix with Sesbania sesban on feed intake, digestibility, body weight change, and carcass parameters of Arsi-Bale sheep fed a basal diet of native grass hay. Trop Anim Health Prod 45(8):1677–1685

Bekele M, Mengistu A, Tamir B (2017) Livestock and feed water productivity in the mixed crop-livestock system. Animal 11:1852–1860. https://doi.org/10.1017/S1751731117000416

Béné C, Arthur R, Norbury H, Allison EH, Beveridge M, Bush SR, Campling L, Leschen W, Little D, Squires D (2016) Contribution of fisheries and aquaculture to food security and poverty reduction: assessing the current evidence. World Dev 2016(79):177–196

Borowiecki KJ (2013) Geographic clustering and productivity: an instrumental variable approach for classical composer. J Urban Econ 73(1):94–110. https://doi.org/10.1016/j.jue.2012.07.004

Bouman B (2007) Rice: feeding the billions. In: Molden D (ed) Water for food, water for life: a comprehensive assessment of water management in agriculture. Earthscan/IWMI, London, pp 515–549

Breman H, Groot JJR, van Keulen H (2001) Resource limitations in Sahelian agriculture. Glob Environ Chang 11(1):59–68

Cohen PJ, Allison EH, Andrew NL, Cinner J, Evans LS, Fabinyi M, Garces LR, Hall SJ, Hicks CC, Hughes TP, Jentoft S, Mills DJ, Masu R, Mbaru EK, Ratner BD (2019) Securing a just space for small-scale fisheries in the blue economy. Front Mar Sci 6:171. https://doi.org/10.3389/fmars.2019.00171

de Fraiture C, Wichelns D, Rockstrom J, Kemp-Benedict E, Eriyagama N, Gordon LJ, Hanjra MA, Hoogeveen J, Huber-Lee A, Karlberg L (2007) Looking ahead to 2050: scenarios of alternative investment approaches. In: Molden D (ed) Water for food, water for life: a comprehensive assessment of water management in agriculture. Earthscan, International Water Management Institute (IWMI), London, Colombo, pp 91–145

de Jong IH, Arif SS, Gollapalli PK, Neelam P, Nofal ER, Reddy KY, Röttcher K, Zohrabi N (2020) Improving agricultural water productivity with a focus on rural transformation. Irrig Drain. Special issue article 70:458. https://doi.org/10.1002/ird.2451

Descheemaeker K, Bunting SW, Bindraban P, Muthuri C, Molden D, Beveridge M, Brakel MV, Herrero M, Clement F, Boelee E, Jarvis DI (2013) Increasing water productivity in agriculture. In: Boelee E (ed) CAB international conference 2013. Managing water and agroecosystems for food security

Diani S, Mohammadi K, Musavi Zadeh MH (2004) Application of geological information system. In: Optimal water management across irrigation and drainage networks, 11th seminar of Iranian National Committee on irrigation and drainage, Tehran

Driedonks N, Rieu I, Vriezen WH (2019) Breeding for plant heat tolerance at vegetative and reproductive stages. Plant Reprod 29(1–2):67–79. https://doi.org/10.1007/s00497-016-0275-9

Driver LJ, Cartwright JM, Knight RR, Wolfe WJ (2020) Species-richness responses to water-withdrawal scenarios and minimum flow levels: evaluating presumptive standards in the tennessee and Cumberland River Basins. Water 12(5):1334. https://doi.org/10.3390/w12051334

Droogers P, Aerts J (2005) Adaptation strategies to climate change and climate variability: a comparative study between seven contrasting river basins. Phys Chem Earth, Parts A/B/C 30(6/7):339–346

Du T, Kang S, Zhang J, Davies WJ (2015) Deficit irrigation and sustainable water-resource strategies in agriculture for China's food security. J Exp Bot 66(8):2253–2269. https://doi.org/10.1093/jxb/erv034

Falkenmark M, Rockström J (2004) Balancing water for humans and nature: the new approach in ecohydrology. Earthscan, Sterling

Fan M, Shen J, Yuan L, Jiang R, Chen X, Davies WJ, Zhang F (2012) Improving crop productivity and resource use efficiency to ensure food security and environmental quality in China. J Exp Bot 63(1):13–24. https://doi.org/10.1093/jxb/err248

Gardner CD, Hartle JC, Garrett RD, Offringa LC, Wasserman AS (2019) Maximizing the intersection of human health and the health of the environment with regard to the amount and type of protein produced and consumed in the United States. Nutr Rev 77(4):197–215. https://doi.org/10.1093/nutrit/nuy073

Garrido A, Martınez-Santos P, Llamas MR (2005) Ground-water irrigation and its implications for water policy in semi-arid countries: the Spanish experience. Hydrogeol J 14(3):340

Gephart JA, Golden CD, Asche F, Belton B, Brugere C, Froehlich HE, Fry JP, Halpern BS, Hicks CC, Jones RC, Klinger DH, Little DC, McCauley DJ, Thilsted SH, Troell M, Allison EH (2021) Scenarios for global aquaculture and its role in human nutrition. Rev Fish Sci Aquacult 29(1):122–138. https://doi.org/10.1080/23308249.2020.1782342

Gerbens-Leenes PW, Mekonnen MM, Hoekstra AY (2013) The water footprint of poultry, pork and beef: a comparative study in different countries and production systems. Water Resour Ind 1–2:25–36

Gichuki F, Cook S, Turral H (2006) Agricultural water productivity: issues, concepts and approaches. In Basin focal project working paper 1 (no. 618-2016-41183)

Glenn Schaible D, Aillery MP (2012) Water conservation in irrigated agriculture: trends and challenges in the face of emerging demands, EIB-99, U.S. Department of Agriculture, Economic Research Service

Gurdeep SM, Kaur M, Kaushik P (2021) Impact of climate change on agriculture and its mitigation strategies: a review. Sustainability 13(3):1318. https://doi.org/10.3390/su13031318

Hasimuna OJ, Maulu S, Nawanzi K, Lundu B, Mphande J, Phiri CJ, Kikamba E, Siankwilimba E, Siavwapa S, Chibesa M (2023) Integrated agriculture-aquaculture as an alternative to improving small-scale fish production in Zambia. Front Sustain Food Syst 7:1161121. https://doi.org/10.3389/fsufs.2023.1161121

Hatfield JL, Dold C (2019) Water-use efficiency: advances and challenges in a changing climate. Front Plant Sci 19(10):103. https://doi.org/10.3389/fpls.2019.00103

He L, Rosa RL (2023) Solutions to agricultural green water scarcity under climate change. PNAS Nexus 2(4):117. https://doi.org/10.1093/pnasnexus/pgad117

Henchion M, Hayes M, Mullen AM, Fenelon M, Tiwari B (2017) Future protein supply and demand: strategies and factors influencing a sustainable equilibrium. Foods 6(7):53. https://doi.org/10.3390/foods6070053

Herrero M, Hugas M, Lele U, Wirakartakusumah A, Torero M (2023) A shift to healthy and sustainable consumption patterns. In: von Braun J, Afsana K, Fresco LO, Hassan MHA (eds) Science and innovations for food systems transformation. Springer, Cham. https://doi.org/10.1007/978-3-031-15703-5_5

Hoekstra AY (2013) The water footprint of modern consumer society. Routledge, New York

Huang Z, Yuan X, Liu X (2021) The key drivers for the changes in global water scarcity: water withdrawal versus water availability. J Hydrol 601:126658. https://doi.org/10.1016/j.jhydrol.2021.126658

Ibba K, Kassout J, Boselli V, Er-Raki S, Oulbi S, Mansouri LE, Bouizgaren A, Sikaoui L, Hadria R (2023) Assessing the impact of deficit irrigation strategies on agronomic and productive parameters of Menara olive cultivar: implications for operational water management. Front Environ Sci 11:1100552. https://doi.org/10.3389/fenvs.2023.1100552

Irmak S (2012) Technology implementations to improve crop water and nutrient productivity. Available at: http://www.worldwaterweek.org/documents/WWW_PDF/2012/Wed/focus-latin-america/Afternoon/Suat-Irmak.pdf

Jägermeyr J (2020) Agriculture's historic twin-challenge toward sustainable water use and food supply for all. Front Sustain Food Syst 4:35. https://doi.org/10.3389/fsufs.2020.00035

Karavolias NG, Horner W, Abugu MN, Evanega SN (2021) Application of gene editing for climate change in agriculture. Front Sustain Food Syst 5:685801. https://doi.org/10.3389/fsufs.2021.685801

Khoshnavaz F, Honar T, Daneshkar AP (2016) Estimation of agriculture water productivity with the help of remote sensing technology, case study: the irrigation network in Qazvin plain, Qazvin. Water Soil Sci 25(1):57–68

Kumar MD, van Dam JC (2013) Drivers of change in agricultural water productivity and its basin-scale improvement in developing economies. Water Int 38(3):312–325

Liu J, Huang Q, Li Z, Liu N, Li J, Huang G (2023) Effect of autumn irrigation on salt leaching under subsurface drainage in an arid irrigation district. Water 15(12):2296. https://doi.org/10.3390/w15122296

Long SP, Ainsworth EA, Leakey ADB, Nösberger J, Ort DR (2006) Food for thought lower-than-expected crop yield stimulation with rising CO2 concentrations. Science 312(5782):1918–1921

Mancosu N, Snyder RL, Kyriakakis G, Spano D (2015) Water scarcity and future challenges for food production. Water 7(3):975–992. https://doi.org/10.3390/w7030975

Matsuno Y, Nakamura K, Masumoto T et al (2006) Prospects for multifunctionality of paddy rice cultivation in Japan and other countries in monsoon Asia. Paddy Water Environ 4:189–197. https://doi.org/10.1007/s10333-006-0048-4

Meena SN, Sharma SK, Singh P, Ram A, Meena BP, Jain D, Singh D, Debnath S, Yadav S, Dhakad U, Verma P, Meena JK, Nandan S (2023) Tillage-based nutrient management practices for sustaining productivity and soil health in the soybean-wheat cropping system in Vertisols of the Indian semi-arid tropics. Front Sustain Food Syst 7:1234344. https://doi.org/10.3389/fsufs.2023.1234344

Mekonnen MM, Hoekstra AY (2012) A global assessment of the water footprint of farm animal products. Ecosystems 15:401–415. https://doi.org/10.1007/s10021-011-9517-8

Molden D, Bin D, Loeve R, Barker R, Tuong TP (2007a) Agricultural water productivity and savings: policy lessons from two diverse sites in China. Water Policy 9(Suppl 1):29–44

Molden D, Oweis TY, Pasquale S, Kijne JW, Hanjra MA, Bindraban PS, Bouman BAM, Cook S, Erenstein O, Farahani H, Hachum A, Hoogeveen J, Mahoo H, Nangia V, Peden D, Sikka A, Silva P, Turral H, Upadhyaya A, Zwart S (2007b) Pathways for increasing agricultural water productivity. In: Molden D (ed) Water for food, water for life: a comprehensive assessment of water management in agriculture. Earthscan, International Water Management Institute (IWMI), London, Colombo, pp 279–310

Molden D, Oweis T, Steduto P, Bindraban P, Hanjra MA, Kijne J (2009) Improving agricultural water productivity: between optimism and caution. Agric Water Manag 97:528. https://doi.org/10.1016/j.agwat.2009.03.023

Mungai LM, Snapp S, Messina JP, Chikowo R, Smith A, Anders E, Richardson RB, Li G (2016) Smallholder farms and the potential for sustainable intensification. Front Plant Sci 7:1720. https://doi.org/10.3389/fpls.2016.01720

Mupaso N, Makombe G, Mugandani R (2023) Smallholder irrigation and poverty reduction in developing countries: a review. Heliyon 9(2):e13341. https://doi.org/10.1016/j.heliyon.2023.e13341

Nhamo L, Mabhaudhi T, Magombeyi M (2016) Improving water sustainability and food security through increased crop water productivity in Malawi. Water 8(9):411. https://doi.org/10.3390/w8090411

Nhemachena C, Nhamo L, Matchaya G, Nhemachena CR, Muchara B, Karuaihe ST, Mpandeli S (2020) Climate change impacts on water and agriculture sectors in southern Africa: threats and opportunities for sustainable development. Water 12(10):2673. https://doi.org/10.3390/w12102673

Peden D et al (2007) Water and livestock for human development. In: Molden D (ed) Water for food, water for life: a comprehensive assessment of water management in agriculture. Earthscan/IWMI, London, Colombo, pp 485–514

PRB (Population Reference Bureau) (2013) United Nations raises projected world population. https://www.prb.org/resources/united-nations-raises-projected-world-population/. Accessed 6 Sept 2023

Pueppke SG, Nurtazin S, Ou W (2020) Water and land as shared resources for agriculture and aquaculture: insights from Asia. Water 12(10):2787. https://doi.org/10.3390/w12102787

Rooks J (2020) Water extraction | What is water extraction & how it fits in water cleanup. Restoration Local Retrieved 8 Nov 2021

Saradadevi R, Palta JA, Siddique KHM (2017) ABA-mediated stomatal response in regulating water use during the development of terminal drought in wheat. Front Plant Sci 8:1251. https://doi.org/10.3389/fpls.2017.01251

Scholz M (2022) Novel water retention and nutrient management technologies and strategies supporting agricultural water management in continental, Pannonian and Boreal regions. Water 14(9):1486. https://doi.org/10.3390/w14091486

Selby J (2005) Oil and water: the contrasting anatomies of resource conflicts. Gov Oppos 40:200–224

Sewilam H, Kimera F, Nasr P (2023) Water energy food nexus model: an integrated aqua-agriculture system to produce tilapia and sweet basil using desalinated water. Environ Sci Pollut Res 30:15975–15990

Sharma B (2012) Developing rainwater management strategies by integrating technologies, institutions and policies for the Blue Nile Basin, Ethiopia. Available at: http://www.worldwaterweek.org/documents/WWW_PDF/2012/Tue/Rainfed-production-undergrowing/Bharat-Sharma.pdf

Sielska A, Nojszewska E (2022) Production function for modeling hospital activities. The case of Polish county hospitals. PLoS One 17(5):e0268350. https://doi.org/10.1371/journal.pone.0268350

Steduto P (1996) Water use efficiency. In: Pereira LS, Feddes LS, Giley RA, Lesaffre B, Sustainability of irrigated agriculture, NATO ASI series E: applied sciences, Kluwer, Dordrecht, pp 193–209

Steduto P, Hsiao TC, Fereres (2007) On the conservative behaviour of biomass water productivity. Irrig Sci 25:189–207

Thornton PK, Rosenstock T, Förch W, Lamanna C, Bell P, Henderson B, Herrero M (2018) A qualitative evaluation of CSA options in mixed crop-livestock systems in developing countries. In: Lipper L, McCarthy N, Zilberman D, Asfaw S, Branca G (eds) Climate smart agriculture. Natural resource management and policy, vol 52. Springer, Cham. https://doi.org/10.1007/978-3-319-61194-5_17

Tzanakakis VA, Paranychianakis NV, Angelakis AN (2020) Water supply and water scarcity. Water 12(9):2347. https://doi.org/10.3390/w12092347

Vadez V, Choudhary S, Kholová J, Hash CT, Srivastava R, Kumar AA, Prandavada A, Anjaiah M (2021) Transpiration efficiency: insights from comparisons of C4 cereal species. J Exp Bot 72(14):5221–5234. https://doi.org/10.1093/jxb/erab251

Verdegem MCJ, Bosma RH, Verreth JAJ (2006) Reducing water use for animal production through aquaculture. Water Resour Dev 22(1):101–113

Wang Q, Zhongjie L, Gui J, Liu J, Ye S, Yuan J, De Silva SS (2018) Paradigm changes in freshwater aquaculture practices in China: moving towards achieving environmental integrity and sustainability. Ambio 2018(47):410–426

Ware T (2023) Irrigation efficiency – definition, types, importance & formula. https://www.twl-irrigation.com/irrigation-efficiency/. Retrieved 6 Sept 2023

Zampaligré N, Dossa LH, Schlecht E (2013) Contribution of browse to ruminant nutrition across three agro-ecological zones of Burkina Faso. J Arid Environ 95:55–64. https://doi.org/10.1016/j.jaridenv.2013.03.011

Zhang W, Xianzeng D, Huang A, Yin H (2019) Analysis and comprehensive evaluation of water use efficiency in China. Water 11(12):2620. https://doi.org/10.3390/w11122620

Zheng Q, Lu H, Huang X, Sun L, Sun G (2020) Effects of urbanization on watershed evapotranspiration and its components in southern China. Water 12(3):645. https://doi.org/10.3390/w12030645

Zwart SJ, Bastiaanssen WGM (2004) Review of measured crop water productivity values for irrigated wheat, rice, cotton and maize. Agric Water Manag 69(2):115–133

Chapter 3
Management and Agricultural Water Productivity Improvement in India and The Ganges Basin

B. Sharma, A. Sikka, J. Lautze, and Tafadzwanashe Mabhaudhi

Abstract The fast transition of India from a water-stressed to a water-scarce country has provided additional impetus for the search for interventions and decision support systems (DSSs) for solutions to problems arising from a mismatch between demand and supply and competing demands of economic and environmental sectors. Agriculture is the largest user of freshwater and increasing water productivity in agriculture is a national challenge requiring urgent attention. Globally, DSSs have gained immense popularity in various domains and, more recently, in agriculture and water resources management but are still limited in developing countries like India. This chapter presents a comprehensive compilation and stocktake of the DSSs developed at various scales in India and specific to the Ganges Basin, aimed at serving as a foundation for future work in this direction. After the criteria-based screening of the literature and reports, each selected DSS was analysed in the context of the aspects covered, key criteria, and the parameters such as relevance, applicability, focus, and scale of application. To reinforce the recommendations, KIIs were held with selected experts and stakeholders. A matrix approach was employed to compile and review the DSS with broader segregation under (i) crop and farm-based decision support systems, (ii) DSS based on artificial intelligence, enhanced machine learning, fuzzy multi-criteria decision making, and knowledge systems, (iii) DSS for real-time operation of micro-irrigation systems, (iv) DSS for management of tanks and reservoirs for water-deficit regions, and (v) DSS for improving water productivity under canal commands and conjunctive management of surface

B. Sharma (✉) · A. Sikka
International Water Management Institute (IWMI), New Delhi, India

J. Lautze
International Water Management Institute (IWMI), Pretoria, South Africa

T. Mabhaudhi
Centre on Climate Change and Planetary Health, London School of Hygiene and Tropical Medicine, London, UK

Centre for Transformative Agricultural and Food Systems, University of KwaZulu-Natal, Carbis Road, Scottsville, Pietermaritzburg, KwaZulu-Natal, South Africa

T. Mabhaudhi et al. (eds.), *Enhancing Water and Food Security Through Improved Agricultural Water Productivity*,
https://doi.org/10.1007/978-981-96-1848-4_3

and groundwater resources. The selected DSSs were then analysed for focus, key findings, relevance, applicability, and scale of application of the DSS.

Keywords Irrigated agriculture · India, decision-support tools · Water-use efficiency · Environmental health

3.1 Introduction

3.1.1 Water Situation in India

Water is an essential resource that sustains life on Earth as it forms part of almost any conceivable economic sector, the lifeline of ecosystems, and the planet's life-supporting system. As per the assessment of the United Nations, among almost 7.7 billion residents of this world, more than 2 billion live in countries undergoing a severe water crisis (Uhlenbrook and Connor 2019). In India, out of the total annual water resources of 1999, 20 billion m^3 (BCM), only about 1122 BCM can be utilized—690 BCM through surface water and 432 BCM through groundwater resources (Bhattacharyya et al. 2015; CGWB 2022). Several basins (Indus, Sabarmati, Krishna, Pennar, etc.) will reach physical water-scarce conditions by 2050, where the utilizable water supply cannot be developed further without severely impacting the environment and riverine water users downstream. The Central Ground Water Board (CGWB) of India estimates that India's groundwater extraction at more than 250 BCM/annum is the largest globally—more than a quarter of the global total. Out of 6881 assessed units, the CGWB classified 17% of groundwater blocks as overexploited (extraction exceeding 100% of the natural replenishment), 5% as critical (extraction reaching about 90–100% of utilizable resources), 14% as semi-critical (extraction in the range of 70–90% of utilizable resources), leaving 63% of groundwater assessment units as 'safe' (CGWB 2022). Water pollution is also a severe problem as almost 70% of its surface water resources and a growing percentage of the groundwater resources are contaminated with biological, organic, inorganic, and toxic pollutants (CGWB 2022).

Agriculture is the largest user of freshwater (>80%), and there are three main reasons why efficient water use in agriculture is crucial. The first is the significant dependency of the Indian economy and rural population on the Indian monsoon. Monsoon rains are vital to agriculture, accounting for about 20% of the national economy and about 50% of employment (Gulati et al. 2019). The second important factor is the scarcity of water resources as the country is moving from being 'water-stressed' (annual per capita water availability of 1441 m^3 in 2015) to a 'water-scarce' country in 2051. The third one relates to the fact that India's water demand patterns are fast changing. In 1960, water withdrawal for the three most water-consuming sectors- agriculture, industry, and domestic was 277 BCM (Shiklomanov and Rodda 2003). This has since increased to about 793 BCM in 2009. Based on

PODIUMSIM analysis, the business-as-usual scenario projects that the total water demand will increase by another 150 BCM or 22% by 2025; and a further 69 BCM or 8% by 2050 (Amarasinghe et al. 2007). These problems are not new but are becoming more widespread and their impacts more devastating, as a large area is experiencing water scarcity. This has provided additional impetus for the search for solutions and decision-support systems for solutions to problems arising from a mismatch between demand and supply and competing demands from other economic and environmental sectors. Increasing water productivity in agriculture has been identified as a national challenge requiring urgent attention.

3.1.2 Water Productivity

The world is witnessing and shall continue to muddle through the increased tensions at the water-food-energy-environment nexus due to burgeoning population growth and the exigencies of climate change (Clothier et al. 2020). The agriculture sector is the single largest user (>70%) of freshwater resources, and the current challenge is to ensure that agricultural water management allows for both reasonable profits to the farmers and enough food, fibre, and biofuel production to meet the demands of the growing population. At the same time, unsustainable costs in energy use and environmental degradation are avoided (Fernandez et al. 2020). Agricultural production systems combine inputs like land, labour, seeds and planting materials, water, fertilizers, energy, and others to produce agricultural outputs. Increases in total production or 'total factor productivity' can be achieved by two approaches: (i) by increasing technical efficiency through more efficient utilization of the production inputs, and (b) by increasing allocative efficiency by producing outputs with the highest returns (e.g. by choice of crops and varieties, the mix of the farming enterprise, etc.). Irrigators are under increasing pressure to ensure food security and long-term environmental sustainability, also increasingly threatened by climate change and the growing population (Mabhaudhi et al. 2018). An analysis of single-factor productivity (say, water) enables us to assess the opportunities for maximizing the returns from using the particular factor. Farmers seeking to maximize benefits will use water and other inputs at levels where the incremental value generated is at least equal to the incremental cost. Under increasing scarcity of resources, performance measures can play an essential role in identifying opportunities to improve the performance of the production system. For this objective, indicators of biophysical water productivity and the economic performance of irrigation are required. Different stakeholders aim to optimize different objectives: while the irrigators try to obtain the highest profitability possible from the farms, orchards, and pastures, environmentalists focus on the importance of preserving current water resources, and policymakers on regulating the demand from different water-consuming sectors (Fernandez et al. 2020). Like land productivity, water productivity is partial-factor productivity that measures how the systems convert water into goods and services (Molden et al. 2010; Nhamo et al. 2016), and is measured as.

$$\text{Water Productivity}\left(\text{WP}\right)=\frac{\text{Output derived from water use}}{\text{Water input}} \tag{3.1}$$

When water input is measured as evapotranspiration (ET), it is called physical water productivity (PWP). When water input is measured as the irrigation water applied, it is called irrigation water productivity (IWP). When the output is transformed into monetary value (for comparison of the options or regions), it is referred to as economic water productivity (EWP). Still, there is a need for the metrics of water productivity and economic benefits to be defined so that water productivity indicators can help improve productivity and profitability. Some of the critical water use indicators and economic analyses endorsed by the *Journal of Agricultural Water Management* are helpful and well accepted for on-farm irrigation decisions (Table 3.1, Clothier et al. 2020; Fernandez et al. 2020).

Table 3.1 Equations and definitions of important water-use efficiency, crop water productivity, and economic water productivity indicators

Indicator	Units	Definition
i. Water-use efficiency and crop water productivity indicators		
$WUE_c=\frac{ETc}{I+P}$	$\frac{m^3\,ha^{-1}}{m^3\,ha^{-1}}$	Crop WUE: ratio between the actual crop evapotranspiration (ETc) and total water applied by irrigation (I) and precipitation (P) (Perry et al. 2009)
$WP_c=\frac{Yield}{ETc}$	$\frac{kg\,ha^{-1}}{m^3\,ha^{-1}}$	Crop water productivity: ratio between the marketable yield produced by a crop and the water consumed by the crop or crop evapotranspiration (ETc). Also referred to as 'physical crop water productivity' (Kijne et al. 2003)
$WP_c=\frac{Yield}{TWU}$	$\frac{kg\,ha^{-1}}{m^3\,ha^{-1}}$	Crop water productivity: ratio between the marketable yield produced by a crop and the total water involved (including special needs) in crop production.
$WP_i=\frac{Yield}{IWU}$	$\frac{kg\,ha^{-1}}{m^3\,ha^{-1}}$	Irrigation water productivity: ratio between the marketable yield produced by a crop during the growing season and the irrigation water applied (IWU) in the same period.
ii. Economic water productivity indicators		
$EWP_c=\frac{Profit}{TWU}$	$\frac{\$\,ha^{-1}}{m^3\,ha^{-1}}$	Economic crop water productivity: ratio between the profit (revenue—variable, fixed, and opportunity costs), in monetary terms ($ or any other currency), by a crop along the growing season and the total amount of water involved in crop production (TWU).
$*EWP_i=\frac{Profit}{IWU}$	$\frac{\$\,ha^{-1}}{m^3\,ha^{-1}}$	Economic irrigation water productivity: ratio between the profit (revenue—variable, fixed, and opportunity costs), in monetary terms ($ or any other currency), by a crop along the growing season and the irrigation water applied (IWU) in the same period.

Source: Fernandez et al. (2020). https://www.sciencedirect.com/science/article/abs/pii/S0378377419317019

*This can be referred to as gross economic irrigation water productivity ($GEWP_i$) when the numerator is gross margin (revenue—variable costs); and net economic irrigation water productivity ($NEWP_i$) when the numerator is calculated as net margin (revenue—variable and fix costs)

Crop yield is a function of crop transpiration as plants use photosynthetically active radiation as the energy in the photosynthesis process to convert CO_2 into biomass using the water transpired by the plant. Evaporation from soil or water surfaces between plants (available through rain or irrigation) is not a crop production function (Ali and Talukdar 2008). The WP concept complemented the existing irrigation or water-use efficiency measures mainly useful for water system managers. These measures ignored the beneficial use of water re-captured and reused in one part of the basin because of the deep percolation and/or runoff losses in other parts of the basin. Another important dimension for estimating the WP is the scale of interest, which may vary from crop plant to field, sub-basin/region/nation, or the world. Crop scale interests crop scientists in selecting water-efficient crops or crop cultivars. Farmers, agronomists, and water specialists are interested in efficient water use at the field or farm scale. Irrigation system managers may address how productive the water available to the irrigation system is. Sub-basin scale is of interest to the planners and river-basin managers for enhancing water productivity at a large scale and catering to other economic sectors of water. River basin planners, development agencies, and researchers assess the water productivity of the renewable water available in the basin as rainfall, surface storage, and groundwater. The basin/region may include multiple production systems, including diversified crop production systems, livestock production systems, silviculture, and fish production. The equation proposed by Cook et al. (2006) can be used to estimate the water productivity of the depleted water in a multiple-use system.

$$\text{Multiple}-\text{use water productivity} = \sum_{j=1}^{p}\sum_{i=1}^{n} Y_{ij}\ A_{ij} \,/ \sum_{j=1}^{p}\sum_{i=1}^{n} W_{ij}\ A_{ij} \tag{3.2}$$

where Y_{ij} is the amount of output for production system j on the field i (kg/ha),

W_{ij} is the amount of water depleted (m^3/ha), A_{ij} is the production area (ha),
p is the number of the production system, and n is the number of fields.

At the basin level, the rationale for improving water productivity is to increase water availability to the users that have limited access to water resources downstream, other sectors of water use like domestic, industries, livelihoods, and the environments whose genuine water needs are marginalized in the basin and enhance total basin-level water benefits through more productive and diversified uses of the available water resources. At the basin level, the productive uses of water can be both tangible and non-tangible. The stakeholders may be the vocal farmers and residents of the habitations, the marginalized and women sections of society, and the silent stakeholders like the environment. As the systems become large and complex, the water benefits shall include economic, social, health, ecosystem, and cultural and spiritual benefits.

Interventions and decisions for improving the water productivity in large basins, like the Ganges Basin, shall no longer be a simple exercise as a given intervention shall lead to multifarious ramifications for different regions, users, and sectors.

Optimal, sustainable, and economically viable decision-making under such complex situations requires high-end technology and suitably designed decision support systems.

3.1.3 Decision Support Systems

Preconceived bias or shortage of time, resources, and funds can lead to less-than-desirable decisions. The current decision-making problems are more complex than in the past, prompting the need for decision support. A decision support system (DSS) is an interactive computer-based system that helps decision-makers use data and models to solve ill-structured, unstructured, or semi-structured problems (Sprague and Carlton 1982). According to Mora et al. (2003), the decision maker employs computer technology to (i) organize the information into problem factors, (ii) attach all the attributes to a model, (iii) use the framework/model to simulate alternatives, and (iv) select the best course of action. The outcomes are reported as parameter conditions, experimental forecasts, and/or recommended actions. A typical architecture of DSS consists of input, processing, and output (Fig. 3.1).

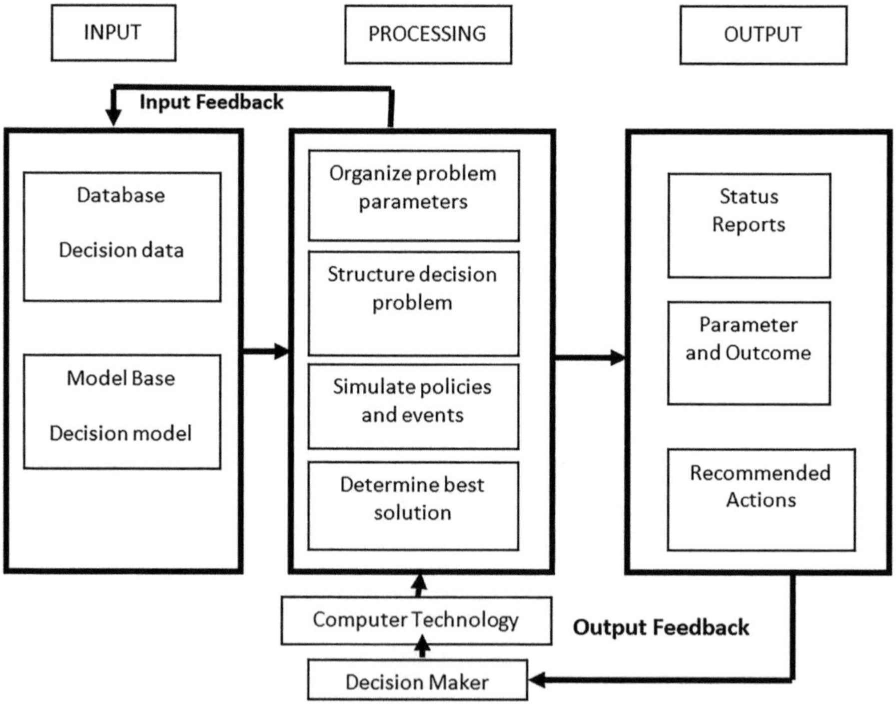

Fig. 3.1 Typical architecture of decision support system. (Adapted from Mora et al. 2003)

DSS facilitates several independent and sequential decisions and generally employs models for analysing problems since modelling enables experimenting with different strategies under different configurations. It is adaptive, flexible, user-friendly, and has a solid graphical user interface. In the absence of a robust decision support system, the growing use of technology and communication systems and the structural complexity of the problems can result in the chain reaction of magnification of errors and associated costs. However, the DSS attempts to support the decision-makers, not replace them, and they have control over all the levels of the process.

DSSs have gained immense popularity in various domains, including manufacturing, business, engineering, health, and more recently in agriculture, water resources management, and estimation and improvement of water productivity (Raman et al. 1992; Arumugam and Mohan 1997; Reddy and Rao 1995; Wellen et al. 2017; Saggi and Jain 2022). Suggested basic components of the spatial decision support system for crop and water productivity management consist of the spatial databases, input attribute database, models and analysis tools, interactive graphical user (Fig. 3.2).

Suggested basic components of the spatial decision support system for crop and water productivity management consist of the spatial databases, input attribute database, models and analysis tools, interactive graphical user interface, and the outputs in the form of regional and field-scale reports, graphs, and maps (Fig. 3.2) (Reddy and Rao 1995).

Globally, DSSs have long been used in agriculture and water management research, service provision, and farm extension. A recent review for improving the decision support systems in irrigated agriculture showed that DSSs in agriculture by

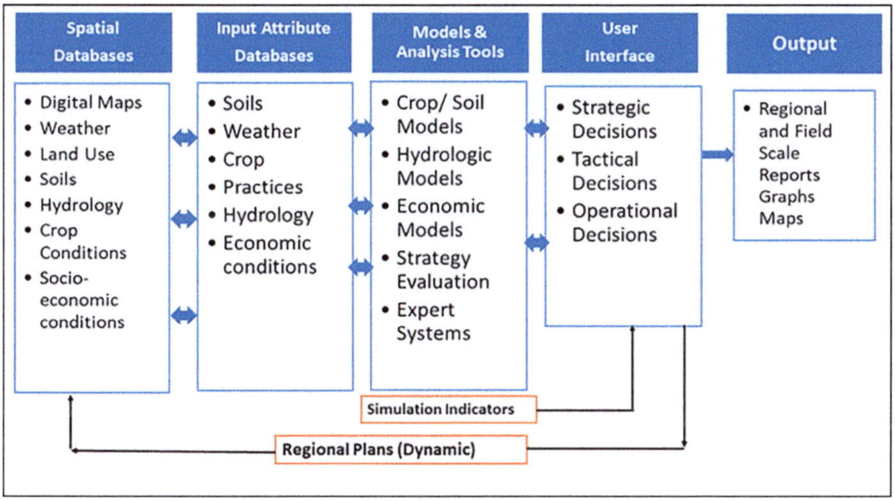

Fig. 3.2 Various components of spatial decision support system for crop and water productivity management. (Modified and redrawn from Reddy and Rao 1995)

water management authorities helps balance water use between different levels and different sectors. Agricultural DSSs can also help decision-makers improve water allocation for enhanced water-use efficiency and productivity, improve nutrient-use efficiency, and minimize environmental pollution. However, fewer DSSs helped in strategic decision-making compared to tactical and operational decisions (Ara et al. 2021). Across regions, most DSSs help users identify more profitable irrigation regimes. Some facilitate a comparison of alternative farming systems and cropping diversity, water productivity, and water-use efficiency, and only a few address the economic aspects of adopting irrigation infrastructure. Most DSSs were designed for use by large farmers, advisors, and service providers at the farm scale, while fewer DSSs helped the decision-making at the regional, catchment, sub-basin, and basin scales. DSSs were generally developed and used locally; these were not publicly available, with insufficient transfer to other regions and users.

For various reasons, the adoption of DSSs by farmers, irrigation departments, development agencies, and service providers is limited, especially in developing countries like India. The key reason is that most DSSs have been developed by researchers and information-technology professionals purely from a 'technology push' perspective instead of the 'end-user pull', meaning these were less demand-driven. The shortcomings of the current agricultural and water use-water productivity DSS (Mackrell et al. 2009) include a lack of understanding of farmers' needs and the decision-making process, which is influenced by various social and economic factors.

Increasing water scarcity, climate change, pressure to provide additional water for increasing domestic and industrial needs, and the environmental flows while maintaining food and nutrition security for a large human population (~1.3 billion) urge Indian farmers and water resource managers to be more water efficient. Sustainable agriculture production and processing have become complex and challenging with the involvement of biological, chemical, physical, and climatic processes and crop management practices increasingly interacting with socio-economic factors and the global markets. Additionally, Indian agriculture has some unique features like the dominance of small and highly diversified agriculture, large dependence of the rural population on agriculture, the uncertainties of monsoon rains for water security, state incentives to farmers in the form of free/cheap energy, fertilizers, and water to support food production, and severe impacts of the climate change in the form of intense heat waves, droughts, and floods. Under these challenging situations, the next generation of agricultural and water professionals using information technologies will play an increasingly important role in agriculture production and natural resource management. This technology allows examining and handling a more comprehensive range of spatial and input attribute databases, such as soils, weather, hydrology, crop condition, and growth. It integrates these with socio-economic and market variables. Simultaneous examination of these and related variables leads to a better understanding of various agricultural processes and their interactions over space and time and then appropriate domains to target new technologies and processes. DSS offers a framework within which such complex systems can be represented in a structured way, allowing them to be more easily understood and helping to add information and new insights for informed decision-making.

Some areas where DSS have been used in agriculture in the Indian context include crop productivity improvement, estimation of crop water requirements, water productivity, and water resource management, including micro-irrigation and irrigation scheduling, and some generic DSS for advisory systems. Recently, there has been an increased focus on using remote sensing-GIS technologies, artificial intelligence, extreme machine learning, and multi-criteria decision-making. However, there is a lack of DSS for decision-making at the higher levels of the basin or sub-basin. There is an urgent need for stock-taking of the application, adoption, and exploring opportunities for improving decision support systems in irrigated agriculture and water productivity for developing the need-based decision support systems. This consultancy report is an effort to contribute to this objective, especially in the Indian context and the Ganges Basin in India.

3.1.4 Approach of the Study

The main deliverable for this report was to develop a comprehensive compilation and stock take of the DSSs developed at various scales in India specific to the Ganges Basin and stratified according to key parameters of relevance, applicability, focus, and scale of application. The next was to assess the selected DSSs to identify the aspects which were covered (and not covered) and detail the important lessons for which the support of key informant interviews with selected development of an innovative, demand-driven, and robust DSS-WP for the Ganges Basin. A conceptual diagram summarizing the steps of the methodology and structure of the DSS-WP report is in Fig. 3.3.

3.1.5 Compilation Methodology and the Stock-Taking of the Decision Support Systems

The methodology for the development of the DSS-WP report consisted of the following main steps:

(i) Literature search: based on the search criteria using relevant keyword searches in Scopus and the grey literature (published informally or non-commercially) was conducted. The search terms found more than 154 entries. The first screening was at the level of article titles to exclude non-relevant topics and non-journal papers (about 100 entries). The second screening was at the article/report abstracts level and making notes of the essential findings. Full-text screening of more than 64 articles/reports was conducted to compile and assess the DSSs included in the report. Further additions to this selection were made

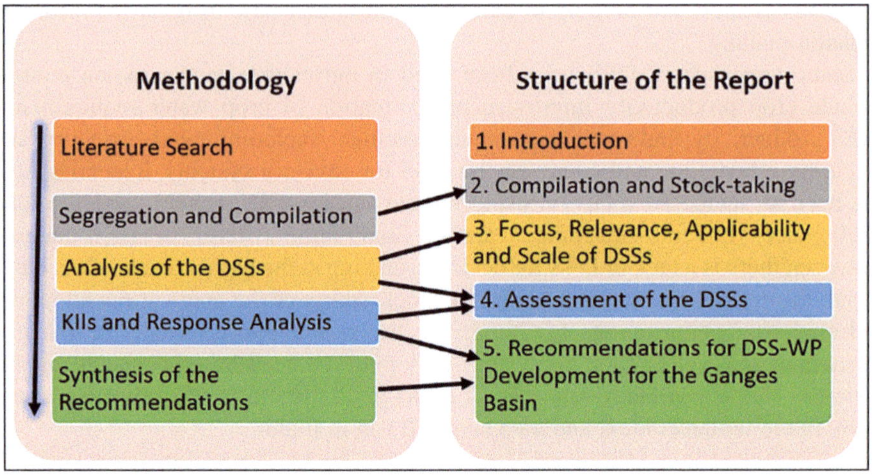

Fig. 3.3 Conceptual elements of the methodology and structure of the DSS-WP report

through many cross-references and additional references suggested by the experts and key-informant stakeholders.

(ii) The selected entries of the DSS review and the additional resources were systematically reviewed and summarized. A reference card system was followed to summarize the highlights of the methodology, geographical location, main findings, recommendations, and any other detail of interest, including specific figures and data. These resources were then segregated under well-defined themes and types of the DSS related to agriculture, water productivity, and water resources management. The segregation and compilation were under the broad themes of (i) crop and farm-based DSS, (ii) DSS based on AI, ELM, Fuzzy MCDM, and knowledge systems, (iii) DSS for real-time operation of micro-irrigation systems, (iv) DSS for operation and management of tanks and reservoirs, and (v) DSS for improving WP under canal commands and conjunctive management of surface and groundwater.

(iii) Next, each DSS was analysed in the context of the aspects covered, key criteria, and the parameters such as relevance, the applicability of the DSS, the focus of the DSS, and the scale of application of the DSS (field, farm, command, sub-basin, region, state, and the basin). The development process was also analysed for its inclusivity, approach (top-down/bottom-up or mixed), economics, and other relevant parameters. This analysis helped identify the gaps (areas/themes not covered) and formulate the detailed lessons learned.

(iv) To gain further insights and practical use of the DSS in India and the Ganges Basin, Key Informant Interviews (KIIs) were held with selected experts/stakeholders to reinforce the analysis and recommendations. All the participants of the Brainstorming Workshop on Water Productivity and Storage in the Ganges Basin (under the Nexus Gains program) were provided with a well-structured instrument (Annexure I), and their responses were gathered electronically (some in person/telephonically) and systematically analysed.

(v) Finally, the recommendations for the development of a useful, interactive, and technology- and demand-driven DSS-WP for the Ganges Basin (and its two selected regions of Ramganga sub-basin, and Bundelkhand region) based on the above compilation, analysis and assessment, and responses and suggestions of the experts in the Ganges Basin were developed.

3.1.6 Stock-Taking of Decision Support Systems

In many parts of the world, including India, the pressure on water resources has intensified and is facing its limits. Moreover, water resource systems are complex and encompass interlinked sub-systems, including technical, agricultural, economic, social, cultural, environmental, and legal aspects. In several instances, increasing pressure on water resources in developing countries like India resulted in a lack of safe and sufficient water for domestic use and inadequate water for economic sectors such as agriculture, industries, energy, and the environment. Such aspects have created public pressure, followed by government and societal responses regarding an increased focus on rational water resources planning and management and enhanced water productivity for all water use sectors. Agriculture, including forestry, is the largest user of freshwater resources (about 80% in India). Still, the extent to which irrigation is employed in any agricultural system is a function of several biophysical, economic, environmental, and social factors. Biophysical factors include crop type, crop water productivity/water-use efficiency, and system diversification; physical factors may include water availability and allocation, irrigation methods, and infrastructure; economic factors include costs, commodity prices, and capital costs. Environmental factors include waterlogging, salinity and land degradation, and social factors may include management associations, aspirations, and gender contexts (Ara et al. 2021). Given this complexity, the users and other decision-makers at different levels face multiple concurrent factors to consider when arriving at prompt and right decisions.

Decision support systems (DSSs) and model codes are among the management instruments that can assist practitioners and the management level of water agencies and other water-related institutional units in arriving at sound, evidence-based decisions. A DSS is an interactive software-based system that helps decision-makers compile useful information from raw data, documents, and personal knowledge to identify and solve problems and optimize decisions (Ara et al. 2021). It is a framework that links together database(s) and the processing environment, an expert knowledge and information system, a modelling and analysis framework, and a communication framework. DSS will help the decision-making process understand the problem and explore various alternative courses of action. It helps the user analyse facts and situations, try out several scenarios, and help select the most appropriate decision. The lifecycle of DSS involves four main stages: (i) knowledge acquisition, (ii) problem structuring and system design, (iii) problem encoding/modelling, and (iv) system testing (Rao and Rajput 2009). DSS provides a customized, flexible, and dedicated management system to assist managers, decision-makers, and policymakers to (i) provide timely, transparent, well-informed, and reproducible answers to essential questions; (ii) quickly and effectively streamline workflow, reduce time, and cost requirements; (iii) transform data and information into knowledge, and (iv) produce understandable results and decisions (Pradhan and Rai 2020). DSSs have transitioned from engineering tools to systems that provide frameworks for stakeholder participation to guide, inform and support decision-making transparently and sustainably (Serrat-Capdevila et al. 2011). Such tools are seamlessly linked and tailored to a context, e.g. estimating water productivity in a region or a basin. A DSS has an open interface that can access models from different disciplines and developers with the help of adapters, which enable the DSS to access prepared input data and model parameters and store relevant model results (GWP 2013).

3.1.7 Use of Artificial Intelligence, Extreme Learning Machines, and Knowledge Engineering for Improved Water Productivity

Artificial intelligence (AI) techniques are excellent for constructing forecasting models and DSS. Several such methods are discussed in the previous sections, but complementary artificial intelligence (AI) paradigms are trending, thanks to rapid advancement in computational capacity and AI theory. In India, the first AI model adopted for ET_0 prediction was the Artificial Neural Network (ANN) (Kumar et al. 2002). Since then, progress has been made in India and elsewhere in the use of machine learning (ML), deep learning, and ridge regression, which help in removing the existing limitations and reduce unimportant components in the forecasting process, which could reduce the computational cost and enhance the accuracy of forecasting procedure.

3.1.7.1 Multi-step Daily Forecasting of ET0 Using Multivariate Complementary Technique

A novel multivariate variational mode decomposition technique (MVMD) integrated with the Ridge Regression (RR) feature selection algorithm and kernel extreme learning machine (KELM) model (i.e. MVMD-RR-KELM) was adopted to multi-step ahead ($t + 3$, and $t + 7$ days) forecasting of daily ET_0 (and thus the water productivity) in different climates of India (Malik et al. 2022). The complementary expert system hybridized with the boosted regression tree (BRT) and extreme gradient boosted (XGBoost) along with the standalone counterpart models (KELM,

BRT, and XGBoost) was examined to validate the robustness of the primary model to forecast the daily ET_0 in three regions of India. Important meteorological inputs, including the maximum and minimum temperatures, relative humidity, wind speed, and solar radiation, were simultaneously decomposed using MVMD pre-processing scheme. Then, the most significant decomposed components were utilized to feed the ML models for multi-step ahead ($t + 3$ and $t + 7$ days) forecasting of the ET_0 in three regions (Hisar in Haryana, and Bathinda and Ludhiana in Punjab state of India; part of the trans-Gangetic basin). The CROPWAT 8.0 software developed by the FAO (built and formulated FAO-56 P-M model) was used to calculate the values of daily ET_0 and considered as the baseline data for assessing AI models in semi-arid and sub-humid climates.

The MVMD is a superior multivariate decomposition tool as it (i) splits the multivariate modulated frequencies that otherwise would not be possible with a multivariate data set, (ii) mode alignment ability of multivariate signals, (iii) robustness to noise, and (iv) quasi-orthogonality across decomposed signals (Rehman and Aftab 2019). Ridge regression overcomes the collinearity issues by penalizing the least-squares loss on the regression coefficients. RR is a successor algorithm of the ordinary least square regression. KELM is an extension of an extreme learning machine (ELM), resulting in a technique with greater stability and accuracy than ELM (Huang 2011). The model was created in MATLAB R2020a, and the stepwise description of model development is given in Malik et al. (2022).

The forecasting accuracy of training and testing results of the hybrid AI model (M_{VMD}-RR-K_{ELM}) for $t + 3$ and $t + 7$ days ahead forecasting of ET_0 at Bathinda and Ludhiana (in Punjab) and Hisar (Haryana) are given in Table 3.2. The performance ranking of this model was better than other hybrid and standalone models, as evidenced by the goodness-of-fit values of Root Mean Square Error (RMSE), Mean Absolute Percentage Error (MAPE), and $U_{95\%}$ (upper endpoint of confidence

Table 3.2 Goodness-of-fit metrics of forecasting ET_0 for three and seven ahead days at Bathinda, Ludhiana, and Hisar stations with MVMD-RR-KELM complementary hybrid AI model

Station	Horizon	Mode	RMSE	MAPE	$U_{95\%}$
Bathinda	$t + 3$	Training	0.535	11.973	1.485
		Testing	0.497	13.537	1.379
	$t + 7$	Training	0.588	13.312	1.632
		Testing	0.555	17.006	1.538
Ludhiana	$t + 3$	Training	0.473	11.176	1.311
		Testing	0.561	14.606	1.535
	$t + 7$	Training	0.545	12.637	1.513
		Testing	0.643	15.085	1.781
Hisar	$t + 3$	Training	0.449	9.504	1.244
		Testing	0.494	12.299	1.364
	$t + 7$	Training	0.542	11.127	1.503
		Testing	0.545	12.132	1.504

Source: Malik et al. (2022)

interval). The climatic parameters data was trained for 1096 days and then tested for 365 days of data. The suggested complementary AI model MVMD-RR-KELM had minimum RMSE, MAPE, and $U_{95\%}$ values.

The goodness-of-fit metrics revealed that the KELM model integrated with MVMD-RR outperformed other standalone models for both horizons ET_0 at the three study locations. The prediction accuracy of the model was good. The use of AI and complementary forecasting models are immensely significant in developing a smart intelligence system for precisely forecasting daily ET0 and accordingly scheduling irrigation plans, thus designing smart DSSs for the sustainable use of water resources in water-scarce regions.

3.1.7.2 DSS to Select Suitable Cropping Pattern Using Fuzzy Multi-criteria Decision-Making

The cropping pattern of a region/basin significantly impacts the use of natural and economic resources and is fundamental to sustainable agriculture. The cropping pattern of a region may be considered a strategic decision that influences the interest of stakeholders, socio-economic effects, sustainability, and degree of automation in farming practices, irrigation methods, farm equipment, and the degree of farming skills. A robust DSS to select the suitable cropping pattern will help all the stakeholders in their cost-effective decision-making for improved crop and water productivity leading to sustainable farming practices and offset any serious concerns (Qureshi et al. 2018).

The fuzzy-based multi-criteria decision-making (MCDM) approach is more effective than the traditional MCDM approach, like the analytical hierarchy process (AHP), wherein weights derived are based on a pairwise criteria comparison. Biasness and vagueness in decision-making may be overcome by employing fuzzy-based methods. The fuzzy technique of order of preference by similarity to the ideal solution (Fuzzy TOPSIS) can be used for MCDM, e.g. in selecting the crop pattern for sustainable agricultural practices. The technique was employed for selecting the eight rabi season crops (wheat, barley, rabi cereals, rabi pulses, sugarcane, spices, and vegetables) using a set of 12 criteria in the framework above (Fig. 3.4).

The framework has the goal at the top and two levels beneath; Level I use 12 criteria (based on expert's opinions and an in-depth literature review), namely, water tariff, crop value, cultivation, crop storage infrastructure, water availability, water quality, soil texture, irrigation method, evapotranspiration, rainfall, and environmental conditions. Level 2 uses eight rabi crops (wheat, barley, gram, rabi cereals, rabi pulses, sugarcane, and spices) to decide the crop pattern. Experts were asked about the importance of each criterion for different crops through a well-structured questionnaire. After establishing the relative importance, the weights were used to judge suitable crop patterns for sustainable agricultural practices. The data obtained from experts' opinions were further analysed using the FTOPSIS methodology (Qureshi et al., 2008). Rankwise fuzzy weights for the different crop criteria were in the order: Water tariff (0.8333) > Crop value (0.800) = Crop demand (0.8000) > Soil

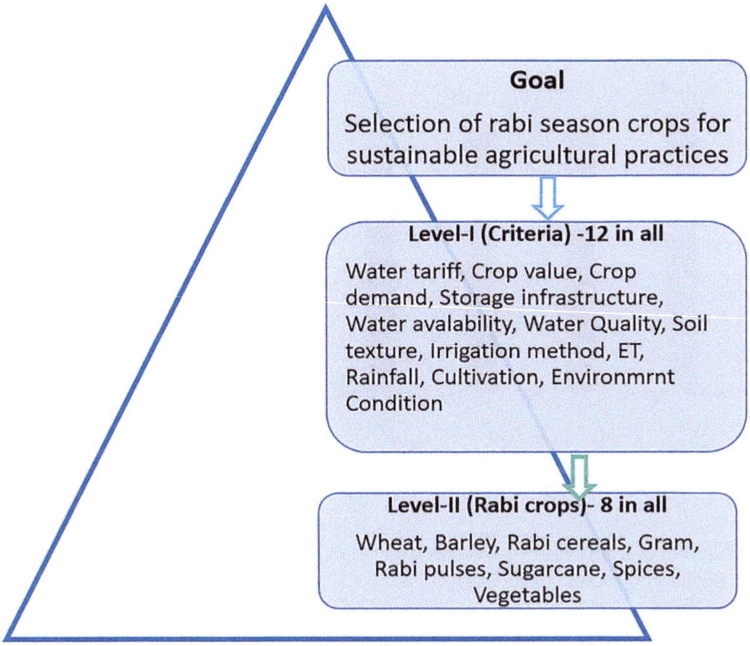

Fig. 3.4 Fuzzy TOPSIS model framework for selecting the suitable cropping pattern under the given criteria to meet the goal of sustainable agricultural practice

texture (0.7933) > Water availability (0.7667) > Water quality (0.7600) = Cultivation (0.7600) > Irrigation method (0.7267) > Environmental condition (0.5000) > Rainfall (0.3800) > Crop storage infrastructure (0.3400). Based on these weights, the closeness coefficients obtained for all the crops represent the ranking of crops for the optimum selection under the given MCDM (Fig. 3.5). The analysis showed a preference for high-value crops like spices, vegetables, sugarcane, and wheat, with high physical and economic water productivity.

The system-based fuzzy environment model will help farmers and agricultural policymakers to formulate a comprehensive DSS for sustainable agriculture practices. It poses many challenges, especially when it is to be derived under the influence of many criteria that influence sustainability and economic viability. The fuzzy TOPSIS methodology has the edge over other methods as it overcomes biases and vagueness to a great extent.

3.1.7.3 CommonKADS Model Framework for the Web-based Agricultural Decision Support System

Agricultural systems are often complex and semi-structured, making the decision support systems helpful tools for the agricultural community. One of the main objectives of agricultural DSSs is to improve productivity, including water

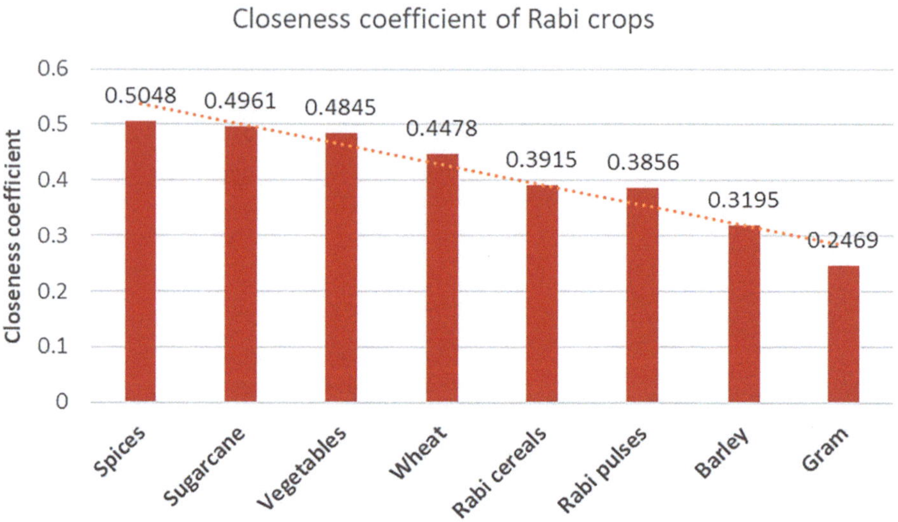

Fig. 3.5 Ranking of Rabi season crops based on the closeness coefficient employing the FTOPSIS technique. (Source: Qureshi et al. 2018)

productivity, sustainability, and profitability of agricultural systems despite differences in variability. DSSs are broadly categorized as DSS in a narrow sense as knowledge-based Intelligent DSS (IDSS) and Web-based DSS (Manos et al. 2004). IDSS and Web-based DSS can be interpreted as a hybrid DSS and Expert System (ES). These can perform diagnostic, advisory, informative, and operational roles in the areas such as irrigation scheduling and estimation of water productivity, farm management, nutrition advisory, and disease forecasting.

The knowledge engineering (KE) modelling approach simplifies the Knowledge-Based Systems (KBS) process by breaking down the problem into smaller tasks called models. Many modelling frameworks are used by KE communities, such as Common KADS (Knowledge Acquisition and Documentation Structure), MIKE, PROTAGE-II, and EXPECT. The Common KADS framework proposes six models in the construction process of KBS, which focus on organization, agents, tasks, communication, knowledge, and design (Schreiber et al. 1994). Figure 3.6 shows the web-based agricultural DSS organization model with each actor's roles and functions (Patel and Bhatt 2014). The knowledge providers are the traditional domain experts. The knowledge engineer elicits domain knowledge from the knowledge providers. Knowledge system developers employ the knowledge system using a suitable software platform. Finally, the farmers and extension agencies are the end user of the knowledge. The final form of the knowledge and advice must be in usable form and should be of practical value to the end user.

Using weather-based modelling, CommonKADS was used for irrigation scheduling (a common practice for improving yield and water productivity). An accurate estimation of ET_0 is significant for implementing weather-based irrigation scheduling. Agent # 1 provides these values, Agent #2 provides information on the soil

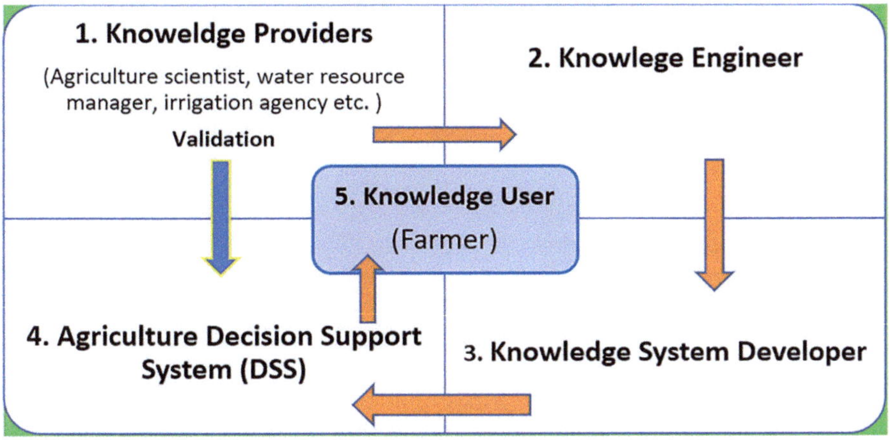

Fig. 3.6 Organization model of a web-based agricultural decision support system

water, and Agent #3 provides information based on crop stage and root zone depth. The communication model structures the knowledge between agents and specifies the details of the information exchanges between the agents. Similarly, the agents take information from the database or the knowledge base. The Knowledge model is an essential aspect of the CommonKADS framework. Finally, the design model suggests the tools for transferring the concept into implementation (Patel and Bhatt 2014). To improve the accessibility of the DSS among the farmers' community, the model may use mobile devices. The modelling framework allows the development of a comprehensive (not crop or task-specific) DSS. Additionally, it has an advantage over the conventional knowledge transfer approach regarding scalability and modularity.

3.1.8 DSS-WP Development for the Ganges Basin, India

The Ganga basin, covering an area of 1.05 million km^2 and traversing 2500 km in length, supports the livelihoods of more than 600 million people. The basin is highly fertile with abundant water resources, allocating a significant share used for agriculture, but with moderate to low agriculture and water productivity (Sharma et al. 2010). During the last six decades, the Ganga basin has undergone substantial hydrogeological and socio-economic changes—while rainfall in the basin has reduced by 11.25%, the evapotranspiration has only reduced by 3.61% with large variations in different regions. Further, any change in the river flow and groundwater abstractions compromises the integrity of the broader ecosystem functions, leading to tremendous pressure on basin water resources (Surinaidu et al. 2020).

Data, time, and resource limitations constrain most studies to model the whole basin, and as such, representative sub-basins are selected per the specific objectives.

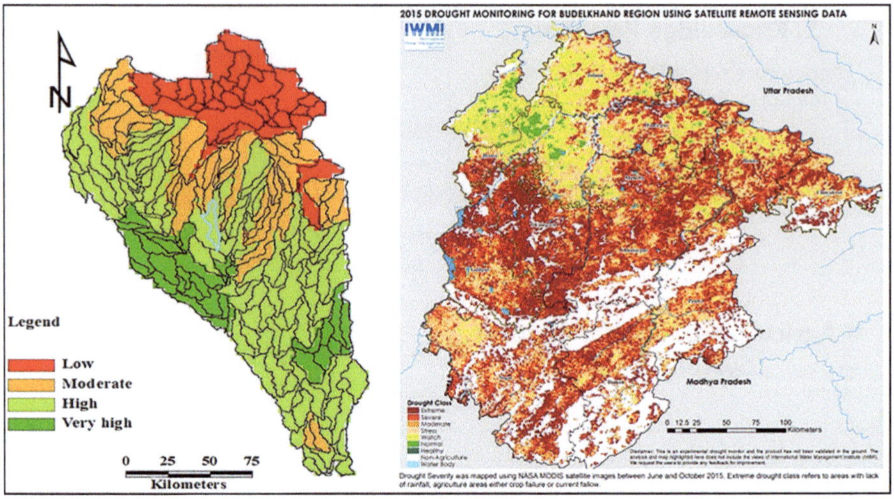

Fig. 3.7 Ramganga sub-basin (left) and Bundelkhand region (right) selected for the DSS-WP studies

For the Nexus Gains WP-2 package on DSS for water productivity, the Ramganga sub-basin and the Bundelkhand region are selected (Fig. 3.7). The two regions are diverse in agricultural, hydrological, and socio-economic aspects. Ramganga sub-basin (32,493 sq. km) is highly regulated through dams, barrages, and canals. The basin's climate is sub-tropical monsoon, and average rainfall varies from 2200 mm in the north to 800 mm near its confluence with Ganga. The average annual pan evaporation rate in the basin is 4.88 mm/day. About 79% of the basin area is under cultivation with rice, wheat, sugarcane, and horticulture.

Bundelkhand region (69,000 sq. km., Yamuna sub-basin) is semi-arid, hot, and drought-prone. Agriculture is mainly rainfed, diverse, complex, under-invested, risky, and vulnerable. Water deficiency, infertility of the land, soil erosion, low yields, and migration are the main features.

3.1.9 Main Elements of the Suggested Framework

The suggested 10-step framework for the development of the DSS-WP shall include the following elements:

3.1.9.1 Development of Water, Land, Energy, Ecosystem, and Socio-economic Database

A comprehensive and easily accessible database on the significant attributes of water, land, agricultural energy use, and socio-economic conditions is essential as the initial step in developing a robust and practical decision support system (DSS). Geodatabases on biophysical and socio-economic resources of the selected sub-basin/region may be developed using ArcGIS (or any other suitable system) to store secondary source maps and data, satellite remote sensing data, and additional GPS field surveys. It may be organized under relevant essential thematic layers, viz., rainfall pattern, canal network with system inflow and characteristics, groundwater resources and quality, location and extent of water bodies, satellite-derived current and past land use, cropping pattern, NDVI, soil texture, terrain, land use pattern, infrastructure, digital cadastral data, and gender-differentiated socio-economic data. Any other unique characteristics of the region, like areas affected by salinity/alkalinity, poor quality groundwater, wastewater resources, flood and/or drought-affected areas, and forest extent and types per the DSS requirement, may be collected. The Geodatabase can be queried for single or multiple attributes/features using the specified criteria such as monsoonal rain, cropping pattern, land use classification, number of women farmers, drought-prone areas, village, and other information. Geodatabase shall also help delineate the areas of low productivity and particular problem areas in the basin (Ramganga) or region (Bundelkhand). All the available data may be built-in for regular updating and change when new data or information becomes available and are thus easy to use by the stakeholders. Once the database is ready, it may be migrated to an open-source platform (Quantum GIS—QGIS 3.28.1 'Firenze) which shall allow easy sharing and distribution of the database and GIS software to the concerned stakeholders for the querying and creating value-added maps, etc. This can be migrated to the appropriate software (Geoserver) for online dissemination and a web map service for multi-thematic layers overlaid with Google Maps for online visualization of the area and identifying the resource constraints and other features at the desired level.

3.1.9.2 Assessment of the Needs and Expectations of the Stakeholders for the DSS-WP

A key reason for the limited success and adoption of the available DSS by the farmers and the related agencies is that most of the DSS in India have been created by researchers, irrigation managers, and software developers purely from a supply rather than a demand perspective. DSS development has proceeded due to a 'technology push' rather than the 'end-user pull' to help solve a problem or improve existing practice. DSS development results from an insufficient understanding of the farmers' and other stakeholders' needs assessment and the decision-making process and thus suffers from a lack of compatibility. At the later stage, low adoption of the DSS may result from the stakeholders' resistance to changing their traditional decision-making process. It shall be worthwhile to allocate sufficient time and resources to understand and document the stakeholders' needs and expectations from the DSS through field surveys, focussed group discussions, well-organized interactive meetings, and reviews. Further, to ensure the sustainability of the DSS, the users need to be supported through hands-on training, continuous and timely support on various queries, and anchoring of the DSS in an agency/organization responsible and responsive to the needs and expectations of the stakeholders.

3.1.9.3 Preparation of Crop Dominance Map, Crop Area, Crop Growth, and Yield Simulations for Grain, Biomass, and Economic Productivity

Crop dominance maps are required to identify the important crops, their spatial coverage, season and duration of growth, and related characteristics for further analysis. This classification can be achieved by using IRS-P6 (Indian Remote Sensing Satellite, series P6 or any other satellite data) and MODIS 8-day time series remote sensing images with a spatial resolution 250 m–500 m for the latest year (Cai et al. 2010, Fig. 3.8). Temporal variations in the NDVI pattern obtained in crop dominance classes also enable differentiation between long-duration crops (sugarcane, agro-forestry, orchards) and short-duration field crops. Crop simulation models like the DSSAT group of models and others can be used to estimate biomass growth and yield patterns. However, crop area, crop health, and yield for the sub-basin/region can be estimated using vegetation indices like LAI and NDVI that can be calculated from LANDSAT and other satellite imageries during various crop growth stages. Synthetic Aperture Radar (SAR) imagery helps collect satisfactory data where cloud cover restricts optical imagery (monsoon season in India), especially for rice

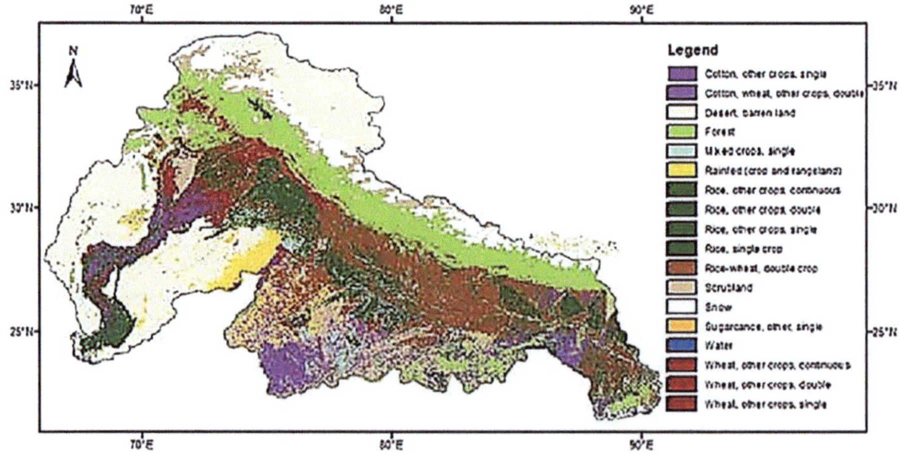

Fig. 3.8 Crop dominance map of the Indo-Gangetic basin with 18 classifications using RS-GIS techniques. (Source: details available in Cai et al. 2010)

and maize crops cultivated during the rainy season. Integrating remote sensing methods with conventional yield simulation techniques substantially improves the accuracy and spatial coverage at reduced costs and efforts.

Conversion of yield/biomass to economic value needs data on market prices of the various commodities. In India, a relatively developed system of 'minimum support price' (a price at which the central government assures the purchase of commodities from the farmers, https://cacp.dacnet.nic.in/content.aspx?pid=32) is available, and the data can be used for the conversion of the yields to their economic value. Alternatively, FAO-declared international prices of the commodities might be used. The performance of DSSAT crop growth models assimilated with remote sensing LAI/NDVI data for different crops was quite satisfactory in obtaining the data on the numerator of the WP equation.

3.1.9.4 Estimation of Agriculture and Ecosystem Water Needs

Traditionally, several models like FAO CROPWAT, AquaCrop, and others are available for estimating the crop water requirements (ET_0) under varying water availability and agroecological conditions. Region-wise water needs of the critical crops in areas containing the Ramganga sub-basin and parts of Bundelkhand (districts in Uttar Pradesh) are available in (Kumari et al. 2017). The methods based on AI, ELM, MCDM, and knowledge engineering are also available, but their use is limited. iCROP model in conjunction with IQQM-River system model and visual MODFLOW were successful in areas where conjunctive use of canal water and groundwater is made (Ramganga sub-basin) to work out the canal rostering and operations, conjunctive use of canal and groundwater and can operate as the main engine of the DSS (Raut et al. 2007).

For larger spatial units like the Ramganga sub-basin and Bundelkhand region of the Ganga basin, the ET_a algorithm SEBAL (Surface Energy Balance Algorithm) is widely used and quite successful in estimating the thermal difference between a 'cool pixel' and 'hot pixel' which has a strong correlation with ET_c. It replaces the use of physical methods of measuring the state of water stress in a single leaf to the area of the pixel and is thus more precise and does not require field water measurements, which can be a severe constraint in developing the DSS for large spatial units.

3.1.9.5 DSS Program for Estimating Water Productivity

The standalone window-based DSS-WP for Estimating and Improving Water Productivity (in the Ganges Basin) may be developed by integrating the database, key modules, crop-water-yield module calibrated with CROPWAT and CSM models to generate and evaluate the 'Best Management Practices (BMP)'-based interventions for resource scenarios under water scarcity, drought, and flooding conditions (to cater towards conditions in Bundelkhand region and Ramganga basins). The developed DSS may contain the main modules of Crop Dominance and

Crop Area, Crop Growth and Yield, Crop Water Demand, Canal Supply, Groundwater, Rainfall and Tank Water Supply, Micro-irrigation, Irrigation Scheduling, Best Management practices-based strategies, and Water Productivity. The DSS's heart can be the pythonSEBAL program for integrating the different modules. Additional modules can support the Database and its regular updating: Farmers' Services, Water Resource Services, Agriculture and Marketing Services, and Help and Queries. The Database module displays the different thematic data for assessing constraints and scenarios. The Crop Growth and Yield Module assess the crop area and dominance, growth during different stages, and final biomass and grain yield (by integrating with the Harvest Index). The Crop Water Demand module computes the crop ET from the optical and thermal remote-sensing data and daily weather data using suitable programs (pythonSEBAL/web-based CROPWAT). The irrigation demand is computed by aggregating the water demand of various crops after subtracting adequate rainfall and adding conveyance and application losses. A crop yield response module for the prevailing environments of the Ramganga sub-basin and Bundelkhand regions (and, for that matter, any other basin or region) like deficit irrigation, waterlogging, salinity, water stress, and pest and disease attack should be developed to predict the relative crop yield loss to generate and recommend innovative BMPs for minimizing yield loss and enhancing land and water productivity. The field demonstration data may validate the module. The additional modules may contain information related to Farmers' and Stakeholders' Queries and Help in the operation of the DSS.

3.1.9.6 Mapping of Location/Extent of 'Hotspots' and 'Bright Spots' of Water Productivity in the Ganges Basin

The DSS should have a routine to delineate and map the '*Hotspots*' *and* '*Bright spots*' of land and water productivity (both in absolute and percentage terms for different crops) and identify their potential causative factors. This will help target the specific interventions and help the affected farms and farming families. Analysis of the existing practices in the 'Bright spots' shall help tailor the available interventions for cross-comparisons and upscaling to other locations in the basins. Analysis of these spots' socio-economic conditions shall also help identify the investment and knowledge constraints and opportunities and thus address the same through appropriate measures.

3.1.9.7 Testing of the Impacts of the Interventions for Enhancing Water Productivity

The next step is to test and validate the impacts of the various interventions on enhancing land and water productivity, saving water, saving energy, allocating more water for the livelihoods of the women and the marginalized, and creating sustainable and healthier ecosystems. This may be achieved by running the scenarios for the selected parts of the basin or the region and then extrapolating (under realistic assumptions) to the whole basin or the state. The selected impacts can be both tangible in the form of enhanced physical and economic productivity and improved livelihoods and intangible in the form of reduced drudgery for women and children and sustainable ecosystems. The criteria for impact evaluation may be selected through focussed discussions and suggestions of the stakeholders and based on national and global best practices.

3.1.9.8 Setting up Farmers' Platforms for Demonstration and Validation

Setting up farmers' platforms and demonstration plots helps build the confidence of the farmers and stakeholder agencies, especially the inclusion of women farmers in these groups, in validating the DSS-generated BMP-based interventions. Several such Farmer Platforms of suitable size (say 10–20 ha size each) can be created to serve different objectives per the location and prevailing conditions. These farmer platforms shall be a formal grassroots institution for active participation and consensus-building in the DSS-WP activities. Continuous feedback from these units shall help in course correction and additional innovations. Farmer platforms also provide a workable solution to the small-size farm holdings of the farmers as the participating farmers may have consensus towards similar cropping patterns, irrigation scheduling, and other agronomic and socio-economic interventions. The government of India also provides financial and capacity-building support in setting up farmer platforms/farmer producer organizations and linking them to both input and output linkages for realizing the benefits across the entire value chain. At a higher level, Platform-of-Platform can be created (PoP, https://enam.gov.in/web/pop-dashboard/trading-platforms/platforms) to address the sub-basin level issues and realize higher economic benefits for enhancing economic water productivity. The field demonstrations through Farmer Platforms shall inspire confidence in stakeholders on DSS-generated interventions and validate the DSS modules.

3.1.9.9 Sharing and Dissemination of Database, DSS Program, and Knowledge to Stakeholder Agencies

Regular and continuous sharing/transfer of the Database, DSS program, and knowledge to stakeholders should form an essential component for the development and sustainability of the DSS. Sufficient human, financial, and other resources must be allocated to avoid the critical lesson from the previous experiences: 'the DSS was not available/implemented after completion of the Project'. For this, the DSS may be anchored in a suitable local organization/institution, for which this might not be an additional assignment. Innovations in knowledge engineering (e.g. CommonKADS) and ICT (mobile and web-based applications of the DSS and its specific modules) help in real-time and easy sharing of the product and its ongoing improvement. Participation of the local NGOs working in the area in the related fields better understand the local problems and sensitivities and help effectively communicate with the stakeholders. This will also help build the DSS-WP project's theory of change, communication material, and success stories.

3.1.9.10 Upscaling of the DSS to Other Basins and Regions

Feasibility assessment for upscaling the DSS may be conducted by hosting it in several institutions/agencies relevant to the objectives of the Project (Farmer Science Centres, ICAR institutions, District Agriculture Offices, Irrigation Command Offices, Water Resources Departments, Regional offices of Central Ground Water Board; IIT and NIH at Roorkee for Ramganga basin, and other development departments, private agencies, and Suitable NGOs). These institutions may be involved right from the inception, during the development and implementation process, and validation to ensure ownership and utility. Hands-on training and capacity building of the concerned persons and the institutions, and an effective and responsive 'Help and Query' module of the DSS shall help in upscaling the DSS to other sub-basins in the Ganges Basin in India, Nepal, and Bangladesh.

3.1.10 Conclusion

A matrix approach was employed to compile and review the DSS with broader segregation under (i) crop and farm-based decision support systems, (ii) DSS based on artificial intelligence, enhanced machine learning, fuzzy multi-criteria decision-making, and knowledge systems, (iii) DSS for real-time operation of micro-irrigation systems, (iv) DSS for management of tanks and reservoirs for water-deficit regions, and (v) DSS for improving water productivity under canal commands and conjunctive management of surface and groundwater resources. The selected DSSs were then analysed for focus, key findings, relevance, applicability, and scale of application of the DSS. The available DSSs assist the decision makers, though on a limited scale, in taking the right decisions based on a comparison of different strategies under various scenarios combining the benefits of geographic information systems, expert systems, and simulation models. Notwithstanding these benefits, the user must understand that there are so many uncertainties associated with natural resources, and even the best models may not be able to understand and code the natural processes of biological systems completely.

References

Ali MH, Talukdar MSU (2008) Increasing water productivity in crop production- a synthesis. Agric Water Manag 95:1201–1213

Amarasinghe UA et al (2007) India's water future to 2025–2050. Business-as-usual scenarios and deviations. International Water Management Institute, Colombo, p 47. IWMI research report 123. Accessed 15 Dec 2022

Ara L et al (2021) Application, adoption and opportunities for improving decision support systems in irrigated agriculture: a review. Agric Water Manag 257:107161

Arumugam N, Mohan S (1997) Integrated decision support system for tank irrigation system operation. J Water Resour Plan Manage 123:266–273
Bhattacharyya A et al (2015) Water resources in India: its demand, degradation and management. Int J Sci Res Publ 5:346–356
Cai X, Sharma BR, Matin MA, Sharma D, Gunasinghe S (2010) An assessment of crop water productivity in the Indus and Ganges river basins: current status and scope for improvement. International Water Management Institute, Colombo, p 30. (IWMI research report 140). https://doi.org/10.5337/2010.232
CGWB (2022) National compilation on dynamic groundwater resources of India, 2017. Central Ground Water Board (CGWB), Ministry of Water Resources, River Development and Ganga Rejuvenation, New Delhi, p 422
Clothier B, Jovanovic N, Zhang X (2020) Reporting on water productivity and economic performance at the water-food nexus. Agric Water Manag 237:106123
Cook S, Gichuki F, Turral H (2006) Agricultural water productivity: issues, concepts, and approaches. Colombo: International Water Management Institute (IWMI). Challenge Program Secretariat. 17p. (Challenge Program on Water & Food, Basin Focal Project Working Paper 1)
Fernandez JE et al (2020) Water use indicators and economic analysis for on-farm irrigation decision: a super high-density olive tree orchard case study. Agric Water Manag 237:106074
Gulati A et al (2019) Getting more from less: story of India's shrinking water resources. NABARD and ICRIER report, Indian Council for Research on International Economic Relations, New Delhi. p 170. Accessed 12 Nov 2022
GWP (2013) The role of decision support systems and models in integrated river basin management. Technical Focus Paper. Global Water Partnership, p 48. Accessed 10 Nov 2022
Huang GB (2011) Extreme learning machine for regression and multiclass classification. IEEE Trans Syst Man Cybern Part B 42:513–529
Kijne JW, Barker R, Molden D (2003) Improving water productivity in agriculture. In: Kijne JW, Barker R, Molden D (eds) Water productivity in agriculture: limits and opportunities for improvement. CABI Publishing, pp xi–xix
Kumar M et al (2002) Estimating evapotranspiration using artificial neural network. J Irrig Drain Eng 128:224–233
Kumari M, Singh OP, Meena DC (2017) Crop water requirement, water productivity and comparative advantage of crop production in different regions of Uttar Pradesh. Int J Curr Microbiol App Sci 6(7):2043–2052
Mabhaudhi T et al (2018) Prospects for improving irrigated agriculture in southern Africa: linking water, energy and food. Water 10:1881
Mackrell D, Kerr D, Von Hellens L (2009) A qualitative case study of the adoption and use of an agricultural decision support system in the Australian cotton industry: the socio-technical view. Decis Support Syst 47:143–153
Malik A et al (2022) Multi-step daily forecasting of reference evapotranspiration for different climates of India: a modern multivariate complementary technique reinforced with ridge regression feature selection. Agric Water Manag 272:107812
Manos B, Ciani A, Bournaris T, Vassiladon I, Papathanasious J (2004) A taxonomy survey of decision support systems in agriculture. Agric Econ Rev 5(2):80–93
Molden D et al (2010) Improving agricultural water productivity: optimism and caution. Agric Water Manag 97:528–535
Mora M, Forgionne G, Gupta J (2003) Decision-making support systems: achievements and challenges for the new decade. Natural Resources Planning, Silva Fennica, Harrisburg
Nhamo L, Mabhaudhi T, Magombeyi M (2016) Improving water sustainability and food security through increased crop water productivity in Malawi. Water 8:411
Patel J, Bhatt C (2014) A CommonKADS model framework for a web-based agricultural decision support system. Int J Food Syst Dyn 5(4):196–203

Perry C, Steduto P, Allen R, Burt C (2009) Increasing productivity in irrigated agriculture: agronomic constraints and hydrological realities. Agric Water Manag 96:1517–1524. Water-use efficiency and productivity: rethinking the basin approach. Available from: https://www.researchgate.net/publication/233068666_Water-Use_Efficiency_and_Productivity_Rethinking_the_Basin_Approach. Accessed 21 Sept 2023

Pradhan A, Rai RK (2020) Role of decision support system for river pollution control. J Water Eng Manag 1(1):1–13

Qureshi MRN, Singh RK, Hasan MA (2018) Decision support model to select crop pattern for sustainable agricultural practices using fuzzy MCDM. Environ Dev Sustain 20:641–659

Raman H, Mohan S, Rangacharya NCV (1992) Decision support for crop planning during droughts. J Irrig Drain Eng 118:229–241

Rao KB, Rajput TBS (2009) Decision support system for efficient water management in canal command areas. Curr Sci 97:90–98

Raut AK, Bansal AK, Verma VK, Marr JA (2007) GIS-based decision support systems for conjunctive irrigation management in India. Draft final report of the Project available at https://swaraup.gov.in

Reddy MN, Rao NH (1995) GIS-based decision support systems in agriculture. ICAR- National Academy of Agricultural Research Management, Hyderabad, p 11. Accessed 9 Dec 2022

Rehman N, Aftab H (2019) Multivariate variational mode decomposition. IEEE Trans Signal Process 99:1. [Accessed through research gate on 28 November 2022]

Saggi MK, Jain S (2022) A survey towards decision support system on smart irrigation scheduling using machine learning approaches. Arch Comput Methods Eng 29:4455

Schreiber A et al (1994) CommonKADS: a comprehensive methodology for KBS development. IEEE Expert December:28–37

Serrat-Capdevila A, Valdes J, Gupta H (2011) Decision support systems in water resources planning and management: stakeholder participation and the sustainable path to science-based decision making. In: Jao C (ed) Efficient decision support systems: practice and challenges from current to future. Intech Open, London. https://doi.org/10.5772/16897

Sharma B et al (2010) The Indus and the Ganges: river basins under extreme pressure. Water Int 35(5):493–521. https://doi.org/10.1080/02508060.2010.512996

Shiklomanov IA, Rodda JC (2003) World water resources at the beginning of the twenty-first century. Cambridge University Press, p 25

Sprague RH, Carlton ED (1982) Building effective decision support systems. Prentice-Hall, Englewood Cliffs

Surinaidu L et al (2020) Assessment of long-term hydrogeological changes and plausible solutions to manage hydrological extremes in the transnational Ganga river basin. H2O Open J 3(1):457–480. https://doi.org/10.2166/h2oj.2020.049

Uhlenbrook S, Connor R (2019) The United Nations world water development report 2019: leaving no one behind. UNESCO, Paris, p 201

Wellen J, Raes D, Tychon B (2017) On the use of decision support tools for improved irrigation management: AquaCrop-based applications. INTECH Open Science

Chapter 4
Remote Sensing Maize Water Stress in Smallholder Farms: A Systematic Review of Progress, Challenges, and the Way Forward Using Earth Observation Data

M. Kapari, M. Sibanda, J. Magidi, L. Nhamo, S. Mpandeli, and Tafadzwanashe Mabhaudhi

Abstract Maize (*Zea mays* L.) is a staple food crop that smallholder farmers mostly cultivate under rain-fed conditions in Southern Africa. Despite significant contributions to food production by smallholder farmers, they face climate change-related

---

M. Kapari (✉)
Water Research Commission of South Africa, Pretoria, South Africa

University of the Western Cape, Bellville, Cape Town, South Africa
e-mail: 4280223@myuwc.ac.za

M. Sibanda
University of the Western Cape, Bellville, Cape Town, South Africa

J. Magidi
Tshwane University of Technology, Pretoria, South Africa

L. Nhamo
Water Research Commission of South Africa, Pretoria, South Africa

Centre for Transformative Agricultural and Food Systems, University of KwaZulu-Natal (UKZN), Pietermaritzburg, South Africa

S. Mpandeli
Department of Environmental, Water and Earth Sciences, Tshwane University of Technology (TUT), Pretoria, South Africa

Faculty of Science, Engineering and Agriculture, University of Venda, Thohoyandou, South Africa

T. Mabhaudhi
Centre on Climate Change and Planetary Health, London School of Hygiene and Tropical Medicine, London, UK

Centre for Transformative Agricultural and Food Systems, University of KwaZulu-Natal, Carbis Road, Scottsville, Pietermaritzburg, KwaZulu-Natal, South Africa

T. Mabhaudhi et al. (eds.), *Enhancing Water and Food Security Through Improved Agricultural Water Productivity*,
https://doi.org/10.1007/978-981-96-1848-4_4

challenges such as drought, resulting in crop water stress and significant yield losses. This is exacerbated by the lack of financial resources, mechanical skills, and sound climate change adaptation strategies, increasing the yield gaps. This could potentially be addressed through technological advancements such as precision farming systems. Remote-sensing systems are sufficient and well equipped to address crop production's complex and technical assessments, such as crop water stress, inexpensively and efficiently. This study sought to systematically review the literature on the progress, emerging gaps, and opportunities in applying remote sensing technologies in quantifying maize water stress. Adhering to the PRISMA guide, 100 peer-reviewed articles were examined from Web of Science, Scopus, Google Scholar, and ScienceDirect. Results significantly increasing research efforts have been exerted from 2002 to the present, with the majority of research articles (37%) being conducted in the United States and the least (12%) in the African continent. Specifically, 17 different Earth observation sensors were used to map maize water stress. Landsat is the most widely utilized sensor, particularly the red and near-infrared regions of the electromagnetic spectrum, along with their derivatives. These Landsat spectral derivatives are used mostly in conjunction with the surface energy model in retrieved literature. However, there is a dearth of literature on remote sensing maize crop water stress in smallholder croplands. This is mainly because these agricultural systems are extremely small (<1 ha) and heterogeneous to be detected by moderate spatial resolution sensors that are freely available. Furthermore, validation mechanisms, data, and fine spatial resolution suitable for these croplands are scanty, if not expensive. Providentially, UAV-based remote sensing technologies, which are relatively cheaper, with ultra-high spatial resolutions, and user-defined acquisition times have emerged as suitable alternatives. In this regard, more research efforts are required to assess the prospects of these technologies, especially in smallholder farms in southern Africa associated with limited resources.

Keywords Remote sensing · Precision agriculture · UAVs · Food security · Maize

4.1 Introduction

Maize (*Zea mays L.*) is one of the most important grains in the world and serves as a carbohydrate source for human nutrition in developing countries and the most critical animal feed in developed countries (Tandzi and Mutengwa 2020). Maize is a staple cereal crop in southern Africa and other regions, with an increased food demand due to population growth (Ndlovu et al. 2021). Therefore, increasing maize production is essential to meet the increasing demand for food and nutrition security (Ekpa et al. 2018). Maize is the most dominant crop smallholder farmers grow, especially in developing regions such as Southern Africa (Ndlovu et al. 2021). Unfortunately, smallholders are also susceptible to climate change shocks, such as reduction and high variability of precipitation and droughts, making it difficult to achieve the desired yields (Chitja et al. 2015). However, smallholders are the most affected by climate change shocks. They lack financial and mechanical skills and are less equipped with infrastructure, agricultural equipment (Ruwanza et al. 2022), and sound adaptation strategies. Song et al. (2019) stated that substantial water stress during the maize's vegetative growth inhibits the crop's plant growth and leaf area, resulting in declined yield. Moisture conditions are essential in plant function, water and energy exchange with the atmosphere, and drought and fire risk (Mirzaie et al. 2014). As such, when assessing the impact of climate change on agriculture and agricultural water management, water availability is an essential factor to consider (Matese et al. 2018). Drought caused by climate change, according to Zhang et al. (2019c), will reduce agricultural water resources, necessitating maximum production per unit of applied irrigation water.

Subsequently, effective monitoring methods of maize water status are essential in implementing informed irrigation strategies while avoiding water waste and crop yield loss (Ihuoma and Madramootoo 2017; Zhang et al. 2019c). Meanwhile, minimizing water use in agriculture is a prerequisite for freeing water resources for other sectors of society in increasing demand (Rossini et al. 2015). As such, numerous studies have investigated the effects of water stress on maize's further growth and development stages (Sreelash et al. 2017; Dong et al. 2021). Maize crop water stress can be detected by assessing and monitoring crop physiological characteristics, soil moisture content, and remote sensing (RS) technology (Ihuoma and Madramootoo 2017). However, conventional maize crop water stress monitoring techniques rely on in-situ soil water measurements and meteorological variables to estimate the amount of water lost from the plant–soil system over a specific period (Elkatoury et al. 2020). Although accurate, methods such as the lysimeter, eddy covariance, and atmometers are point-based (Padilla et al. 2011); hence, they lack spatial representativeness. Furthermore, these methods are time-consuming, labour-intensive, and costly and do not consider the spatial variability of soil and plants (Maes et al. 2018). With a large amount of data available from various earth observation platforms, research in monitoring crop water stress using satellite and aerial RS techniques has emerged as the most spatially explicit method (Maes et al. 2018). RS is recognized as a fast, non-destructive technique widely used to quantify biophysical and biochemical parameters related to maize crop water stress at various scales (Mirzaie et al. 2014). RS allows for acquiring very high spatial, spectral, and temporal resolution data as input for precision agriculture (Zhang et al. 2019c). RS is one of the most essential tools for monitoring crops (Ahmad et al. 2021). Canopy reflectance measurements by RS technology facilitate easy, non-destructive and low labour-intensity data collection (Elmetwalli et al. 2021). The electromagnetic energy plants reflect depends on the crop's physiological and structural conditions and the environment (Wijesingha et al. 2021). Developing RS data acquisition and analysis techniques helps obtain accurate models for estimating crop parameters (Gerhards et al. 2019). Various earth observation sensor platforms (ground instruments, unmanned aerial vehicles (UAVs), airborne and satellite) have been deployed to acquire remotely sensed data on croplands for estimating plant growth and health parameters through various modelling approaches (Rossini et al. 2013; Zhang et al. 2019c). These include multispectral (Mariapaola Ambrosone et al. 2020), thermal infrared (Mangus et al. 2016), hyperspectral (Rossini et al. 2013), red-edge (Shao et al. 2021), visible (Genc et al. 2013), and radar (van Emmerik et al. 2015) sensors application on different platforms to detect these different reflected energies.

However, crop monitoring applications, including maize, are very demanding regarding the temporal and spatial resolution requirements. Until recent technological advancements, it has often been challenging to meet the monitoring needs of plant water stress, even at local scales. To meet the demands of farm-scale monitoring, RS techniques must have at least 1–3 days of repetition of data acquisition and spatial resolution of less than 10 m (Zhang et al. 2016). As such, increased adoption of RS use at a farm-scale/local scale has primarily been due to the widespread use of new technologies that integrate high-resolution cameras onboard unmanned aerial vehicles (UAVs) (Messina and Modica 2020). In addition, recently, the significance, capability, and possibilities of using RS data in monitoring crop water stress are well established; this includes using RS data in machine learning algorithms (Virnodkar et al. 2020).

Nonetheless, aerial RS has the disadvantage of being difficult and expensive to operate. The advantages of low cost, simple structure, convenient transportation, high flexibility, short operating cycles, and high spatiotemporal resolution allow UAV systems to collect crop information with the desired spatiotemporal resolution (Zhang et al. 2019c). This makes UAVs more suitable for rapidly and effectively monitoring crop moisture stress on an agricultural scale (Zhang et al. 2019c, 2022). To devise efficient irrigation strategies for smallholder croplands, it is crucial to determine suitable indicators for monitoring the water status of maize crops at the individual farm level (Alvino and Marino 2017).

Much work has been carried out in reviewing the literature on the utility of remotely sensed data in estimating crop moisture stress parameters. For instance, leaf area index (LAI) (Matese et al. 2018; Wijesingha et al. 2021), leaf water content (LWC) (Song et al. 2021); crop water stress index (CWSI) (Testi et al. 2008; Matese et al. 2018; Alordzinu et al. 2021), and canopy water content (CWC). Alvino and Marino (2017) reviewed the literature on the relationship between surface temperature and remotely-sensed vegetation indices in relation to water-use efficiency (WUE) and evapotranspiration, while Yan et al. (2006) reviewed the literature on the advances in RS applications in mapping soil moisture, with a focus on methodological details and discussed issues related to estimating soil moisture from remotely sensed data. Greco et al. (2012) limited their literature review to utilizing ground-based thermal remotely sensed data, excluding satellite and airborne techniques. Meanwhile, recent literature reviews focus on the utility of UAV remotely sensed data in estimating crop water stress (Awais et al. 2021); this includes the use of UAV remotely sensed thermal data for mapping crop water stress (Messina and Modica 2020; Ahmad et al. 2021). Although a review on cereal crops' water footprint estimation using RS and other methods (Feng et al. 2021) included maize, little to no research efforts have been exerted towards reviewing the literature on utilizing RS data in estimating maize crop water stress at local scales. This lack of summarized knowledge limits the realization of efficient, reliable, responsive, practical, and appropriate RS options to estimate maize water stress for smallholder croplands.

Therefore, in light of the dearth of knowledge on RS application on maize water stress, this study conducted a systematic review to assess the literature on the progress, challenges and the way forward on using RS to monitor maize water stress. Understanding the progress made in RS will provide insights into the needs and areas of expansion required in maize water stress detection, such as various methods available. This will benefit role players in the agricultural and water sector research to effectively determine maize needs and further direct irrigation plans. The study aims to achieve this by establishing advances in monitoring maize water stress using RS technologies and identifying challenges and opportunities associated with using RS technologies to monitor maize water stress.

4.2 Methods

This study intended to conduct a systematic review of the use of RS in assessing maize water stress. The Preferred Reporting Items for Systematic Reviews (PRISMA) guidelines were followed to produce this review (Albeha et al. 2020). PRISMA provides a guideline checklist that is peer-accepted and thus will be followed in this paper. The review is structured into two sections. The first section outlines progress in adopting RS for estimating maize water stress; the second section outlines the challenges and the way forward in applying RS in monitoring maize water stress. The literature search and analysis were conducted in four stages to address these sections. These stages describe the article selection criteria and search strategy, eligibility selection criteria, data extraction and analysis procedures.

Stage 1: Literature Search
Four electronic databases, Scopus, Web of Science, ScienceDirect, and Google Scholar, were used for a systematic search with no limitations to the year of publication. The keywords combination used to obtain data across SCOPUS, Web of Science, ScienceDirect, and Google Scholar databases were searched using the string in Table 4.1.

Stage 2: Screening and Eligibility Criteria
The articles eligible for the analysis had to meet the following criteria:

1. Peer-reviewed articles in accredited journals
2. Studies written in English
3. Studies focusing on other crops' water stress and/or maize
4. Studies based on RS techniques in detecting maize water

The bibliographic information of the articles was retrieved and compiled on Mendeley's desktop for screening preparations.

About 95, 180, 50, and 14 articles were retrieved from Scopus, Web of Science, Google Scholar, and ScienceDirect, respectively (Table 4.1). Titles and abstracts were reviewed using the abovementioned criteria to determine the study's eligibility to be included in the study. The first screening process comprised the removal of duplicates, resulting in a total of 169. After that, irrelevant articles were removed ($n = 17$), including articles not written in English, resulting in 152 articles. Included

Table 4.1 Key search words that were used in this study

Search database	Search criterion	Total no. of articles	No. of articles
SCOPUS	TITLE-ABS-KEY (("Maize" AND "Water stress") AND "RS" OR "GIS")	114	95
Web of Science	("Maize" AND "Water stress") AND ("RS" OR "GIS"); ("Corn" AND "Water stress") AND ("RS" OR "GIS"); ("Maize" AND "Water stress") AND "RS" OR "GIS" AND ("smallholder farms" OR "smallholder agriculture") (All Fields)	392	180
Google Scholar	("smallholder farms" AND ("maize" AND "RS" AND "water stress" OR "Leaf moisture" OR "stress" OR "water content*" OR "moisture content") AND "unmanned aerial vehicles*" OR "drones") and (("maize" AND "water stress") AND "RS" OR "GIS" AND "smallholder farms" OR "smallholder agriculture")	1092	50
ScienceDirect	("maize water stress" OR "corn water content" AND "RS" AND "farmers")	574	14
Articles considered before screening after removing duplicates = 169			

Source: Own data

and excluded articles were recorded per PRISMA and meta-analysis statement (Fig. 4.1). All studies unavailable in portable document format (pdf) were excluded, resulting in 138 studies. The pdf information of the articles was retrieved and compiled on *Mendeley desktop* for further screening preparations. Full-length articles of the selected articles were then downloaded, and the number of retained articles after the screening was recorded at 100. After that, a Microsoft Excel spreadsheet was created to capture each study's details and was used for quantitative assessment.

Stage 3: Data Extraction

Bibliometric information of the selected articles, such as the author's names, the title of the article, the year of publication, the name of the journal, keywords, and the abstract digital object identifier (DOI), uniform resource locator (URL) was exported from Mendeley to a Microsoft Excel spreadsheet. Furthermore, information on the study was conducted (country), the type of farm studied, the type of platforms, sensors, and instruments used, the type of RS method used, the sensors' spectrum coverage, various vegetation indices, type of a water stress indicator studied, and prediction method used were also captured after going through each article. In addition, the coefficient of determination (R^2) in each study, the source of water for the studied maize, and validation methods were recorded. The categorical data were then converted into numerical variables of zeros (No) and ones (Yes) to prepare for data analysis.

Stage 4: Data Analysis

Data extracted from the retrieved articles were subjected to qualitative and quantitative analysis. First, bibliometric analysis was performed to visualize the literature's

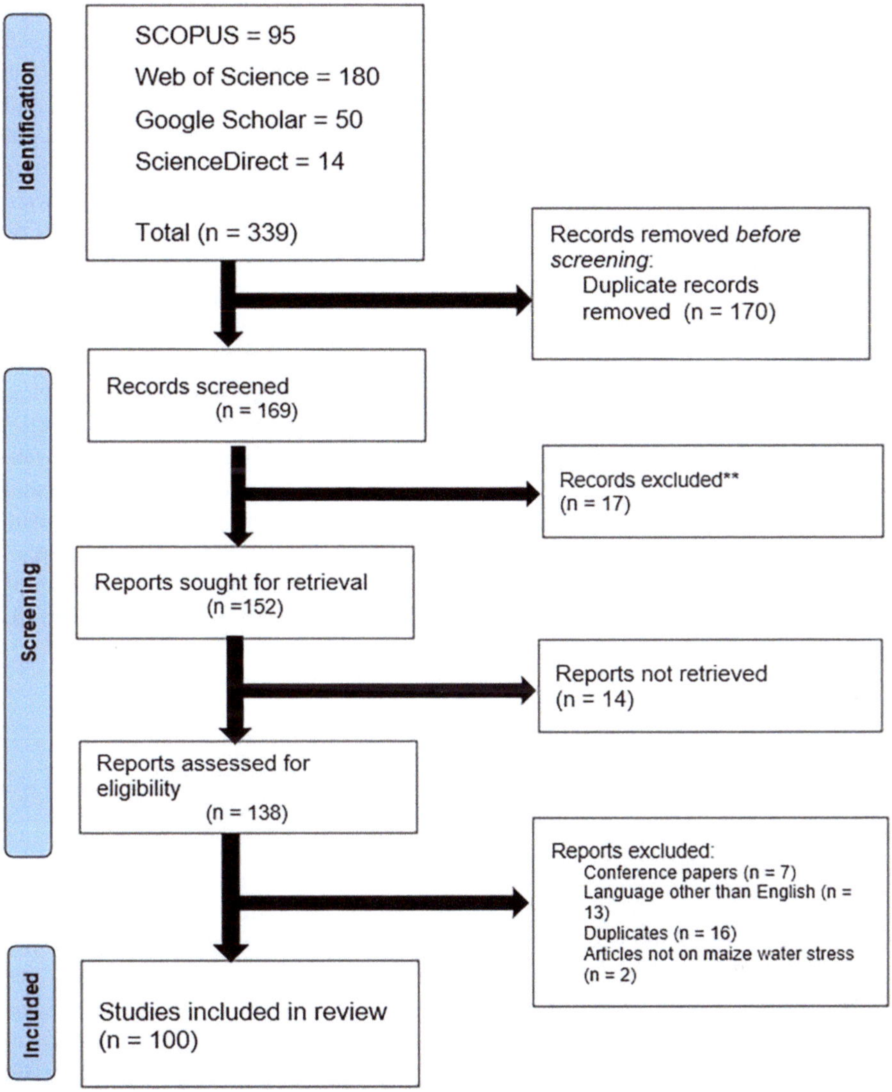

Fig. 4.1 Results and filtering process of articles that comply with the search keywords. (Source: Own data)

occurrence and co-occurrence networks of key terms. Bibliometric analysis is a quantitative method for evaluating published articles useful for evaluating peer-reviewed studies in a specific field (Han et al. 2020). The analysis used *VOSviewer* software (van Eck and Waltman 2010). *VOSviewer* provides network visualization of binding terms in terms of linked clusters. Creating a map using the VOSviewer included selecting a counting method, selecting the minimum number of occurrences for a term, calculating the relevance score for the co-occurrence terms and, based on this score, displaying the most relevant items, displaying a map depending on the selected items, consecutively (Masenyama et al. 2022). The features of *VOSviewer* for bibliometric mapping are described in detail by van Eck and Waltman (2010). The final database's titles and abstract were divided into two groups of studies using UAVs and studied using satellite and airborne (collectively) were used as input text data in *VOSviewer* to generate graphical visualizations that rely on critical terms for monitoring and mapping the occurrence and co-occurrence of maize water status (van Eck and Waltman 2010) by different sensors. Titles and abstracts of the final database articles were used in the *VOSviewer* software to investigate the trend of concepts and topics in mapping and monitoring maize water stress.

Graphs were created using *Microsoft Excel* for each historical study's characteristics and are reported in the Sect. 4.3. The review was then divided into two sections, which serve the purpose of the study. The first section examines previous maize moisture stress mapping advances and modelling using RS data. This section presents and details trends in the literature for quantitatively assessing maize water stress. The final stage outlines and discusses the challenges and way forward associated with knowledge generation in maize water stress mapping and modelling using remotely sensed data. Both stages considered literature search characteristics, plant biophysical parameters, earth observation sensors, sensor platforms, algorithms, and previously used optimal spectral settings in the discussion. Not only were frequencies assessed in this study, but bias tests were also not conducted.

4.3 Results

4.3.1 Literature Search Characteristics

In evaluating the evolution of topical concepts in predicting and mapping maize crop water stress on VOSviewer based on the information from titles only, the network map in Fig. 4.2 categorized the identified concepts into four clusters: red, green, blue and yellow. The first cluster in red has words "day", "rmse" (root mean square error), "stage", "water", "eta" (actual evapotranspiration), "wue" (water use efficiency), "irrigation", "accuracy", "reflectance", and "water deficit". This cluster links the application of airborne and satellite data in maize water stress detection; maize is grown at different "irrigation" and "water deficit" levels, and the spectral "reflectance" is measured during the "day" in all the different phenological stages. After that, the airborne and satellite data is analysed to model maize "wue" The model is then validated by comparing with ground data using "rmse" as a statistical measurement indicator. The second cluster in green comprises the words "impact", "relationship", "drought stress", and "growth". The words "impact" and "relationship" denote the purpose of the study, which determines "drought stress" on maize "growth". The third cluster in blue has key terms "soil", "ndvi" (normalized difference vegetation index), "vegetation indices", and "vwc" (vegetation water content).

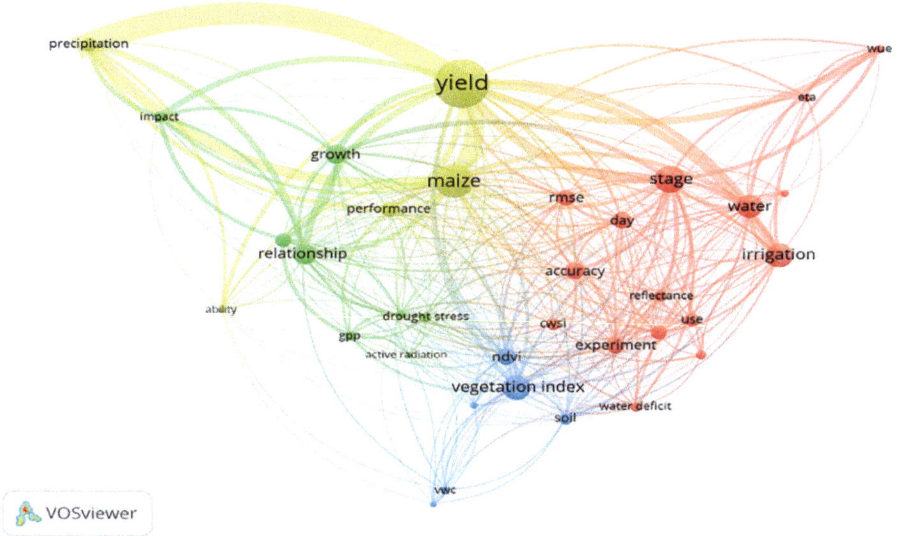

Fig. 4.2 Concepts of satellites and airborne sensors application in maize water stress detection from studies' abstracts

This cluster links the input data in the models predicting maize water stress, "soil" reflectance or moisture, "ndvi" (normalized vegetation index), and other "vegetation indices" and "vwc" measurements. The last cluster in yellow has key terms "ability", "precipitation", "maize", and "yield". This particular cluster denotes the "ability" of satellite and airborne sensors to estimate "maize" "yield" for rainfed maize under "precipitation".

Similarly, results from the co-occurrence of concepts in studies utilizing UAVs to monitor and assess maize water stress in Fig. 4.3 show four cluster categories. The first cluster, in red, has the concepts "temperature", "UAV", "cwsi" (crop water stress index), "lst" (land surface temperature), "rmse", "value", and "ground". This cluster signifies "temperature" data obtained using "UAV" has been utilized to estimate "lst" and "cwsi"; the relationship between UAV data and "ground" acquired "values" is compared and presented in "rmse" values. The second cluster (green) has the terms "data", "crop", "ndvi", and "season". This particular cluster denotes the use of maize "crop" "reflectance" data has been utilized to develop "ndvi" for the growing "season" of the crop. The third cluster (blue) connects "chlorophyll fluorescence" and "pri" as the indicators of "maize" and "water stress". The last cluster links the potential of UAVs to "model" maize water stress at a "field" scale using "cdi" (crop drought index) as a proxy.

In assessing the titles and abstracts, the co-occurrence analysis resulted in four clusters similar to the only abstract analysis (Fig. 4.2) shown in Fig. 4.4. Cluster 1 (Red) has key terms "soil", "vegetation index", "vwc", and "rmse". This cluster denotes the relationship between measured and estimated "vegetation index" and "vwc" has been determined using "rmse" values. The second cluster (green) has

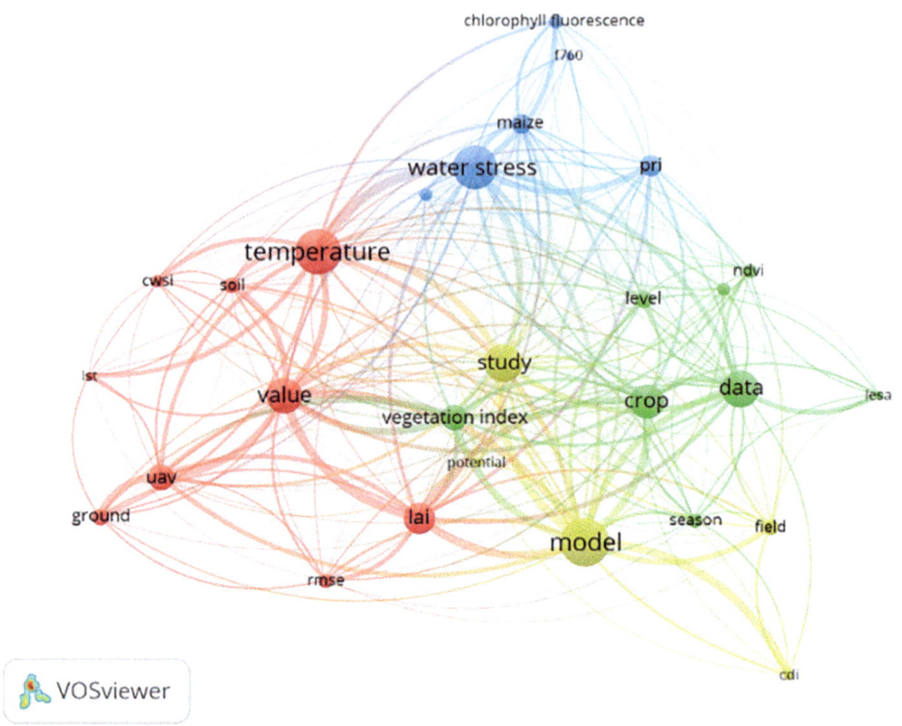

Fig. 4.3 Concepts of UAV application in maize water stress detection from studies' abstracts

terms such as "satellite", "drought", "growth", "impact", "relationship", and "response". This cluster links the prospects of utilizing "satellite" remotely sensed data in assessing the "response" of maize "growth" to "drought" stress. Also, including the words "impact" and "relationship" represent the purpose of the study, thus informing the remote sensing model applied. The third cluster (blue) is "wue", "eta", and "irrigation". This particular cluster denotes the measurements of maize "wue" and "eta" at different irrigation levels during the changing phenological "stages". Cluster 4 has "precipitation", "maize", and "yield"; this cluster links "precipitation" levels with the increase or decrease of "maize" "yield".

The UAV used in monitoring and analysing maize water stress in Fig. 4.5 shows four cluster categories. The first cluster in Red has key terms "cwsi", "lst", "soil", "value", "lai", "vegetation index", "ground", and "rmse". It links the indicators of maize water stress "soil" moisture, "lst", "lai", and "cwsi" and other "vegetation indices"; they are compared with "ground" measurement using "rmse" for accuracy assessment. The second cluster in green has the keywords; "data", "season", "ndvi", "crop", and "level". This cluster denotes that "ndvi" data collection procedures, "data" is collected during maize "crop" growing "season" at different water stress "levels". The third cluster in blue has key terms "pri", "maize", "canopy temperature", and "water stress detection". This cluster links the proxy used to determine

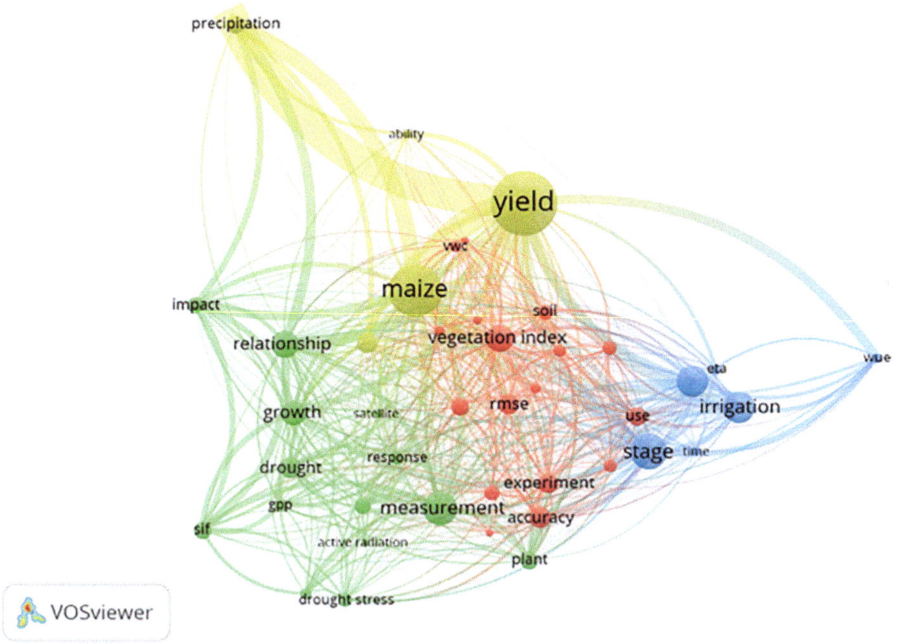

Fig. 4.4 Concepts of satellites and airborne sensors application in maize water stress detection from studies' titles and abstracts

"maize" and "water stress detection", which include "canopy temperature" reflectance and "pri". The last cluster in yellow links the potential of UAVs to "model" maize water stress at a "field" scale.

4.4 Trends in Publications

Recently, much has been focused on assessing and modelling maize water stress using remotely sensed data. This is demonstrated by the constant rise in research that evaluated and modelled maize water stress using RS techniques. The study results reveal that the earliest publication of maize water stress was in 2002 (Fig. 4.6). Since then, there has been a constant publication rate with an increase from 2011. The year 2011 shows a drastic increase of 15 more publications compared to the beginning of 2002. Overall, results show that RS applications for understanding maize water stress significantly increased from 2011 to date. As a result, 100 articles were published from 2002 to mid-2022. Overall, 38 countries around the world remotely sensed maize water stress. Regarding the spatial distribution of these studies, results show that the United States (37) is leading, globally, in the utilization of remotely sensed maize water stress, followed by China (Fig. 4.7).

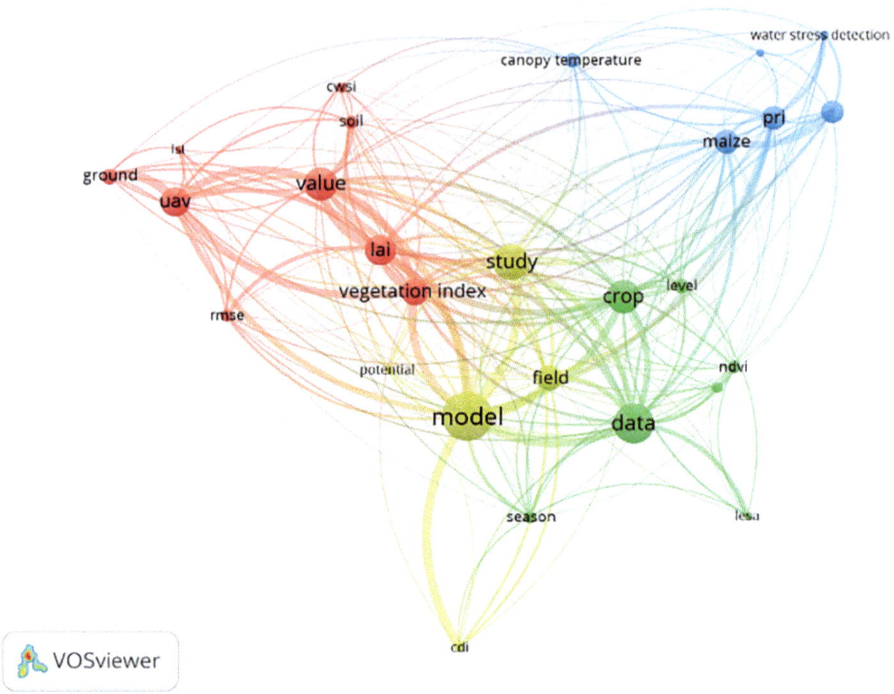

Fig. 4.5 Concepts of UAV application in maize water stress detection from studies' titles and abstracts

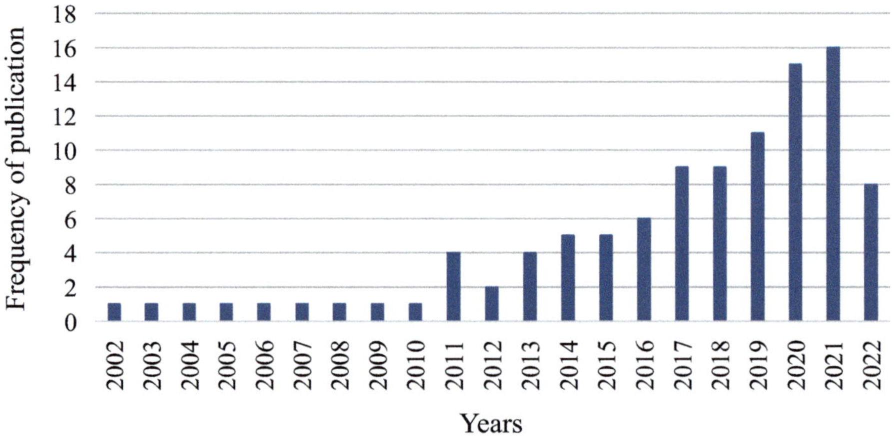

Fig. 4.6 The number of studies that utilized remote sensing to assess maize water stress

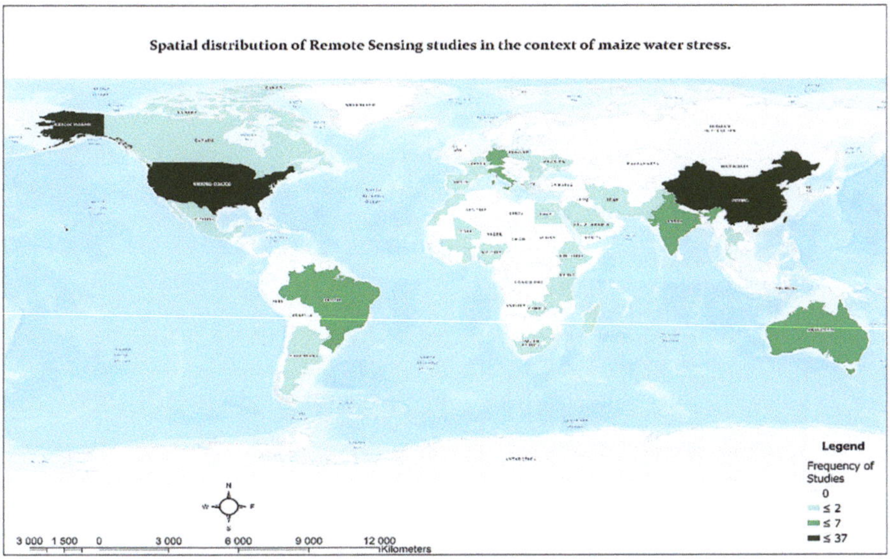

Fig. 4.7 The spatial distribution of Remote Sensing studies in the context of maize water stress

Nonetheless, fewer research efforts were observed from Africa, contributing only 12% of the studies, and South Africa contributed only 1%.

4.5 Maize Water Stress Indicators

Four groups of maize water stress indicators were captured in this study (Fig. 4.8). This study recorded these as plant, soil, temperature, and water-based indicators. Out of four identified groups of maize water stress indicators, the highest frequency of publication was noted on plant-based indicators (56%), water-based indicators (24%), temperature-based indicators (12%), and minor soil-based indicators (8%) in that order. In total, 25 indicators were identified in the literature, with soil water content being the most frequently studied indicator (16%). This could be because, in the case of water stress, variables, including the quantity and timing of precipitation, the physical characteristics of the soil, and the soil horizonation, all impact the amount of soil water available to plants (Brogi et al. 2020). The second indicator with the most frequencies is the leaf area index (LAI) (14%) (Fig. 4.8a). Canopy temperature also has a notable frequency of publication record of (12%) (Fig. 4.8c), followed by evapotranspiration (11%) (Fig. 4.8b).

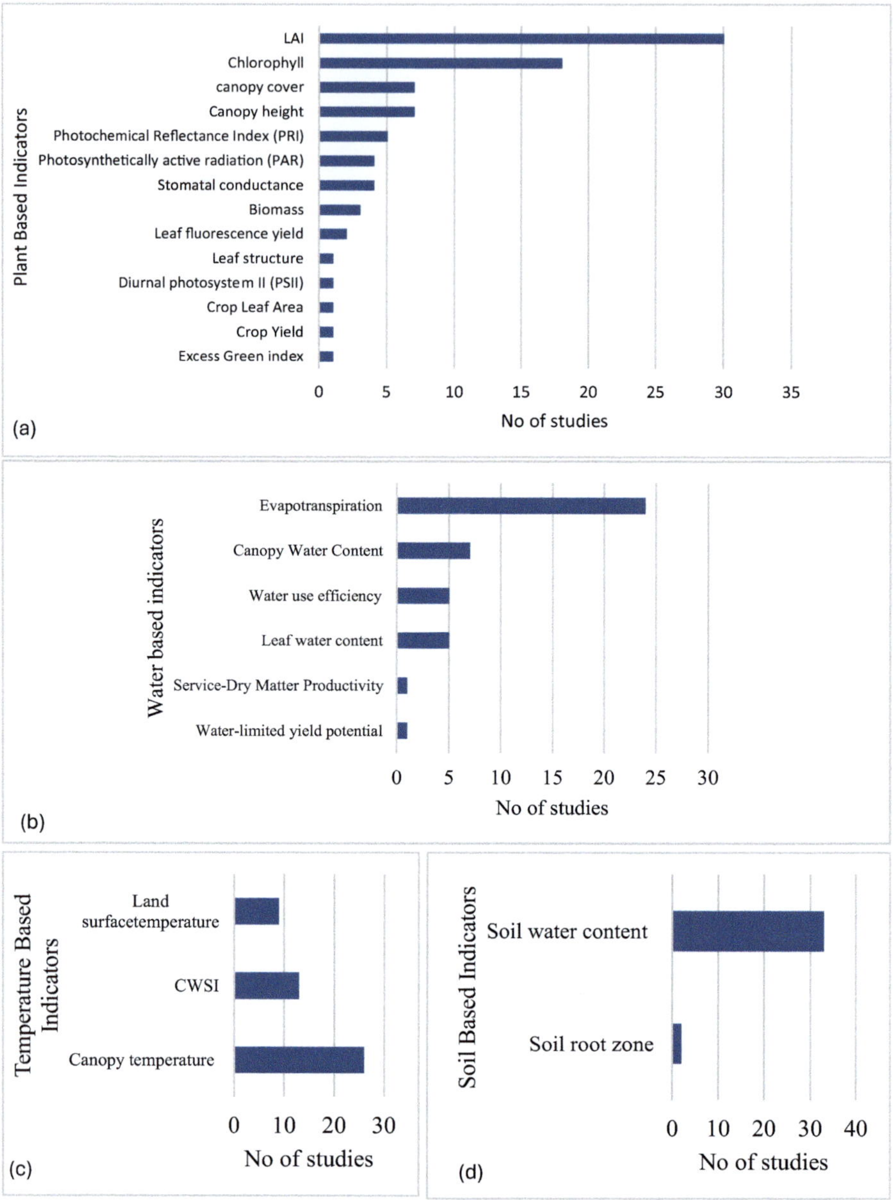

Fig. 4.8 Plant (**a**), water (**b**), temperature (**c**), and soil (**d**) based maize water stress proxies utilized in studies to determine maize water stress

4.6 Platforms and Sensors

Overall, 17 satellite and airborne sensors were collectively recorded from the literature (Fig. 4.9). Moderate resolution imaging spectroradiometer (MODIS) has the highest frequency of publication (n = 21), followed by Landsat 7 ETM (n = 11), then Landsat 8 (n = 8), and after that Landsat 5 (n = 6). Meanwhile, six different UAV models were recorded, and 11 studies explored the potential of drones in mapping maize crop water stress (Fig. 4.10). Studies that utilized UAV-captured data for maize water stress mapping are significantly lower than those that use satellite and airborne captured data. Of the 85 studies that utilized satellite, airborne, and drone-captured data, only 18.8% used drone-captured data.

Meanwhile, results show that 16 different spectroradiometers and infrared thermal radiometers (IRTs) have collectively been utilized for measuring spectral and canopy temperature reflectance (Fig. 4.11a). ASD Field spectroradiometer is the single most frequently utilized among the 16 (38.89%), followed by S1–11 and S1–121 IRT sensors (8.33% each) (Fig. 4.11a), low frequently utilized spectrometers are QE65Pro and the type of spectroradiometers which were not specified in the literature with the frequency of 8.33% each. The rest of the spectrometers and IRTs have the lowest frequency of 2.78% each. This high number of spectroradiometers could account for the results in Fig. 4.17 that energy balance models are the most widely used RS models in maize water stress since they measure RED and near-infrared (NIR) surface reflectance data, which is used as input data for the models.

Generally, FLIR Duo Pro, followed by MicaSense, and then Canon cameras were the dominant UAV sensors recorded in this study (Fig. 4.11b). The high usage of MicanSense could be attributed to the MicaSense RedEdge multispectral sensor covering the Red, Green and Blue (RGB), RedEdge, NIR, and thermal infrared sections of the electromagnetic spectrum. Except for the thermal infrared range, this sensor's spectral settings are comparable to those of the well-known Sentinel 2

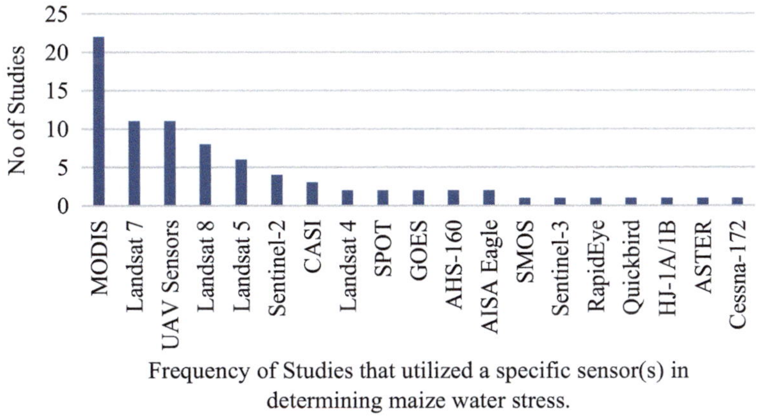

Fig. 4.9 UAV, airborne, and satellite sensors used in studies estimating maize water stress

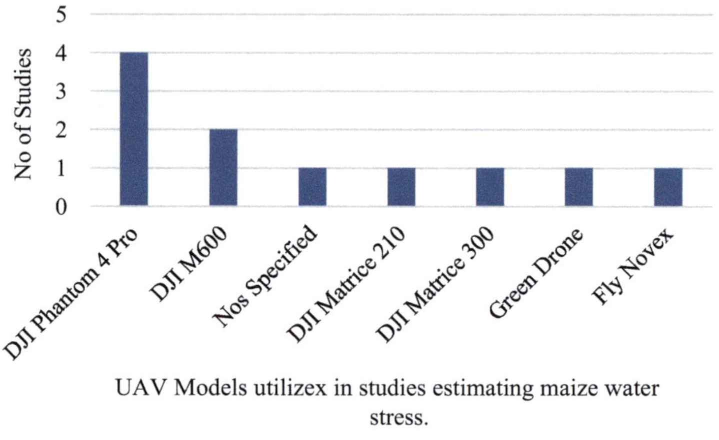

Fig. 4.10 Unmanned aerial vehicles (UAVs) model type

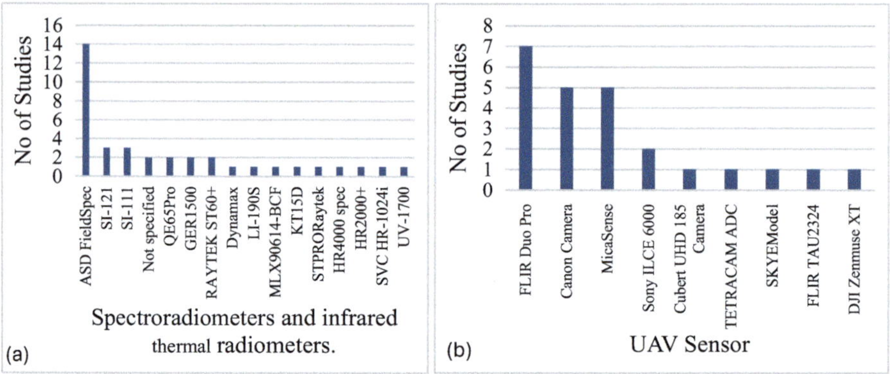

Fig. 4.11 Spectroradiometer and infrared thermal radiometers (**a**) and UAV sensors (**b**) used in studies using remote sensing to determine maize water stress

multispectral instrument, which covers nearly the same spectral regions (Sibanda et al. 2021). On the contrary, Hyperspec VNIR (37.5%) and Tetracam Multi-Camera Array (MCA) (25%) are the most frequently used airborne sensors found in this study (Fig. 4.12a), while the least utilized airborne sensors are FLIR A320 (12.5%), Hyperspec Fluorescence (12.5%), and AVIRIS (12.5%). Furthermore, some studies utilized camera sensors without embedding them on a sensor platform; Tamarisk 320 has been utilized in more than one study, whereas other cameras have only been utilized in one study each (Fig. 4.12b).

Upon assessing whether there were significant differences in the performance (R^2) of these sensors in estimating maize crop water stress, the findings showed that all satellite and airborne sensors exhibited significant differences (p = 0.05) (Fig. 4.13). The highest R^2 averaged value was obtained from QUICKBIRD with an average prediction accuracy of 91%, followed by Cessna-172 aircraft (88%).

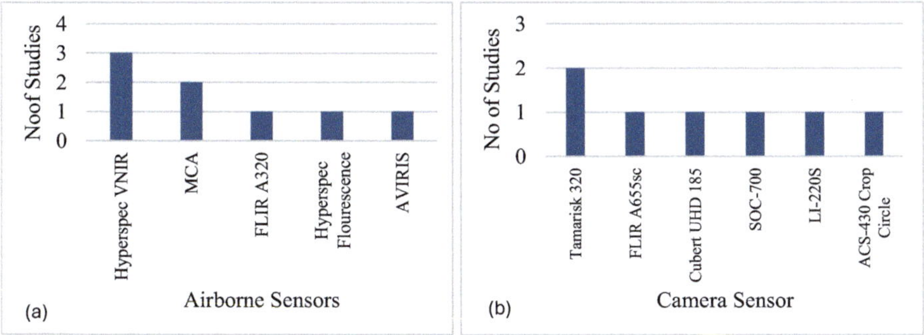

Fig. 4.12 Airborne (**a**) and camera ground sensors (**b**) utilized in studies determining maize water stress

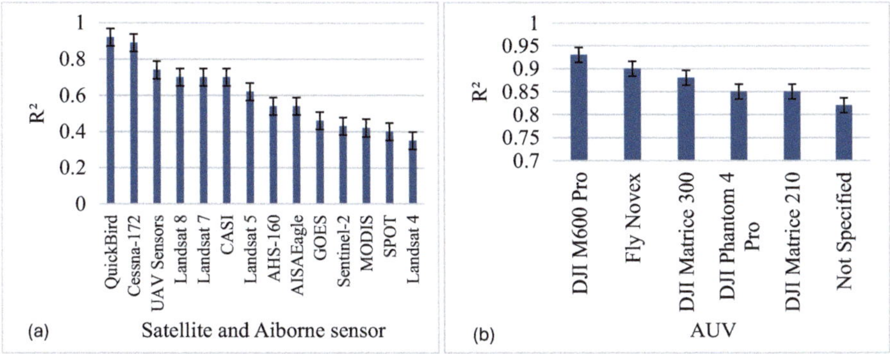

Fig. 4.13 Average correlation coefficient values for Earth observation sensors used in the maize water stress studies

Meanwhile, the freely available and commonly utilized sensors, including Landsat 8, 7, 5 and 4, have average R^2 of 70%, 70%, 63% and 34%, respectively. In addition, Sentinel-2, MODIS, GOES and SPOT have yielded low average prediction accuracies (R^2) of 46%, 44%, 47%, and 41%, respectively. On the contrary, the utility of UAVs reveals high average prediction accuracies with R^2 ranging from 82% to 93%. But similarly to Quickbird, UAVs' average R^2 is high.

4.7 Vegetation Indices and Spectral Characteristics

Numerous vegetation indices (VIs) are derived from satellite and UAV-borne sensors, field spectrometers, and cameras (incomplete sentence). Although numerous VI were identified in the literature, only those utilized in more than two studies were considered in this review (Table 4.2). As a result, 14 different VIs were recorded in this study. The most widely used VI is the Normalized difference vegetation index

Table 4.2 Spectral indices commonly utilized for dictating maize crop water stress

Indices	Acronym	Equation	References
Normalized difference vegetation index	NDVI	(R800 – R670)/(R800 + R670)	Zhao et al. (2018)
Optimized soil-adjusted vegetation index	OSAVI	(1 + 0.16) * (r800 – r670)/(r800 + r670 + 0.16)	Zhou et al. (2022)
Green NDVI	GNDVI	(R800 – R550)/(R800 + R550)	Zhao et al. (2018)
Normalized difference red-edge index	NDRE	(R750 – R705)/(R750 – R705)	Liu et al. (2018)
Enhanced vegetation index	EVI	2.5 × (R800 – R690)/(R800 + 6.0 × R690–7.5 × R490)	Liu and Huete (1995)
Transformed chemical absorption reflectance index	TCARI	3 ((R700 – R670) – 0.2 (R700 – R550) R700/R670)	Liu et al. (2010)
Simple ratio	SR	r900/r680	Gitelson and Merzlyak (1996)
Soil-adjusted vegetation index	SAVI	(1 + L) NIR – RED/NIR + RED + L	Huete (1988)
Photochemical reflectance index	PRI	(R570 – R530)/(R570 + R530)	Suárez et al. (2009)
Renormalized difference vegetation index	RDVI	(R800 – R670)/($\sqrt{}$R800 = R670)	Zhang et al. (2019c)
Difference vegetation index	DVI	r900 – r680	Li et al. (2021b)
Normalized difference water index	NDWI	(R860 – R1240)/(R860 + R1240)	Zhou et al. (2022)
Leaf chlorophyll index	LCI	(R850 – R710)/(R850 + R680)	Ramachandiran and Pazhanivelan (2017)
Vegetation temperature condition index	VTCI	LSTmax(NDVI) – LST(NDVI)/LSTmax(NDVI) – LSTmin(NDVI)	Wang et al. (2018a)

Source: Own Data

(NDVI) with a frequency of 41% (Fig. 4.14). NDVI was followed by OSAVI and GNDVI, both utilized in 8% of the retrieved literature. Other indices included TCARI (7%), SR (7% each), EVI (6%), SAVI, PRI, NDRE (4% each), NDWI and RDVI (3%), LCI, DVI, VTCI (2% each). Even though SAVI is among the least used VIs, it has the highest average r^2 value. The most commonly used sections of the electromagnetic spectrum in deriving these VIs were red (R) and near-infrared (NIR). Results show that analysis of maize water stress has been done using the provided wavelength spectra: the visible wavelength spectra in the blue range

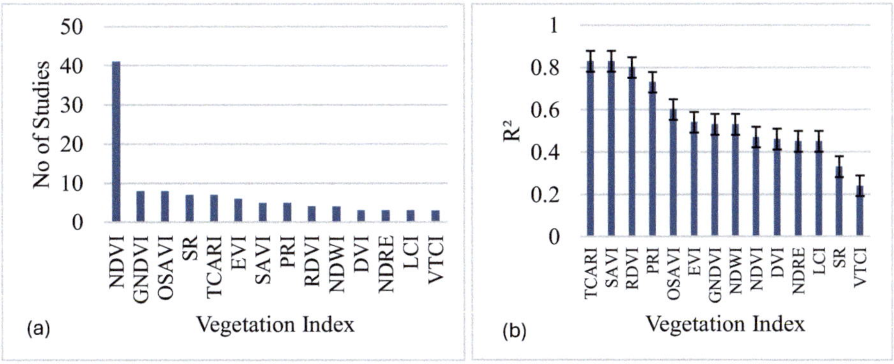

Fig. 4.14 Vegetation indices used in studies analysing maize water stress using remote sensing data (**a**) and average correlation coefficient values (**b**)

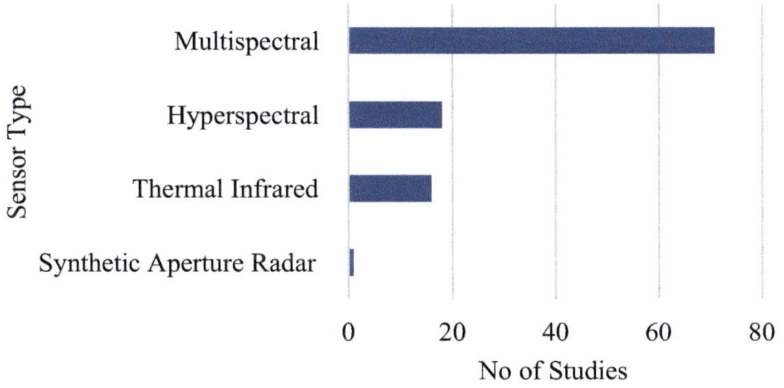

Fig. 4.15 Spectrum coverage of sensor technologies used in remote sensing

(450–495 nm), the green range (495–570 nm), the red range (620–750 nm), and the near-infrared wavelength spectra (850–1700 nm). Moreover, there is also the use of red-edge band reflectance.

Results show that the most explored sensors (67%) that cover NIR and Red electromagnetic spectrum regions have been multispectral sensors (Fig. 4.15). This is concurrent with the use of satellite sensors with freely available data, such as Landsat and MODIS, to study maize water stress (Fig. 4.8). Even some of the UAV sensors that have caught researchers' attention are multispectral; for instance, the MicaSense family of multispectral cameras capture data at a high spatial resolution not only in the visible spectrum but also in the red edge and near-infrared regions of the electromagnetic spectrum. Hyperspectral sensors are the second most utilized sensors (17%), followed by thermal infrared sensors (16%) (Fig. 4.15). SAR is the least utilized sensor type.

4.8 Machine Learning Algorithms Utilized in Remote Sensing for Maize Water Stress

Nine algorithms were utilized in maize water stress detection (Fig. 4.16). Linear regression was the most frequently utilized algorithm (30%), followed by least squares regression (20%) and random (11%). Meanwhile, the support vector machine was among the least adopted algorithms (6%). Figure 4.17 shows the average coefficient determination accuracies derived using different machine learning algorithms ranging between 97% and 30%. Multi-layer perceptron (MLP) neural network exhibited the highest prediction accuracy, followed by a simple algorithm (88%).

4.9 Models Utilized in RS Maize Crop Water Stress

Only 39% of the studies followed a specific method for estimating or modelling maize crop water stress. Specifically, surface energy models were utilized in 25.64% of the studies, followed by FAO-56 (20.51%), AquaCrop (7.69%), and CERES (5.14%). Each of the remaining 16 models was adopted in 2.56% of the studies. These results show that surface energy models (Fig. 4.18) are the most utilized in maize water stress, which could mean they are best suited for determining maize water stress. However, WOrld FOod Studies (WOFOST), Soil-Canopy-Observation of Photosynthesis and Energy fluxes (SCOPE), and Mapping Evapotranspiration at

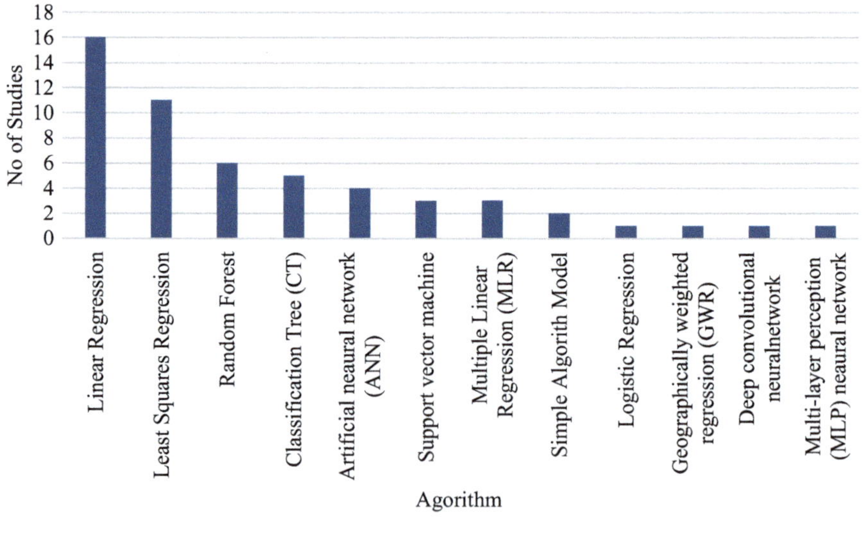

Fig. 4.16 Machine learning algorithms and regression models utilized in studies using the application of remote sensing in maize water stress

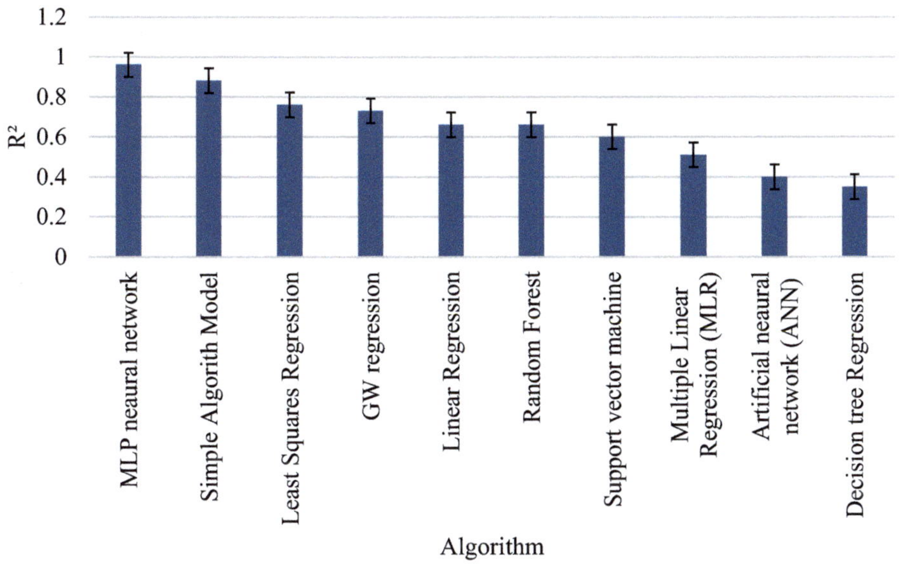

Fig. 4.17 Average coefficient of determination (R^2) values produced by machine learning algorithms applied in the studies

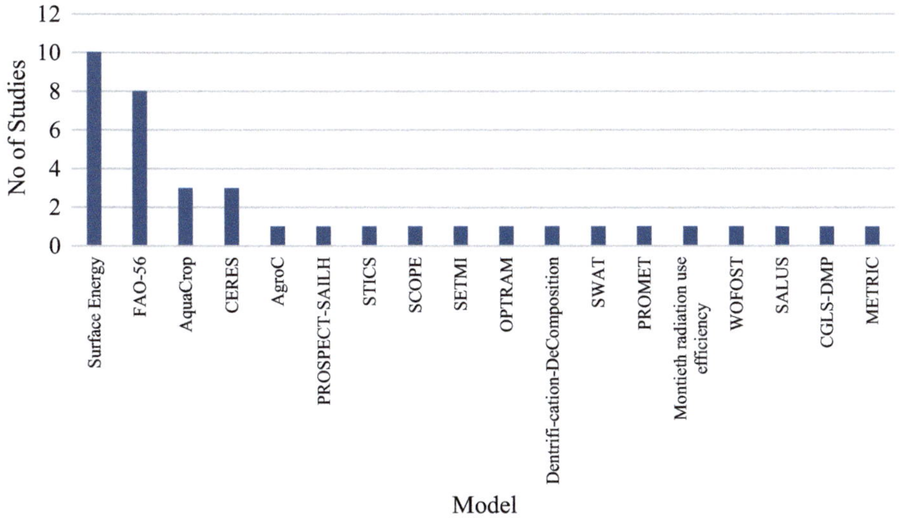

Fig. 4.18 Models followed in determining maize water stress

high Resolution using Internalized Calibration (METRIC) model (s) were least utilized in retrieved literature.

Statistical accuracy assessment models such as the root mean square error and coefficient of determination are the most frequently used model validation methods

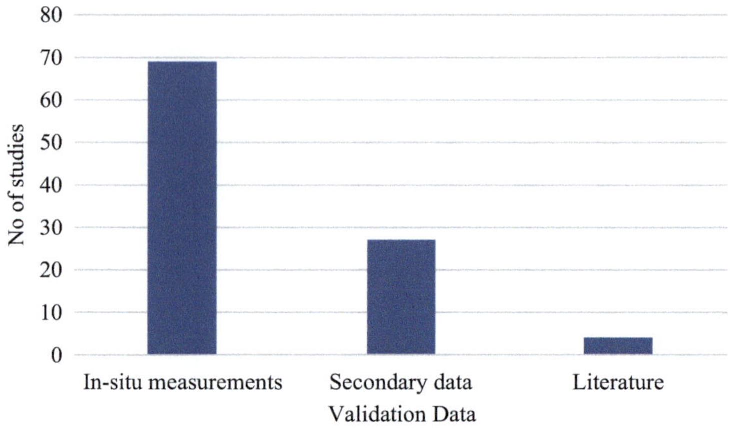

Fig. 4.19 Validation data for the studies using remote sensing in determining maize water stress

in analysing maize water stress from retrieved literature. Results show that 69% of the reviewed studies have relied on statistical accuracy assessments such as root mean square error (RMSE), the coefficient of determination (R^2), the mean absolute error (MAE), and normalized root mean square error (NRMSE) for model validation (Fig. 4.19). While many researchers are leaning towards these statistical validations, meaning that they have to acquire ground-measured data to compare with estimated, few have preferred datasets from another data source (26%). Very few have used data from the literature referencing work that has already been published (5%).

4.10 Discussion

4.10.1 Critical Terms in the Literature

4.10.1.1 Changes in the Critical Terms in the Literature

In terms of assessing the evolution of key terms from literature, results revealed that utilized UAV captured data have been utilized to determine maize water stress from 2014 and 2015 until recently (Figs. 4.20 and 4.22). "Chlorophyll fluorescence" and "pri" (photochemical reflectance index) are the earliest proxies of maize water stress determination, followed by "temperature" from 2017 (Fig. 4.20). Figure 4.22 reveals similar results; however, the "canopy temperature" focus here is shown to have begun in 2015. Both figures also show that the recognition of the UAVs' "potential" to monitor maize water stress gained popularity following the years 2019. Concurrently, "ground" measured data usage increased in the same years to validate UAV-captured data using "rmse" (Figs. 4.20 and 4.22). In addition, Fig. 4.22

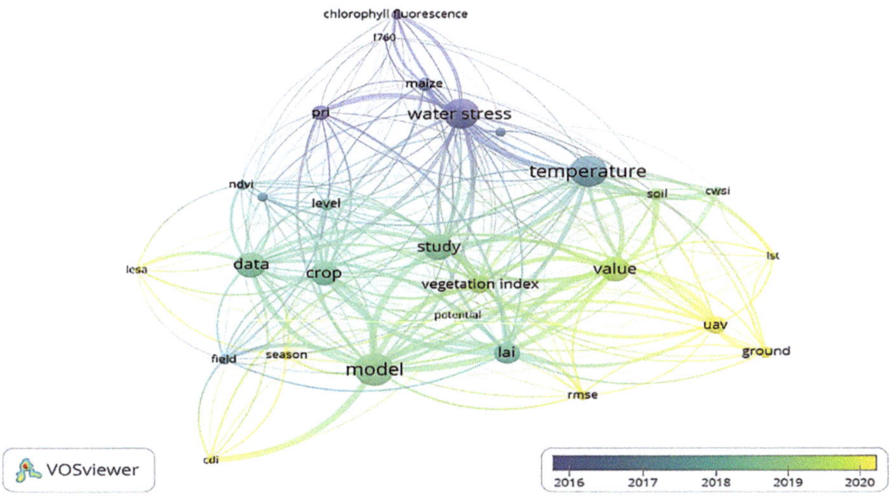

Fig. 4.20 Topical concepts in mapping maize water stress using UAV sensors from studies' abstracts

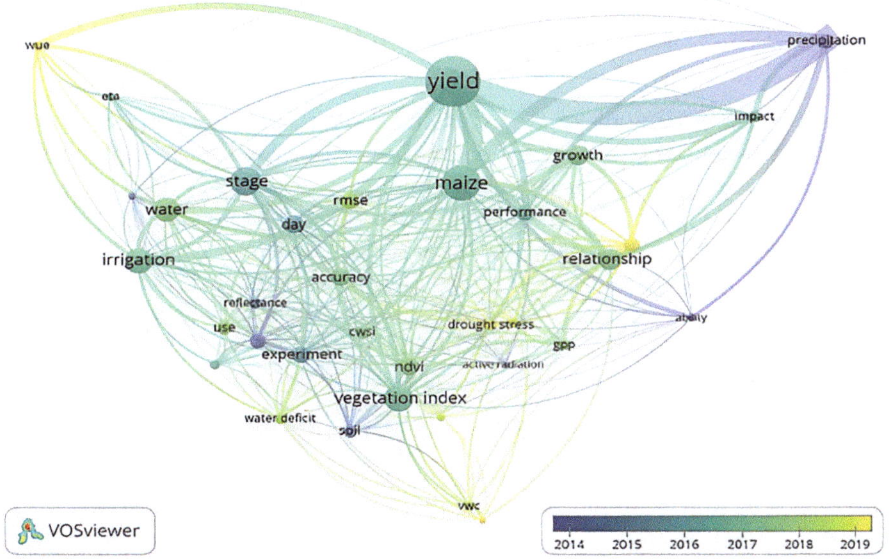

Fig. 4.21 Topical concepts in mapping maize water stress using satellite and airborne sensors from studies' abstracts

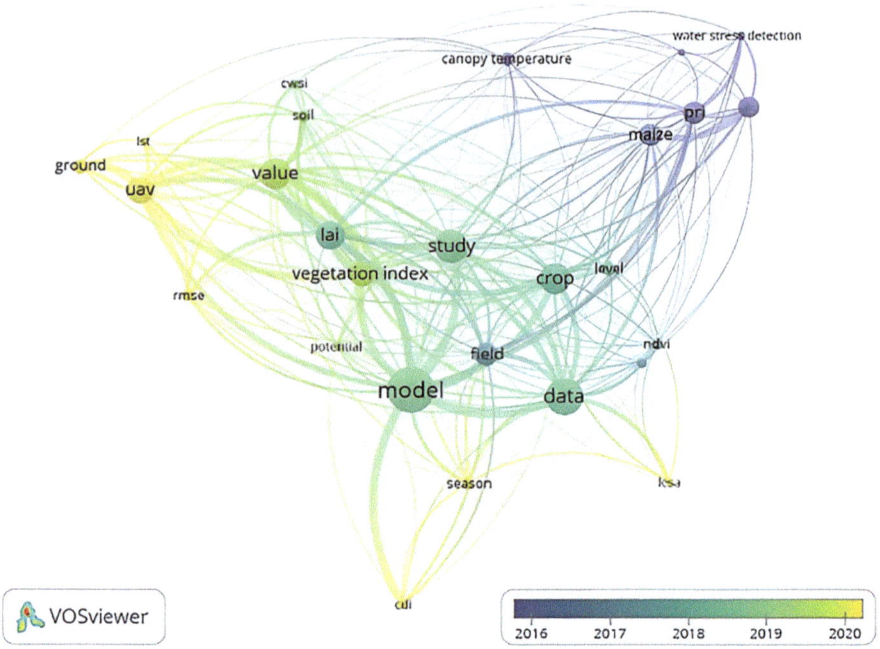

Fig. 4.22 Topical concepts in mapping maize water stress using UAVs from studies' titles and abstracts

reveals that the recognition of UAVs' ability to determine maize water stress at "field" increased from 2017. As early as 2013 and 2014, satellites and airborne sensors had already been increasingly utilized in maize water stress (Figs. 4.21, 4.22 and 4.23). Both Figures reveal that from 2016, the research community became increasingly aware that maize spectral "reflectance" changes with different phenological stages during its "growth". Following then, "experiments" became increasingly important in monitoring maize water stress at different "irrigation" levels and "drought stress" (Fig. 4.21). From 2019 until recently, "wue" and "vwc" have become the most widely adopted maize water stress indicators using satellites and airborne captured data.

4.10.1.2 The Role of Remote Sensing Platforms in Determining Maize Water Stress

Generally, the use of RS in determining maize water stress has considerably increased in recent years (Fig. 4.6). In reality, increasing RS use in maize water stress detection has been necessitated by reducing agricultural water losses to transfer water resources to other uses in a society where demand is rising (Rossini et al. 2015). Additionally, many academics are concentrating on discovering alternatives

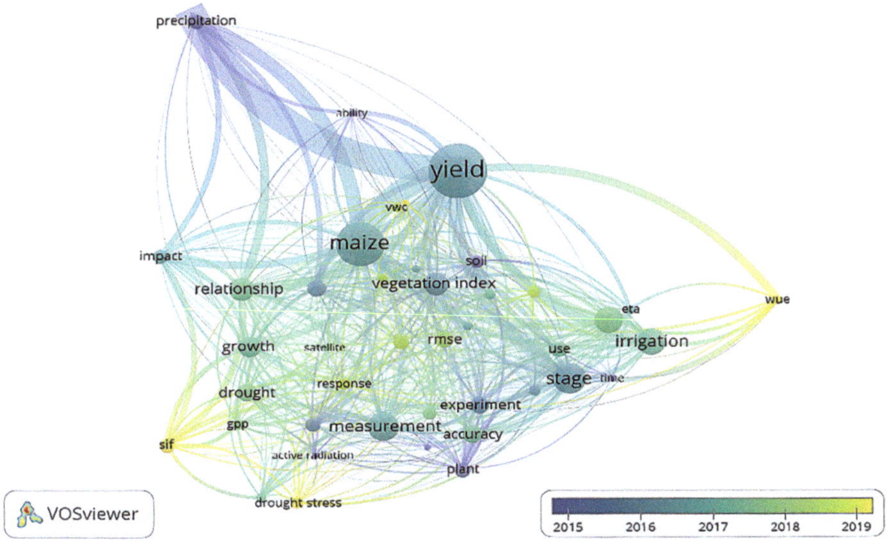

Fig. 4.23 Topical concepts in mapping maize water stress using satellite and airborne sensors from studies' titles and abstracts

and methods to boost agricultural productivity while using less irrigation water due to the knowledge that the world's water deficit is worsening alongside rising food demand. Especially for a commodity like maize, which is intensively farmed and calls for much irrigation (Rossini et al. 2015). Results revealed that the first article on RS of maize crop water stress from retrieved literature was published in 2002. Then after, one article per year was published until 2010. Then from 2011, the number of published articles increased significantly to above 15 only by 2020 (Fig. 4.6). This shows that what was initially described as the potential and benefits of using RS to study crop characteristics, particularly for maize, has evolved into a well-established tool to study water status, primarily due to better access to remotely sensed data and advancement in RS technology. The capacity to examine the relationship between geographical variability in crop development and water stresses is provided by the analysis of crop phenology using remotely sensed datasets (Wang et al. 2016).

Results also revealed that more than 17 satellite and airborne platforms and 7 UAV types were utilized to acquire remotely sensed data at varied geographical, spatial, and temporal resolutions for estimating maize crop water stress (Figs. 4.9 and 4.10). The fact that some satellite images are freely available adds an advantage to their agricultural utilization (Kyere et al. 2019). Landsat alone is the oldest earth monitoring platform running in space, thus offering over a 47-year collection of satellite data of the world at 30 m resolution (Kyere et al. 2019). Results revealed that Landsat is the most utilized RS data source for monitoring maize water stress. In particular, the Landsat 7 Enhanced Thematic Mapper Plus (ETM+) sensor, the most utilized Landsat sensor, has been commercially available since 1999 and freely available since 2008, followed by Landsat 8, which was launched in 2013 (USGS 2013). Lastly, Landsat 4 and 5 (Thematic Mapper), launched in 1982 and 1984 (Butcher et al. 2019), have also been used in studies determining maize water stress. For instance, Wang et al. (2016) utilized Landsat 5 (Thematic Mapper) data to detect spatial-temporal patterns of maize leaf growth under climate stress for the St. Joseph River watershed, United States, using NDVI retrieved from 2000 to 2010. While all Landsat missions provide data at 30 m resolution, the launch of each mission has been to provide better temporal and spectral coverage and spatial and radiometric resolution than its predecessor (USGS 2013). For instance, Thematic Mapper (TM) provides data at 30 m resolution and has seven spectral bands, including the thermal infrared band (120 m resolution) (Masina et al. 2020). Compared to its predecessor, ETM+ has improved thermal band resolution from 120 to 60 m, which can be utilized to estimate surface temperature as a crop water stress variable (Masina et al. 2020). When Reyes-González et al. (2019) utilized the 60 m thermal band from ETM+ to estimate surface temperature, they found a good correlation between estimated and measured surface temperature with R^2 of 0.87. Padilla et al. (2011) used a red and near-infrared band of TM and ETM+ to compute soil-adjusted vegetation index (SAVI) as an input in the FAO-56 model for the estimation of maize evapotranspiration (ET). With the r^2 of 0.89, there is a good comparison between daily estimated and measured ET. The significant increase in the utilization of Landsat 8 (OLI and TIRS) can be attributed to the fact that compared to its predecessor, it has improved radiometric performance to 12-bit (Butcher et al. 2019) and thus has a high sensitivity to a very slight difference in electromagnetic energy. In a nutshell, Landsat, with a new mission better than its predecessor, provides consistently improved remotely sensed data quality available in archives from 1972 (USGS 2013).

Unlike Landsat, MODIS was the second most widely used sensor platform in mapping maize water stress. Unlike Landsat, MODIS has coarse resolution at a scale of ≥250 m, thus limiting its application over heterogeneous agricultural systems since crops and fields are frequently mixed (Wulder et al. 2019). When MODIS and Landsat were compared, results revealed that due to MODIS' low spatial and temporal resolution (Berni et al. 2009), the Landsat-derived map better preserves spatial detail at field scale (Ren et al. 2021). Nonetheless, according to Ren et al. (2021), an excellent indicator of irrigation and non-irrigation, for instance, is the time-series NDVI produced from MODIS. MODIS provides a variety of phenology products that authors have used to assess maize water status; leaf area index (Ines et al. 2013; Wang et al. 2018a), enhanced vegetation index (Holzman et al. 2018), vegetation temperature condition index (Wang et al. 2018a). Moreover, evapotranspiration, vegetation, albedo, and land cover products provided by MODIS (Javadian et al. 2020) have contributed to the study of maize water stress. This variety of data products from MODIS determines maize water stress (Fig. 4.8), giving MODIS a high frequency of publications (Fig. 4.9). In particular, Wan et al. (2021) utilized Terra MODIS 8-day global land surface temperature (LST)/emissivity products (MOD11A2) with a spatial resolution of 1 km to calculate the LST and Terra MODIS 8-day global surface reflectance products (MOD09A1) at 500 m spatial resolution to calculate the vegetation indices (including NDVI and EVI). Whereas

Li et al. (2021b) assessed the ability of MODIS LST/Emissivity product (MOD11A1 and MYD11A1) and MODIS/Aqua LST/3-Band Emissivity product (MYD21A1) to estimate daytime LST and found that MOD11A1 and MYD11A1 products underestimated daytime LST with a RMSE <2.9 °C. whereas MYD21A1 LST achieved an overall RMSE <3.6 °C regarding daytime LST. Even though MODIS offers many data products including LST data its low spatial and temporal resolution makes it infeasible for application in small holder plots which are small in size.

On the other hand, Sentinel, a still-emerging platform, gives users access to high-resolution images with pixels as large as 10 m (Rivera-Marin et al. 2022). They also offer an opportunity to improve the assessment of several variables at a higher spatial and temporal resolution level when figuring out the water status of maize. The program's introduction in 2015 (Mananze et al. 2018) may have influenced the increase in publications following that year (Fig. 4.6). Weiss and Jacob (2019) state that Sentinel-2 can provide spectral sampling to identify crop types and varieties. Furthermore, Sentinel usage's ability to assess spatial-temporal variability of crop water status has been widely recognized in the literature. For instance, Ambrosone et al. (2020) used Sentinel-2 data to estimate soil water content on a spatially heterogeneous agricultural landscape through the Optical Trapezoid Model (OPTRAM) (Fig. 4.18). They found that estimations by OPTRAM are close to the in situ measurement value. Also, Mananze et al. (2018) used Sentinel-2 images to retrieve maize LAI at small-scale fields (<5 ha) in Mozambique and found that Sentinel-2 can accurately estimate LAI at field scale. Perhaps Sentinel's ability to account for spatial heterogeneity, thus giving a viable option of remotely sensed data applied in small-scale fields, will increase the RS data usage in determining maize water status.

While the medium resolution of Landsat (30 m) to coarse resolution of MODIS (250–1000 m) has restricted their use at the small farm scale, UAVs have become cutting-edge field phenotyping platforms to provide spatial data in agriculture (Zhang et al. 2019c). Even though satellite platforms like Sentinel provide the highest spatial-temporal resolution of the freely available spaceborne data, UAVs can offer higher-quality data at the user-determined scale and time. For instance, Cheng et al. (2022) used UAV data obtained at the height of 30 m and 6-day temporal resolution. In contrast, Zhang et al. (2019c) obtained UAV multispectral data at a flight height of 70 m with a spatial resolution of 4.7 cm and 14 flights during the study period (2017.06.26–08.29). Because UAV platforms can hover over an area of interest and acquire imagery at lower altitudes, allowing for a finer ground sampling distance, they have been deemed suitable for better quantifying maize water stress at a field scale. As such, it was noted from the results that research on using UAVs for mapping and monitoring maize water stress is growing considerably. For instance, in their study, Zhang et al. (2019b) mapped the maize crop coefficient (Kc) using UAV-obtained data under various deficit irrigation levels. They found that when compared to on-site measurement, Kc calculated from UAV data had a significant ability to assess field variability of crops and soil. When also estimating Kc, similarly to Zhang et al. (2019b), Shao et al. (2021) found that UAV-based VIs effectively correlate with ground-based data ($R^2 = 0.65$). In addition, while estimating maize water stress index (CWSI), Zhang et al. (2019c) found that CWSI values from UAV data retrieved by VI-CWSI regression models had more capabilities to evaluate soil and crop field variability. The fact that UAV platforms can produce datasets with incredibly high spatial and temporal resolutions when compared to satellite imagery and, therefore, perform better in predicting maize water stress at local scales can attest to its growing adoption in the agriculture research field.

RS data and techniques are relevant for many crops' water requirements, inventory and monitoring requirements; however, these are not limited to satellite and airborne platforms. Literature reveals that radiometric ground-based remotely sensed data has also proven to be a reliable RS data acquisition method. Camera sensors have been used as sources of ground remotely sensed data, especially in field experiments (DeJonge et al. 2016); even though other authors have found it challenging to obtain maize phenology measurement for the entire growth cycle, it has been deemed a possible source of remotely sensed data. Tsakmakis et al. (2021) captured the green canopy cover (CC) using an RGB camera mounted on a selfie stick from roughly 3 m height aboveground; images were further processed via GIMP and PhotoShop software. In addition, An et al. (2019) captured maize drought stress RGB images in the field and found that for the total dataset, classification and identification accuracy of drought stress from the captured images was 95.95% and 98.14%, respectively. Even the feasibility of thermal cameras to quantify high-resolution spatial canopy temperatures relative to soil moisture has been examined (Mangus et al. 2016). In their study, Mangus et al. (2016) found that while compensating for changing ambient greenhouse conditions, the thermal infrared system kept a measurement accuracy of ±0.62 °C.

4.10.1.3 The Basis of Maize Water Stress Estimation Using Remote Sensing Models

Empirical approaches, called "regressions", calibrate a numerical relationship between ground-measured biophysical variables and the remotely sensed spectral reflectance (Weiss and Jacob 2019). For instance, linear or nonlinear relationships were established between VIs and LAI (Rossini et al. 2015; Shao et al. 2021), chlorophyll content (Behmann et al. 2014), or water content (Zhou et al. 2022). Multivariate analysis techniques, particularly linear regression, were probably the most used algorithms because they are simple and easy to implement (Fig. 4.16). However, a significant limitation of linear regression is that a nonlinear relationship between response variables and environmental variables cannot be detected (Xie et al. 2021). Therefore, results reveal that more advanced techniques of machine learning, such as support vector machine, random forest (Filgueiras et al. 2020), partial least squares (PLS) (Elmetwalli et al. 2021) and neural networks have been utilized as complementary tools in maize water stress studies (Fig. 4.16).

The success of machine learning algorithms can be explained by the fact that they are non-parametric and thus do not rely on any assumptions about data distribution (Barrett et al. 2014). Also, machine learning techniques have a simple structure, fitting solid ability, and high estimation accuracy and can reveal the linear and nonlinear relationships between response and environmental variables (Guo et al. 2015; Deng et al. 2021). This provides a convenient and new method for evaluating and estimating maize crop physiological parameters. Algorithms such as PLS can reduce spectral response complexity and multi-collinearity by performing simple vector space projection operations, thereby reducing over-fitting (Krishna et al. 2019). PLS has been successfully implemented with spectral data to estimate leaf stomatal conductance (Sobejano-Paz et al. 2020), LAI (Elmetwalli et al. 2021), and water content (Mirzaie et al. 2014). Also, the random forest (RF) model can model highly nonlinear dimensional relationships, resistance to 'overfitting', relative robustness in the presence of noise in the data, the establishment of an impartial measure of the error rate, and the ability to determine the relevance of the variables used (Xie et al. 2021). It has therefore been successfully applied to (i) predicted agricultural data such as maize crop coefficient for crop evapotranspiration assimilation (Shao et al. 2021), and (ii) classified irrigated and non-irrigated maize (Ren et al. 2021). Nonetheless, machine learning and multivariate algorithms can estimate crop water stress parameters, thus their usage in maize water stress studies.

In addition, models that diagnose current maize water status and predict possible impacts for imminent management methods have also been utilized in maize water stress monitoring (Fig. 4.18). Results from these methods can be used to inform recommendations about irrigation scheduling to address maize water status. Results reveal the widely used models comprise the surface energy, Acquacrop, FAO-56, and CERES model, which is included in the Decision Support System for Agrotechnology Transfer (DSSAT) (Ines et al. 2013; Ahmad et al. 2018), the Simulateur mulTIdisciplinaire pour les Cultures Standards (STICS) model (Jégo et al. 2012), the WOrld FOod Studies (WOFOST) model (Guissard et al. 2005) (Fig. 4.18). Remote sensing techniques for estimating ET from satellites have been developed based on energy balances, yielding estimates of actual evapotranspiration. University of Idaho (USA) created the Mapping EvapoTranspiration with High Resolution and Internalized Calibration (METRIC) model to estimate ET, which is based on the Surface Energy Balance Algorithms from the Land (SEBAL) model (Alvino and Marino 2017). METRIC calculates evapotranspiration from Landsat imagery by calculating the available energy using the earth surface temperature (Ts) derived from satellite imagery thermal bands to constrain the heat flux for one or more layers (canopy and soil) and then computing the latent heat as a residual to the surface energy balance (Alvino and Marino 2017). In fact, Reyes-González et al. (2019) found that the METRIC model is a good estimator of surface temperature with an r^2 of 0.87.

The METRIC model has also been found to be adopted in studies using remote sensing data for maize water stress modelling (Fig. 4.18). According to Claverie et al. (2012), crop models were initially developed to simulate crop growth on agricultural fields with well-known soil, climate, and agricultural practices that were spatially homogeneous. They have been used in a variety of agricultural and environmental issues. STICS, for example, simulates nitrogen, water, and energy, among other things, whereas WOFOST focuses on carbon and water while ignoring nitrogen balance. Some agricultural applications may also necessitate the description of crop growth over large areas, necessitating the execution of numerous simulations in a multilocal, spatially distributed manner. This necessitates the development of simple methods with fewer parameters, such as AQUACROP (Steduto et al. 2009), which examine crop biomass production in relation to water availability. STICS and CERES are complex models that simulate various agro-environmental variables by describing numerous coupled phenological and physiological processes like photosynthesis, respiration, evapotranspiration, and nitrogen uptake. However, these models require a large number of input data and parameters. This information may be available during scientific experiments or from some farmers on a local scale. Still, it is not generally available across large areas, meaning they best perform in small-scale fields.

4.10.1.4 The Use of Wavebands and Selection of Spectral Indices in Determining Maize Water Stress

Application of RS in determining maize water stress has focused on the optical reflective domain of the electromagnetic spectrum (0.4–2.5 μm), with results revealing a variety of available multi- and hyperspectral sensors: ground level (e.g. ASD FieldSpec), airborne (e.g. AVIRIS, Aisa), drones (e.g. DJI Matrice), and spaceborne (e.g. MODIS, Sentinel, Landsat, SPOT) (Fig. 4.9). These optical sensors have been adopted for spectral reflectance properties, particularly the canopy, leaves, and underlying soil (Gerhards et al. 2019). The opportunity to develop narrowband vegetation indices has been opened, especially by the available remotely sensed data, by simply interpreting vegetation reflectance signatures and plant physiology and structure, such as photosynthetic activity, greenness or fractional vegetation cover, and canopy water content. Therefore, VIs have been considered indicators of maize water stress. In particular, results revealed that the most commonly used VI, the NDVI (Fig. 4.14), relies on near-infrared and red bands. NDVI serves as a quantitative gauge of the greenness of the vegetation (Ihuoma and Madramootoo 2017). Since chlorophyll absorption during photosynthesis results is low in the blue and red regions of the spectrum, the green region peaks, giving rise to green vegetation colour (Yue et al. 2018). However, since the effectiveness of photosynthesis is reduced with the increase in leaf senescence during the reproductive stage, the correlation between red-NIR-based VIs (such as NDVI) and measured maize water stress indicator is reduced during this stage (Yue et al. 2018). Sobejano-Paz et al. (2020) further elaborate that NDVI tends to saturate in high vegetation conditions. Because of this, in their study, Chakraborty et al. (2020) added new VIs in their analysis, GNDVI and NDRE and discovered that the crop vigour grew in the early development phases, peaked in the mid-growth stages and then fell in the late growth stages. Furthermore, Haboudane et al. (2002) empirically determined that NDVI products are unstable due to their dependence on soil reflectance and sun view geometry. Nevertheless, NVI is employed in irrigation studies for crop cover mapping as a way to calculate crop coefficients (Kc) for use in the standard FAO-56 approach (Fig. 4.18) (Rossini et al. 2013) to track changes in LAI (Suárez et al. 2009; Rossini et al. 2015; Sobejano-Paz et al. 2020), among others, to inform irrigation planning. The shortcomings of NDVI are remedied by adopting chlorophyll and structure indices based on visible (RGB), NIR and red-edge bands. OSAVI to minimize soil background effects on surface reflectance readings (Rossini et al. 2013; Costa-Filho et al. 2020; Sobejano-Paz et al. 2020; Elmetwalli et al. 2021); SAVI, TCARI, because of its sensitivity to chlorophyll content (Rossini et al. 2015; Zhang et al. 2019c) and RDVI. These indices aim to reduce the spectral noise caused by various non-photosynthetic materials and minimize soil brightness influences from spectral vegetation indices involving red and Near-Infrared. OSAVI, GNDVI, NDRE, SAVI, RDVI, and TCARI were all recorded in the findings of the study, with OSAVI being the most frequently published among them all.

On the contrary, due to short-term changes in xanthophyll pigments when induced with water stress, PRI is directly linked to the photosynthetic process (Gerhards et al. 2019). Calculated at reflective wavelengths 530 and 570 nm (Suárez et al. 2009), PRI is the pre-visual index for water-stress detection (Gerhards et al. 2019). Panigada et al. (2014) found that contrary to traditional greenness indices (e.g. NDVI, OSAVI), PRI was able to track the development of plant water stress from when only the plant physiology is affected in the early phase to the later stages when the plant structure starts to be affected, therefore presenting the higher potential for early detection. Similar to NDVI, concerns over PRI's sensitivity to pigment levels, structural changes, illumination effects, soil background, and viewing angles have led to its improvements to ensure more effective results (M Rossini et al. 2015). The resulting modified PRI (570–515 nm) has overcome the PRI's demerit and thus proved more effective in detecting water stress (Sobejano-Paz et al. 2020). Furthermore, modified RI has been found to best relate with other maize water stress indicators: canopy temperature, relative water content, and photosynthetic efficiency (Rossini et al. 2013). Nonetheless, results reveal that PRI's ability to determine maize water stress has been recognized and utilized in the literature.

The Normalized Difference Water Index (NDWI) tracks changes in the water content of leaves using NIR and Short Wave Infrared (SWIF) at wavelengths of about 860 nm and 1240 nm (Zhang and Zhou 2019; Ndlovu et al. 2021; Zhou et al. 2022). While the NIR reflectance is influenced by internal leaf structure and dry matter content but not by water content, the SWIR reflectance reflects changes in the plant's water content and the spongy mesophyll structure in vegetation canopies. The combination of NIR and SWIF improves the accuracy of determining the water content of vegetation by removing variations brought on by changes in leaf internal structure and dry matter content (Ihuoma and Madramootoo 2017). Studies have found NDWI a dependable estimator of water content since it presented the highest sensitivity to CWC (Zhang and Zhou 2019). Similarly, Zhou et al. (2022) demonstrated that maize water content was better estimated using the NDWI than the NDVI and OSAVI.

4.10.2 Challenges and Way Forward Use RS Technologies to Map and Monitor Maize Water Stress

The number and distribution of research utilizing RS to examine maize water stress varied significantly around the globe. The ability of RS application in monitoring maize water stress has been adequately exploited by developed nations such as the United States and China. Specifically, America had the most publications, followed by Asia, Europe, and Africa. Over 74 countries have examined maize water stress using RS, with the United States contributing most research efforts and studies (Fig. 4.7). This illustrates an imbalance in the global distribution of maize water stress studies employing RS data, which may skew the global community's comprehension of maize water requirements and dynamics. Consequently, the capacity to increase production while conserving water, policies to plan irrigation technologies, and timely scheduling to address maize water irrigation requirements globally may also be skewed.

Meanwhile, fewer than five published studies sought to evaluate the utility of remotely sensed data in Africa and the global south (Fig. 4.7). Surprisingly, the countries with the lowest numbers of publications are those where maize is a staple crop grown commercially and in smallholder croplands that are more susceptible to crop water stress. Despite the widely acknowledged potential of RS in mapping and monitoring crops using freely accessible earth observation-data products, complex interactions between weather and geography, particularly the highly diverse rainfall patterns and the lack of ancillary data, make it difficult to apply RS techniques, especially those primarily designed for the global north.

Since small-scale farmers predominate in Africa and grow rainfed maize, no data supports and validates these models. In brief, the dominant rainfed crop monitoring technologies suitable for African cropping systems are severely hampered by excessive cloud cover during the rainy season, which restricts cloud-free remotely sensed data (Sun et al. 2019). Most freely available remotely sensed data from high spatial resolution sensors have inadequate temporal resolutions for recording significant maize phenological changes. They are compounded by frequent cloud cover, which leads to missing data (Chivasa et al. 2017). In their study, Javadian et al. (2020) articulated that MODIS ET product coverage is limited to North Africa; thus, the region was excluded from their study. Additionally, to acquire 8-day images composite with 70% agricultural area under precise sky circumstances in sub-Saharan Africa, a revisit period of 1–3 days in August is needed (Bégué et al. 2020).

Perhaps the challenge faced by African countries in accessing high resolution freely available data is reinforced by the technological resources strategy that is mainly driven by the Global North institutions or funded through international organizations, thus resulting in data and products that only partially meet the African demand (Bégué et al. 2020). Therefore, data processing methods are often limited for the applicant in African agricultural systems, which are more diversified yet much less documented than agricultural systems in industrialized countries. Thus, research on applying RS data to estimate maize water stress is urgently required in

Southern Africa and Africa. Fortunately, current advances in sensor technology have opened the opportunity for obtaining remotely sensed vegetation data using UAV technology with high spatial resolution at a user-determined temporal resolution. However, its use in small-scale farmers of Africa still needs to be further examined. Using UAV technology to obtain data is crucial in exploring the heterogeneous nature of agricultural systems practised in Southern Africa. It can contribute to the detailed testing of this available high-resolution sensor capabilities to discriminate maize from other crops and improve water stress estimates' accuracy.

Apart from water stress, different stresses that are rare but sometimes accounted for, like pest and disease infestation, crop development stage, local weather, edaphic factors, and landscape, affect maize reflectance properties (Chivasa et al. 2017). While using maize canopy temperature measurements, Kullberg et al. (2017) acknowledged that other stresses that may have contributed to higher canopy temperature, such as diseases, may exist. Therefore, to remedy this, their study used the Degrees Above Canopy Threshold (DACT) method. Other studies did not consider the influence of factors that might have altered the spectral reflectance of maize other than water stress (Rojas 2007; Rattalino Edreira et al. 2018). However, it suffices to note that a considerable amount of publications considered a change in spectral reflectance brought by different stages of maize growth and, therefore, obtained data measurements at different phenological stages (Bahir et al. 2017; Masina et al. 2020; Ndlovu et al. 2021; Shuai and Basso 2022). Nevertheless, it is significant to note that despite these limitations, remotely sensed data can still play a significant role in determining maize water stress accurately in fragmented agricultural systems, which can be achieved through the utilization of high spatial, spectral, and temporal resolution such as UAVs together with suitable analysis techniques, assisted by training the model using ground-truthing data.

Particularly in Africa, where susceptibility to shocks from climate variability and unpredictable precipitation is significant, research activities need to be encouraged to investigate the value of RS in monitoring maize water stress. The shortage of data that might restrict the use of remotely sensed data in maize water stress monitoring in these regions presents the opportunity to use UAVs to gather spatial data. UAVs are becoming a cutting-edge source of near-real-time spatial data for mapping and monitoring crop water requirements in light of the fourth industrial revolution, which will increase the productivity of the agricultural sector. Compared to the widely adopted satellite data shown in the Sect. 4.6, UAV data offer a strong chance of delivering accurate, swift, and spatially explicit data models for identifying maize water requirements and irrigation scheduling.

4.11 Conclusion

Remote sensing systems can be applied in target water stress identification. Other than applications such as crop growth assessment, irrigation levels, and crop yield, digital image techniques are performed for leaf and canopy phenotypic classification to detect crop water stress with the help of digital imagery data. Efficient ground-based sensors and UAV systems are becoming essential to advance image collection. Different symptoms are significant to immediately estimate crop water stress, which cannot be estimated using only a visible and thermal infrared image system and red-edge. Yet the utilization of remotely sensed data in predicting maize water stress in Africa is scant. The review further established that applying RS in maize water stress estimation should not overlook the intrinsic limitations caused by low- and medium-resolution sensors in transferring these methodologies to fragmented agricultural systems. Spatial patterns in these heterogeneous agricultural systems, like field size and shape, define the appropriate spatial, temporal, and spectral resolution to use. Mixed pixel problems remain challenging with low-to-medium resolution sensors in fragmented agricultural systems. However, significant improvements in water stress estimation are expected with high-resolution sensors like UAVs, whose pixel sizes are several times smaller than the field sizes prevalent in heterogeneous cropping systems. Findings from studies related to the detection of crop water stress using remote–sensing systems further upgrade the scope of remote sensing technology, management, and techniques and open up new perspectives for research on crop water stress management, especially in African countries. Therefore, further studies are necessary to investigate using high-resolution multispectral RS in estimating maize water stress in heterogeneous agricultural systems to increase our understanding of African agriculture and improve food security through early warning systems.

References

Ahmad I et al (2018) Yield forecasting of spring maize using remote sensing and crop modeling in Faisalabad-Punjab Pakistan. J Indian Soc Remote Sens 46(10):1701–1711. https://doi.org/10.1007/s12524-018-0825-8

Ahmad U, Alvino A, Marino S (2021) A review of crop water stress assessment using remote sensing. Remote Sens 13(20). https://doi.org/10.3390/rs13204155

Albeha M et al (2020) Graduate employability and competence development in higher education—a systematic literature review using PRISMA. Sustainability 12(15):5900

Alordzinu KE et al (2021) Rapid estimation of crop water stress index on tomato growth. Sensors 21(15). https://doi.org/10.3390/s21155142

Alvino A, Marino S (2017) Remote sensing for irrigation of horticultural crops. Horticulture 3(2). https://doi.org/10.3390/horticulturae3020040

Ambrosone M et al (2020) Retrieving soil moisture in rainfed and irrigated fields using Sentinel-2 observations and a modified OPTRAM approach. Int J Appl Earth Obs Geoinf 89:102113. https://doi.org/10.1016/j.jag.2020.102113

An J et al (2019) Identification and classification of maize drought stress using deep convolutional neural network. Symmetry 11(2). https://doi.org/10.3390/sym11020256

Awais M et al (2021) Remotely sensed identification of canopy characteristics using UAV-based imagery under unstable environmental conditions. Environ Technol Innov 22:101465. https://doi.org/10.1016/j.eti.2021.101465

Bahir M et al (2017) Evaluation and aggregation properties of thermal Infra-Red-based evapotranspiration algorithms from 100 m to the km scale over a semi-arid irrigated agricultural area. Remote Sens 9(11). https://doi.org/10.3390/rs9111178

Barrett B et al (2014) Assessment of multi-temporal, multi-sensor radar and ancillary spatial data for grasslands monitoring in Ireland using machine learning approaches. Remote Sens Environ 152(529):109–124. https://doi.org/10.1016/j.rse.2014.05.018

Bégué A et al (2020) Remote sensing products and services in support of agricultural public policies in Africa: overview and challenges. Front Sustain Food Syst 4. https://doi.org/10.3389/fsufs.2020.00058

Behmann J, Steinrücken J, Plümer L (2014) Detection of early plant stress responses in hyperspectral images. ISPRS J Photogramm Remote Sens 93:98–111. https://doi.org/10.1016/j.isprsjprs.2014.03.016

Berni JAJ et al (2009) Mapping canopy conductance and CWSI in olive orchards using high resolution thermal remote sensing imagery. Remote Sens Environ 113(11):2380–2388. https://doi.org/10.1016/j.rse.2009.06.018

Brogi C et al (2020) Simulation of spatial variability in crop leaf area index and yield using agroecosystem modelling and geophysics-based quantitative soil information. Vadose Zone J 19(1). https://doi.org/10.1002/vzj2.20009

Butcher G, Owen L, Barnes C (2019) Landsat: the cornerstone of global land imaging. GIM Int 33:1–7

Chakraborty M, Khot LR, Peters RT (2020) Assessing suitability of modified center pivot irrigation systems in corn production using low altitude aerial imaging techniques. Inf Process Agric 7(1):41–49. https://doi.org/10.1016/j.inpa.2019.06.001

Cheng M et al (2022) Estimation of soil moisture content under high maize canopy coverage from UAV multimodal data and machine learning. Agric Water Manag 264. https://doi.org/10.1016/j.agwat.2022.107530

Chitja J et al (2015) Empowerment of women through water use security, land use security and knowledge generation for improved household food security and sustainable rural livelihoods in selected areas of amongst others the Eastern Cape Province. Pretoria. Water Research Commission

Chivasa W, Mutanga O, Biradar C (2017) Application of remote sensing in estimating maize grain yield in heterogeneous African agricultural landscapes: a review. Int J Remote Sens 38(23):6816–6845. https://doi.org/10.1080/01431161.2017.1365390

Claverie M et al (2012) Maize and sunflower biomass estimation in southwest France using high spatial and temporal resolution remote sensing data. Remote Sens Environ 124:844–857. https://doi.org/10.1016/j.rse.2012.04.005

Costa-Filho E, Chávez JL, Comas L (2020) Determining maize water stress through a remote sensing-based surface energy balance approach. Irrig Sci 38(5–6):501–518. https://doi.org/10.1007/s00271-020-00668-1

DeJonge KC, Mefford BS, Chávez JL (2016) Assessing corn water stress using spectral reflectance. Int J Remote Sens 37(10):2294–2312. https://doi.org/10.1080/01431161.2016.1171929

Deng W et al (2021) A data mining based system for transaction fraud detection. In: 2021 IEEE international conference on consumer electronics and computer engineering (ICCECE). IEEE, pp 542–545. https://doi.org/10.1109/ICCECE51280.2021.9342376

Dong X et al (2021) Leaf water potential of field crops estimated using NDVI in ground-based remote sensing—opportunities to increase prediction precision. PeerJ 9. https://doi.org/10.7717/peerj.12005

Ekpa O et al (2018) Sub-Saharan African maize-based foods: technological perspectives to increase the food and nutrition security impacts of maize breeding programmes. Glob Food Sec 17:48–56. https://doi.org/10.1016/j.gfs.2018.03.007

Elkatoury A, Alazba AA, Abdelbary A (2020) Evaluating the performance of two SEB models for estimating ET based on satellite images in arid regions. Arab J Geosci 13(2). https://doi.org/10.1007/s12517-019-5019-3

Elmetwalli AH et al (2021) Integration of radiometric ground-based data and high-resolution quickbird imagery with multivariate modeling to estimate maize traits in the nile delta of Egypt. Sensors 21(11). https://doi.org/10.3390/s21113915

Feng B et al (2021) A quantitative review of water footprint accounting and crop production simulation based on publications from 2002–2018. Ecol Indic 120. https://doi.org/10.1016/J.ECOLIND.2020.106962

Filgueiras R et al (2020) Soil water content and actual evapotranspiration predictions using regression algorithms and remote sensing data. Agric Water Manag 241:106346. https://doi.org/10.1016/j.agwat.2020.106346

Genc L et al (2013) Paprastojo kukuruzo (Zea mays L.) dregmes streso nustatymas, taikant spektrini atspindi ir klasifikavimo medžio metoda. Zemdirbyste 100(1):81–90. https://doi.org/10.13080/z-a.2013.100.011

Gerhards M et al (2019) Challenges and future perspectives of multi-/hyperspectral thermal infrared remote sensing for crop water-stress detection: a review. Remote Sens 11(10). https://doi.org/10.3390/rs11101240

Gitelson AA, Merzlyak MN (1996) Signature analysis of leaf reflectance spectra: algorithm development for remote sensing of chlorophyll. J Plant Physiol 148(3–4):494–500

Greco M et al (2012) Estimating evapotranspiration and drought stress in DNA In Posidonia oceanica cadmium induces changes with ground-based thermal patterning remote sensing in methylation and chromatin agriculture: a review. J Exp Bot 63(13):4671–4712. https://doi.org/10.1093/jxb/err313

Guissard V, Lucau-Danila C, Defourny P (2005) Crop specific LAI retrieval using optical and radar satellite data for regional crop growth monitoring and modelling. Proc SPIE – Int Soc Opt Eng. https://doi.org/10.1117/12.627495

Guo PT et al (2015) Digital mapping of soil organic matter for rubber plantation at regional scale: an application of random forest plus residuals kriging approach. Geoderma 237–238:49–59. https://doi.org/10.1016/j.geoderma.2014.08.009

Haboudane D et al (2002) Integrated narrowband vegetation indices for prediction of crop chlorophyll content for application to precision agriculture. Remote Sens Environ 81(2–3):416–426. https://doi.org/10.1016/S0034-4257(02)00018-4

Han J et al (2020) Mapping the intellectual structure of research on surgery with mixed reality: bibliometric network analysis (2000–2019). J Biomed Inform 109:103516. https://doi.org/10.1016/j.jbi.2020.103516

Holzman ME et al (2018) Early assessment of crop yield from remotely sensed water stress and solar radiation data. ISPRS J Photogramm Remote Sens 145:297–308. https://doi.org/10.1016/j.isprsjprs.2018.03.014

Huete AR (1988) A soil-adjusted vegetation index (SAVI). Remote Sens Environ 25(3):295–309

Ihuoma SO, Madramootoo CA (2017) Recent advances in crop water stress detection. Comput Electron Agric. Elsevier B.V.:267–275. https://doi.org/10.1016/j.compag.2017.07.026

Ines AVM et al (2013) Assimilation of remotely sensed soil moisture and vegetation with a crop simulation model for maize yield prediction. Remote Sens Environ 138:149–164. https://doi.org/10.1016/j.rse.2013.07.018

Javadian M et al (2020) Global trends in evapotranspiration dominated by increases across large cropland regions. Remote Sens 12(7). https://doi.org/10.3390/rs12071221

Jégo G, Pattey E, Liu J (2012) Using Leaf Area Index, retrieved from optical imagery, in the STICS crop model for predicting yield and biomass of field crops. Field Crop Res 131:63–74. https://doi.org/10.1016/j.fcr.2012.02.012

Krishna G et al (2019) Comparison of various modelling approaches for water deficit stress monitoring in rice crop through hyperspectral remote sensing. Agric Water Manag 213:231–244. https://doi.org/10.1016/j.agwat.2018.08.029

Kullberg EG, DeJonge KC, Chávez JL (2017) Evaluation of thermal remote sensing indices to estimate crop evapotranspiration coefficients. Agric Water Manag 179:64–73. https://doi.org/10.1016/j.agwat.2016.07.007

Kyere I et al (2019) Multi-temporal agricultural land-cover mapping using single-year and multi-year models based on landsat imagery and IACS data. Agronomy 9(6):13–15. https://doi.org/10.3390/agronomy9060309
Li K et al (2021a) Evaluation of four new land surface temperature (LST) products in the U.S. Corn Belt: ECOSTRESS, GOES-R, Landsat, and Sentinel-3. IEEE J Sel Top Appl Earth Obs Remote Sens 14:9931–9945. https://doi.org/10.1109/JSTARS.2021.3114613
Li M et al (2021b) Hyperspectral characteristics and scale effects of leaf and canopy of summer maize under continuous water stresses. Agriculture (Switzerland) 11(12). https://doi.org/10.3390/agriculture11121180
Liu, Huete (1995) A feedback based modification of the NDVI to minimize canopy background and atmospheric noise. IEEE Trans Geosci Remote Sens 33(2):457–465
Liu J et al (2010) Estimating crop stresses, above-ground dry biomass and yield of corn using multi-temporal optical data combined with a radiation use efficiency model. Remote Sens Environ 114(6):1167–1177. https://doi.org/10.1016/j.rse.2010.01.004
Liu J, Chen P, Xu X (2018) Estimating wheat coverage using multispectral images collected by unmanned aerial vehicles and a new sensor. Agro-geoinformatics (Agro-geoinformatics) (pp. 1–5). IEEE.
Maes et al (2018) Can UAV-based infrared thermography be used to study plant-parasite interactions between mistletoe and Eucalypt trees? Remote Sens 10(12):1–16. https://doi.org/10.3390/rs10122062
Mananze S, Pôças I, Cunha M (2018) Retrieval of maize leaf area index using hyperspectral and multispectral data. Remote Sens 10(12). https://doi.org/10.3390/rs10121942
Mangus DL, Sharda A, Zhang N (2016) Development and evaluation of thermal infrared imaging system for high spatial and temporal resolution crop water stress monitoring of corn within a greenhouse. Comput Electron Agric 121:149–159. https://doi.org/10.1016/j.compag.2015.12.007
Masenyama A et al (2022) A systematic review on the use of remote sensing technologies in quantifying grasslands ecosystem services. GIScі Remote Sens 59(1):1000–1025. https://doi.org/10.1080/15481603.2022.2088652
Masina M et al (2020) Remote sensing analysis of surface temperature from heterogeneous data in a maize field and related water stress. Remote Sens 12(15). https://doi.org/10.3390/RS12152506
Matese A et al (2018) Estimation of water stress in grapevines using proximal and remote sensing methods. Remote Sens 10(1):1–16. https://doi.org/10.3390/rs10010114
Messina G, Modica G (2020) Applications of UAV thermal imagery in precision agriculture: state of the art and future research outlook. Remote Sens 12(9). https://doi.org/10.3390/RS12091491
Mirzaie M et al (2014) Comparative analysis of different uni- and multivariate methods for estimation of vegetation water content using hyper-spectral measurements. Int J Appl Earth Obs Geoinf 26(1):1–11. https://doi.org/10.1016/j.jag.2013.04.004
Ndlovu HS et al (2021) A comparative estimation of maize leaf water content using machine learning techniques and unmanned aerial vehicle (Uav)-based proximal and remotely sensed data. Remote Sens 13(20). https://doi.org/10.3390/rs13204091
Padilla et al (2011) Integration of vegetation indices into a water balance model to estimate evapotranspiration of wheat and corn. Hydrol Earth Syst Sci 15(4):1213–1225. https://doi.org/10.5194/hess-15-1213-2011
Panigada C et al (2014) Fluorescence, PRI and canopy temperature for water stress detection in cereal crops. Int J Appl Earth Obs Geoinf 30(1):167–178. https://doi.org/10.1016/j.jag.2014.02.002
Ramachandiran K, Pazhanivelan S (2017) Relationship between spectral vegetation indices and crop parameters of maize under nitrogen and water stress. J Agrometeorology 19(2):114–119
Rattalino Edreira JI et al (2018) Water productivity of rainfed maize and wheat: a local to global perspective. Agric For Meteorol 259:364–373. https://doi.org/10.1016/j.agrformet.2018.05.019

Ren J et al (2021) A two-step mapping of irrigated corn with multi-temporal MODIS and Landsat analysis ready data. ISPRS J Photogramm Remote Sens 176:69–82. https://doi.org/10.1016/j.isprsjprs.2021.04.007

Reyes-González A et al (2019) Comparison of leaf area index, surface temperature, and actual evapotranspiration estimated using the METRIC model and in situ measurements. Sensors (Switzerland) 19(8). https://doi.org/10.3390/s19081857

Rivera-Marin D, Dash J, Ogutu B (2022) The use of remote sensing for desertification studies: a review. J Arid Environ 206:104829. https://doi.org/10.1016/j.jaridenv.2022.104829

Rojas O (2007) Operational maize yield model development and validation based on remote sensing and agro-meteorological data in Kenya. Int J Remote Sens 28(17):3775–3793. https://doi.org/10.1080/01431160601075608

Rossini M et al (2013) Assessing canopy PRI from airborne imagery to map water stress in maize. ISPRS J Photogramm Remote Sens 86:168–177. https://doi.org/10.1016/j.isprsjprs.2013.10.002

Rossini M et al (2015) Discriminating irrigated and rainfed maize with diurnal fluorescence and canopy temperature airborne maps. ISPRS Int J Geo Inf 4(2):626–646. https://doi.org/10.3390/ijgi4020626

Ruwanza S, Thondhlana G, Falayi M (2022) Research progress and conceptual insights on drought impacts and responses among smallholder farmers in South Africa. Land 11(2):159

Shao G et al (2021) Mapping maize crop coefficient Kc using random forest algorithm based on leaf area index and UAV-based multispectral vegetation indices. Agric Water Manag 252:106906. https://doi.org/10.1016/j.agwat.2021.106906

Shuai G, Basso B (2022) Subfield maize yield prediction improves when in-season crop water deficit is included in remote sensing imagery-based models. Remote Sens Environ 272:112938. https://doi.org/10.1016/j.rse.2022.112938

Sibanda M et al (2021) Application of drone technologies in surface water resources monitoring and assessment: a systematic review of progress, challenges, and opportunities in the global south. Drones 5(3):1–21. https://doi.org/10.3390/drones5030084

Sobejano-Paz V et al (2020) Hyperspectral and thermal sensing of stomatal conductance, transpiration, and photosynthesis for soybean and maize under drought. Remote Sens 12(19):1–32. https://doi.org/10.3390/rs12193182

Song L, Jin J, He J (2019) Effects of severe water stress on maize growth processes in the field. Sustainability 11(18). https://doi.org/10.3390/su11185086

Song X, Zhou G, He Q (2021) Critical leaf water content for maize photosynthesis under drought stress and its response to rewatering. Sustainability (Switzerland) 13(13):1–14. https://doi.org/10.3390/su13137218

Sreelash K et al (2017) Estimation of available water capacity components of two-layered soils using crop model inversion: effect of crop type and water regime. J Hydrol 546:166–178. https://doi.org/10.1016/j.jhydrol.2016.12.049

Steduto P, Hsiao TC, Raes D, Fereres E (2009) AquaCrop—The FAO crop model to simulate yield response to water: I. Concepts and underlying principles. J. Agron 101(3), pp. 426–437

Suárez L et al (2009) Modelling PRI for water stress detection using radiative transfer models. Remote Sens Environ 113(4):730–744. https://doi.org/10.1016/j.rse.2008.12.001

Sun C et al (2019) Using of multi-source and multi-temporal remote sensing data improves crop-type mapping in the subtropical agriculture region. Sensors 19(10):1–23. https://doi.org/10.3390/s19102401

Tandzi LN, Mutengwa CS (2020) Estimation of maize (Zea mays L.) yield per harvest area: appropriate methods. Agronomy 10(1):1–18. https://doi.org/10.3390/agronomy10010029

Testi L et al (2008) Crop water stress index is a sensitive water stress indicator in pistachio trees. Irrig Sci 26(5):395–405. https://doi.org/10.1007/s00271-008-0104-5

Tsakmakis ID, Gikas GD, Sylaios GK (2021) Integration of Sentinel-derived NDVI to reduce uncertainties in the operational field monitoring of maize. Agric Water Manag 255. https://doi.org/10.1016/j.agwat.2021.106998

USGS (2013) Landsat — a global land-imaging mission. United Stated Geological Survey

van Eck NJ, Waltman L (2010) Software survey: VOSviewer, a computer program for bibliometric mapping. Scientometrics 84(2):523–538. https://doi.org/10.1007/s11192-009-0146-3
van Emmerik T et al (2015) Impact of diurnal variation in vegetation water content on radar backscatter from maize during water stress. IEEE Trans Geosci Remote Sens 53(7):3855–3869. https://doi.org/10.1109/TGRS.2014.2386142
Virnodkar SS, Pachghare VK, Patil VC, Jha SK (2020) Remote sensing and machine learning for crop water stress determination in various crops: a critical review. Precis Agric 21(5), pp.1121–1155.
Wan W et al (2021) Drought monitoring of the maize planting areas in Northeast and North China Plain. Agric Water Manag 245(2):106636. https://doi.org/10.1016/j.agwat.2020.106636
Wang R, Cherkauer K, Bowling L (2016) Corn response to climate stress detected with satellite-based NDVI time series. Remote Sens 8(4). https://doi.org/10.3390/rs8040269
Wang L et al (2018a) Developing an integrated indicator for monitoring maize growth condition using remotely sensed vegetation temperature condition index and leaf area index. Comput Electron Agric 152(17):340–349. https://doi.org/10.1016/j.compag.2018.07.026
Wang P et al (2018b) Monitoring growth condition of spring maize in Northeast China using a process-based model. Int J Appl Earth Obs Geoinf 66:27–36. https://doi.org/10.1016/j.jag.2017.11.001
Weiss M, Jacob F (2019) Remote sensing for agricultural applications: a meta-review. https://doi.org/10.1016/j.rse.2019.111402
Wijesingha J et al (2021) Comparison of spaceborne and uav-borne remote sensing spectral data for estimating monsoon crop vegetation parameters. Sensors 21(8). https://doi.org/10.3390/s21082886
Wulder MA et al (2019) Current status of Landsat program, science, and applications. Remote Sens Environ 225:127–147. https://doi.org/10.1016/j.rse.2019.02.015
Xie X et al (2021) Comparison of random forest and multiple linear regression models for estimation of soil extracellular enzyme activities in agricultural reclaimed coastal saline land. Ecol Indic 120:106925. https://doi.org/10.1016/j.ecolind.2020.106925
Yan F et al (2006) Progress in soil moisture estimation from remote sensing data for agricultural drought monitoring. https://doi.org/10.1117/12.689309
Yue J et al (2018) A comparison of regression techniques for estimation of above-ground winter wheat biomass using near-surface spectroscopy. Remote Sens 10(1). https://doi.org/10.3390/rs10010066
Zhang F, Zhou G (2019) Estimation of vegetation water content using hyperspectral vegetation indices: a comparison of crop water indicators in response to water stress treatments for summer maize. BMC Ecol 19(1):1–12. https://doi.org/10.1186/s12898-019-0233-0
Zhang J et al (2016) Using satellite multispectral imagery for damage mapping of armyworm (Spodoptera frugiperda) in maize at a regional scale. Pest Manag Sci 72(2):335–348. https://doi.org/10.1002/ps.4003
Zhang L et al (2019a) Maize canopy temperature extracted from UAV thermal and RGB imagery and its application in water stress monitoring. Front Plant Sci 10:1–18. https://doi.org/10.3389/fpls.2019.01270
Zhang Y et al (2019b) Maize crop coefficient estimated from UAV-measured multispectral vegetation indices. Sensors 19(23):1–17. https://doi.org/10.3390/s19235250
Zhang L et al (2019c) Mapping maize water stress based on UAV multispectral remote sensing. Remote Sens 11(6):1–24. https://doi.org/10.3390/rs11060605
Zhang X et al (2022) Combining spectral and texture features of UAS-based multispectral images for maize leaf area index estimation. Remote Sens 14(2). https://doi.org/10.3390/rs14020331
Zhao L et al (2018) Retrieving the diurnal FPAR of a maize canopy from the jointing stage to the tasseling stage with vegetation indices under different water stresses and light conditions. Sensors 18(11). https://doi.org/10.3390/s18113965
Zhou H et al (2022) Dynamic characteristics of canopy and vegetation water content during an entire maize growing season in relation to spectral-based indices. Remote Sens 14(3). https://doi.org/10.3390/rs14030584

Chapter 5
Application of Endophytes in Bioremediation, Biotransformation, and Water Disinfection for Irrigation Systems

Linda U. Obi, Muiz O. Akinyemi, Kazeem A. Alayande, and Rasheed A. Adeleke

Abstract Globally, freshwater is insufficiently distributed to meet all present and future water demands. As a result, agricultural water demands must be satisfied by creative methods such as wastewater treatment and recycling. Several physical and chemical remediation procedures have been utilised to degrade or remove environmental pollutants in wastewater. Some techniques produce toxic metabolites, are ecologically unfriendly and expensive, or entail the relocation of contaminants rather than reducing them. It is, therefore, critical to incorporate biological approaches that use organisms' metabolic activities to break down or change toxic contaminants into less dangerous intermediates or products. Employing plant symbiont microorganisms known as endophytes has been shown as a sustainable and eco-friendly approach for decontaminating water and enhancing agricultural output. In this chapter, we highlight the potential of endophytes as wastewater bioremediation, biotransformation, and disinfection agents.

Keywords Agricultural product · Endophytes · Irrigation · Wastewater · Eco-friendliness

L. U. Obi · K. A. Alayande · R. A. Adeleke (✉)
North-West University, Potchefstroom, South Africa

M. O. Akinyemi
North-West University, Potchefstroom, South Africa

National Centre for Food Manufacturing, University of Lincoln, Holbeach, United Kingdom
e-mail: Rasheed.adeleke@nwu.ac.za

T. Mabhaudhi et al. (eds.), *Enhancing Water and Food Security Through Improved Agricultural Water Productivity*,
https://doi.org/10.1007/978-981-96-1848-4_5

5.1 Introduction

Agriculture water demand is hardly met, relying solely on rainfall and soil water accessible to the root capillarity. It is common knowledge that sustainable agricultural practices face problems of water shortages and inefficient irrigation water systems in most parts of the world. The current climate change and adverse weather events are exacerbating the situation, affecting virtually all regions of our planet. Irrigation is an essential strategy for addressing the world's growing food insecurity. It provides a reliable water supply to sustain agricultural farms, and almost one-half of the available irrigation water is supplied from groundwater originating from shallow and deep-water tables (He et al. 2020a, b). Ground and surface water are the largest and most important sources of freshwater for irrigation (Siebert 2010). While available freshwater is sufficient to meet all current and future water demands, its spatial and temporal distributions are insufficient.

Many regions, particularly the arid and semi-arid regions, such as South Africa, have inadequate water resources to meet requirements (Mengistu et al. 2021). In these water-scarce areas, the reuse of domestic wastewater (DWW) (e.g., artificial ponds from surface runoff) for irrigation is a reasonable alternative agriculture practice to achieve sustainable water management (Patience et al. 2021; Phakathi et al. 2021). In all these regions of the earth with shallow levels of rainfall, there is little available quality water to support human and plant activities due to low levels of precipitation and high rate of evapotranspiration coupled with prolonged dry periods and extreme weather conditions (Bortolini et al. 2018).

Reports and literature reviews originating in the countries on the quality of various aquatic sources in recent decades suggest the presence of inorganic (Agunbiade and Moodley 2014; Rimayi et al. 2018; Gani et al. 2021), organic (Vasseghian et al. 2021), and biological (Verlicchi and Grillini 2020) contaminants including water used for irrigation of crops. Often, these contaminants occur in effluents from anthropogenic sources such as domestic sewage, wastewater treatment plants, agricultural land use, and industrial and mining operations, which may be discharged into the water system through numerous paths. Using irrigation water containing even small amounts of these pollutants puts biodiversity, consumer health, and food quality in danger immediately and over time. These dangers include but are not limited to food poisoning, spoilage and wastage, malnutrition, infertility, stillbirths, decreased effectiveness of antibiotics, loss of agricultural products, rejection of products for export abroad, and even human and animal fatalities.

Moreover, numerous studies have listed various physical (like filtration) and chemical (such as nanoparticles, charcoal) techniques that can be used to clean wastewater for agricultural use; however, there are few studies of a similar nature describing biological methods, particularly the employment of endophytic microbes in irrigation water purification. This chapter focuses on the biological application of endophyte as a tool for bioremediation, biotransformation, and water disinfection for irrigation systems. Plants have symbionts called endophytes; these microbes confer growth benefits to the plant host by mediating activities such as nitrogen fixation, phosphate solubilization, and biological control of plant diseases (Raimi and Adeleke 2021). Numerous studies have shown that endophytes can sweep up, detoxify, and even destroy pollutants in wastewater. Therefore, we postulate that employing these isolates alone or combined to clean irrigation water would boost farm product output, safety and quality.

5.2 Application of Endophytes in Bioremediation of Contaminated Aqueous System

The dawn of microbial biotechnology has established the beneficial functions of endophytes in different sectors, such as agriculture, medicine, and industry (Fig. 5.1). The application of endophytes for environmental remediation of pollutants cannot be overemphasized due to their efficiency, eco-friendliness, and sustainability. Their ability to degrade complex organic compounds makes them significant in the bioremediation of contaminated aqueous environments. Endophytes are endosymbionts, often bacteria or fungi, that exist in the internal tissues of healthy plants (Raimi and Adeleke 2021). Endophytes have been employed in beneficial metabolic activities, from improving plant growth and development to producing bioactive and/or metabolic substances of medicinal and ecological importance (Fadiji and Babalola 2020). They are ubiquitous, and different plant-associated endophytes have been linked with critical biological processes such as biocontrol

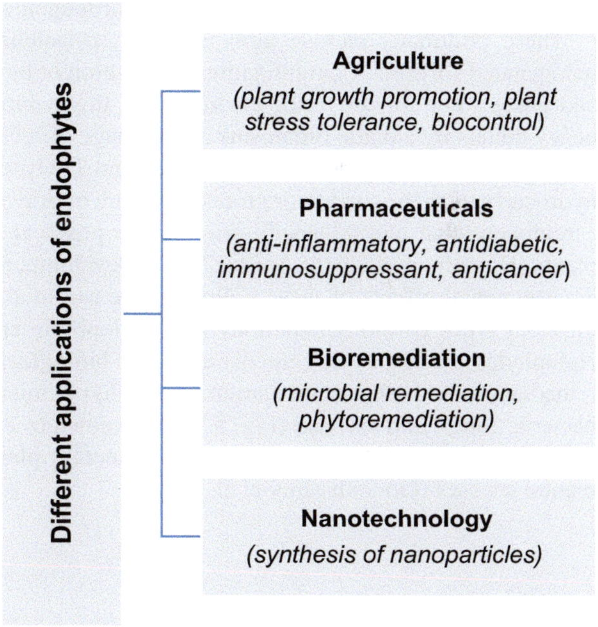

Fig. 5.1 Beneficial applications of endophytes

and bioremediation (Rodriguez et al. 2009; Anyasi and Atagana 2019). This is due to their potential to produce active intermediates or products by altering complex compounds to less-toxic counterparts (Wang and Dai 2011).

Different physical and chemical remediation techniques have been used to degrade or remove environmental pollutants. Some of these techniques are toxic metabolite producers, environmentally unfriendly, cost-intensive, or involve the relocation of contaminants instead of degrading them (Chlebek and Hupert-Kocurek 2019; Khan et al. 2019). Incorporating biological techniques that utilize the metabolic functions of organisms to degrade or transform these toxic pollutants into less harmful intermediates or products is imperative (Akhtar and Mannan, 2020). Such techniques, including bioremediation with endophytic microorganisms, have been recorded as sustainable and eco-friendly means of environmental decontamination (Singh et al. 2020). Production of relevant biological compounds by endophytes enhances their application as a bioremediation agent for contaminated aquatic ecosystems.

Contaminated aquatic systems refer to bodies of water contaminated by various pollutants, leading to detrimental effects on the ecosystem and potentially endangering human health. Contamination of aquatic systems could be due to natural disasters or anthropogenic activities. The discharge from mining, petrochemical, and textile or dye industries comprises harmful substances that present a health risk to humans. Notably, soil and water often exhibit a significant accumulation of heavy metals at high concentrations (Saravanan 2021). The preservation and restoration of these ecosystems are crucial for maintaining the health and sustainability of our water resources. Furthermore, aquatic ecosystems such as wetlands are potential water sources for economic purposes; however, these ecosystems are usually contaminated with different pollutants from anthropogenic sources.

These pollutants include agrochemicals, petroleum hydrocarbons as well as halogenated solvents. Contamination or pollution of the aquatic ecosystem is severe and poses some threats to the environment; this comprises possible absorption of heavy metals by aquatic organisms and seepage of recalcitrant and toxic pollutants into the underlying groundwater (Kumar and Dwivedi 2021). Heavy metals and hydrocarbons in the surface or groundwater are of grave concern as many are human carcinogens that can adversely affect some physical and metabolic processes in plants (Hussain et al. 2018; Anitha 2022). Developmental strategies to mitigate the environmental effects of these pollutants are essential, hence the incorporation of different types of bioremediations with endophytic entities. Bioremediation with endophytes encompasses remediation with biological entities such as microbial remediation and phytoremediation. Endophytic microorganisms, encompassing bacteria, fungi, and archaea (Fig. 5.2), are known to engage in symbiotic associations with various plant components and protect the plants against different environmental stresses (Govindasamy et al. 2018).

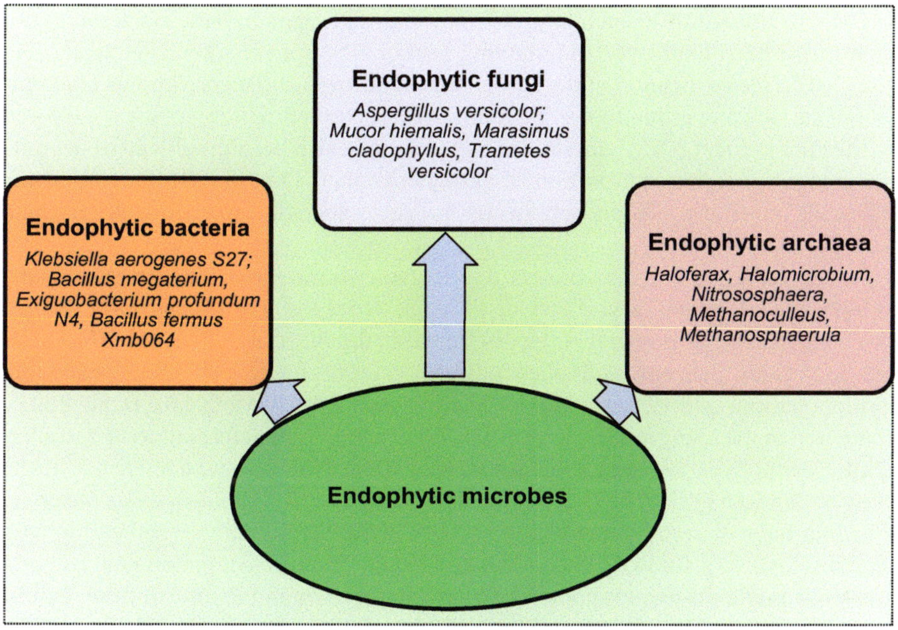

Fig. 5.2 Endophytic microbes used in bioremediation of aqueous systems

5.3 Application of Endophytic Microbes in the Decontamination of Aquatic Systems

5.3.1 Bacterial Endophytes as Remediation Agents for Aquatic Systems

Bacteria are effective biodegradation and bioremediation agents due to their ubiquity and versatility. Given the right conditions, bacteria have few universal toxins, making it likely that they can break down any given substrate. Bacteria-mediated processes have been employed to reduce the toxicity of environmental contaminants such as heavy metals. Heavy metal stress can be more easily tolerated and treated by endophytic bacteria as biodegradation, particularly in live cells, is the primary elimination route for endophytic bacteria (Sim et al. 2019). Removal of environmental pollutants by bacterial endophytic entities is essential and promising. The ability of these bacterial strains to reduce heavy metals concentration in plant species and enhance plant growth was demonstrated by Ali et al. (2017).

Metals are more readily available to plants, thanks to endophytic bacteria, which facilitate their movement through plant tissues. The potential of endophytic bacteria *Klebsiella aerogenes* S27 to completely decolorize a carcinogenic chemical, malachite green (MG), in wastewater was examined by Shang et al. (2019), where the endophyte facilitated transformation via actions of novel enzymatic oxidoreductase

KaTMR to transform malachite green into less-toxic metabolites. Bacterial endophytes have been implicated in removing xenobiotics from contaminated water (Ho et al. 2012). Detoxification of the agrochemical atrazine, which is a known contaminant of surface and groundwater by endophytic bacteria Streptomyces was explored by Mesquini et al. (2015). *Bacillus megaterium* has also been employed to degrade and detoxify a significant portion of agrochemicals (Liu et al. 2014). Endophytes utilize the enzyme organophosphorus hydrolase to degrade agrochemicals (Barman et al. 2014).

Azo dyes are stable xenobiotics due to their stable molecular structures, and endophytes are potential purifiers of effluents from the dye (Shang et al. 2019). *Exiguobacterium profundum* strain N4 converts the diazo dye Reactive Black 5 to less hazardous benzene and phenanthrene via enzymatic oxidation, reduction, desulphonation, and demethylation (Sharma and Roy 2015; Karaś et al. 2021). Reduction in the cytogenetic effect of DB-14 dye due to transformation by the alkaliphilic endophytic bacteria *Bacillus fermus* strain Xmb064 was observed in the study of Sandesh et al. (2019). Absolute degradation of the DB-14 dye was observed after four days of aerobic incubation at 37 °C with the aforementioned endophytic bacteria. Also, in a study by Hussain et al. (2018), 20 strains of endophytic bacteria isolated from the roots and shoots of *Typha domingensis* and *Pistia stratiotes* exhibited credible potential in biodegrading heavily contaminated textile effluent. All the bacterial test strains showcased significant reduction in the chemical and biological oxygen demand, coloration intensity, toxicity level, organic nutrients, and heavy metals.

A study by Ijaz et al. (2015) showed the efficiency of a relatively recent technique, floating treatment wetlands (FTWs). The technique involves the cultivation of plants on buoyant mats without soil. In this setup, the submerged plant biomass freely hangs in the water column beneath the mat. This innovative approach, characterized by its simplicity and easy establishment used with a pollutant-degrading endophytic bacteria, has been reported to be highly effective in enhancing sewage effluent quality. The introduced endophytes exhibited a significant level of colonization in both the root and shoot of the plants, as well as in the wastewater itself (Ijaz et al. 2015).

5.3.2 Endophytes as Mycoremediation Agents for Aquatic Systems

Fungi are known for rapid growth, quick adaptability, and accumulation of high concentrations of heavy metals in water environments when compared with bacteria. They remediate polluted aquatic ecosystems through bioaccumulation, biomineralization, and adsorption of the available pollutants (Dang et al. 2018; Kumar and Dwivedi 2021). Using fungal entities to remediate or rid the polluted aquatic ecosystems of contaminants is called mycoremediation. A possible means of responding to heavy metal stress by endophytes is through the production of a defence mechanism known as extracellular polymeric substances (EPS), which binds/immobilizes heavy metal ions. These secreted EPS play a critical role in enhancing the heavy metal tolerance ability of most endophytes at different concentrations of several heavy metals (Dang et al. 2018). Endophytic fungi (EF) are abundant in plants and can display their high metal tolerance capacity by proliferating in a heavy metal-polluted environment (Mohammadian et al. 2017). Different EF have been employed in different mycoremediation strategies, and most of them have demonstrated their pollutant remediation strategies through enzymatic detoxification, degradation, filtration, immobilization, etc. (Kumar and Dwivedi 2021). Mycoremediation of hydrocarbons contaminated aquatic ecosystem by the EF *Pleurotus ostreatus* exhibited the ability of the ligninolytic enzyme, manganese peroxidase, in breaking down polycyclic aromatic hydrocarbons (PAHs) (Arfaeinia et al. 2019).

Mycoremediation of agrochemicals such as glyphosate, paraquat, endosulfan, and aldrin is an efficient and environmentally friendly means of remediating agrochemical-contaminated water as some conventional methods will either incur cost or generate toxic intermediates (Eman et al. 2013). Degradation of environmental contaminants such as agrochemicals has been demonstrated by the characteristic ability of the EF, *Trametes versicolor* (Bhadouria et al. 2020). *Trametes versicolor* portrayed its fast adsorptive and, subsequently, high degradation ability on hydrophobic synthetic pesticides, chlorpyrifos, dicofol, and cypermethrin, according to Hu et al. (2020). Several studies have demonstrated the degradation capacities of EF in removing agrochemicals from wastewater. Amongst the different fungi analyzed for their potential to metabolize the herbicide Granstar (tribenuron-methyl), *Aspergillus versicolor* exhibited a higher ability to degrade tribenuron-methyl to different metabolites within 35 days of study (AI-Jawhari and AI-Sead 2016).

Pharmaceuticals are one of the most common xenobiotics in wastewater, surface, and groundwater. This threatens public health, hence the need for their removal. The study of Esterhuizen-Londt et al. (2016) investigated the potential of the EF *Mucor hiemalis* to take up and degrade acetaminophen in wastewater. They also observed the ability of the fungi to exhibit no oxidative stress that could have resulted from exposure to acetaminophen; the bioremediation process produced no harmful metabolites. Another EF identified in the mycoremediation of pharmaceuticals, including opiates and nonsteroidal anti-inflammatory drugs (NSAIDs), is *Trametes versicolor* (Marco-Urrea et al. 2010; Asif et al. 2017). Using their ligninolytic enzymes, *T. versicolor* has likewise been employed in the degradation of antibiotics component of pharmaceutical effluents, and the fungal entity is known to be effective in the degradation of antibiotics such as ciprofloxacin, ofloxacin, and sulfamethazine (Prieto et al. 2011; García-Galán et al. 2011). Degradation of ofloxacin and norfloxacin using the manganese peroxidase enzymes of the EF *Irpex lacteus* has also been reported (Čvančarová et al. 2015). The nonspecificity of the ligninolytic enzymes of EF enhances their ability to degrade different xenobiotics in contaminated water. These enzymes include laccases, peroxidases, catalases, hydrolases, proteases, pectinases, amylases, phosphatases, lipases, cellulases, and tyrosinases (Yang et al. 2017; Rana 2019; Choudhary et al. 2021).

Endophytic fungi have been known to be efficient bioremediation due to the nongeneration of toxic intermediates during degradation. They have been utilized in the mycoremediation of various wastewater contaminants, including synthetic dyes. The studies by Ngieng et al. (2013) and Yang et al. (2017) evaluated the degradation ability of *Marasmius cladophyllus* and *T. versicolor* concerning different synthetic dyes. They observed the potential of the aforementioned fungal entities to break down these dyes using their ligninolytic enzymes to generate no toxic by-products. Triphenylmethane is a component of effluents from textile industries. Gao et al. (2020) demonstrated that the EF *Bjerkandera adusta* possesses the capability to efficiently degrade a high molecular weight dye, triphenylmethane, through the oxidation pathway using enzymes such as manganese peroxidase and lignin peroxidase.

Endophytes naturally produce a surface-active protein compound, hydrophobins, which enhances the solubility of hydrophobic compounds and subsequently improves bioavailability and possible biodegradation by prospective enzymes (Kulkarni et al. 2017). Endophytes, especially endophytic fungi (EF), secrete an extensive range of extracellular enzymes which catalyze different metabolic activities, including the degradation of low-molecular-weight and high-molecular-weight hydrocarbons (Akhtar and Mannan 2020; Bhadra et al. 2022). Several endophytes have been portrayed in the degradation of hydrocarbons, including the degradation of long chain hydrocarbons and BTEX by EF Aspergillus and Mucor, as well as exploiting the degradation potential of different species of Fusarium in degrading different hydrocarbons have also been exploited (Asemoloye et al. 2020; AI-Jawhari 2022). These fungi are also capable of producing active substances such as cis-4-acetoxyoxymellein and 8-deoxy-6-hydroxy-cis-4-acetoxyoxymellein which have been employed in algicidal activities of eradicating toxic algal blooms from wastewater. These fungi utilize the active substances to degrade the algal cells' nuclear membrane, leading to possible algal cell deaths (Hussain et al. 2015).

Mycofiltration, a form of mycoremediation, is an eco-friendly wastewater treatment technique that utilizes the filtration abilities of the mycelium of certain fungal entities to capture and degrade environmental contaminants in water (Mnkandla and Otomo 2021). Different studies have reported on the high biosorptive capacity of EF as an efficient strategy for eradicating heavy metals in the environment. The potential of fungal mycelia to filter and improve water quality in a built ecosystem for fish culture was investigated by Chandra et al. (2022). Their study suggested the capacity of fungal mycelium to take up heavy metals, degrade contaminants, and significantly reduce coliform bacteria's microbial load in the built ecosystem. A significant reduction in assayed content of heavy metals (Pb, Zn, Fe, Cu) was observed as the concentration of these metals was within the acceptable range for the proliferation of aquatic organisms.

The study by Pini and Geddes (2020) also portrayed the feasibility of mycofiltration in remediating biological contaminants in water. They exploited the mycelia of *Pleurotus ostreatus* to trap and degrade *E. coli* in river water. *E. coli* are pathogenic fecal coliform that is of public health concern hence the need for their removal from the environment. Besides the removal of biotic pathogens like *E. coli*, *Pleurotus ostreatus* has been employed in the removal of abiotic environmental contaminants such as hydrocarbons, heavy metals, and polychlorinated biphenyls (PCBs) (Kumar and Dwivedi 2021).

5.3.3 Endophytic Archaea in the Remediation of Aquatic Systems

Endophytes are traditionally recognized as bacteria or fungi; however, recent classification has expanded to include archaea as endophytes. Notably, archaeal phyla such as Thaumarchaeota, Crenarchaeota, and Euryarchaeota have been identified in this role. Endophytic archaeal genera encompass a variety of species, including Methanoculleus, Thermoplasma, Methanococcus, Methanosphaerula, Methanospirillum, Nitrosopumilus, Nitrososphaera, Halomicrobium, Halogeometricum, Haloferax, Thermoplasma, and more (Fadiji et al. 2020). The understanding of these archaea's distribution, significance, function, and activity in relation to their host plants is still speculative, as mentioned by Müller et al. (2015). Not many studies have been conducted on the potential of endophytic archaea to decontaminate aquatic systems (Chow et al. 2022); however, the plant-growth-promoting abilities of endophytic archaea cannot be overemphasized.

Endophytic archaea play a vital role in sustainable agriculture as they possess specific genes that facilitate various plant metabolic functions, including nutrient cycling, siderophore production, and modulation of plant hormones. Moreover, endophytic archaea augment the plants' capacity to withstand and endure abiotic stress (Jung et al. 2020; Chow et al. 2022). Although susceptible to salinity, endophytic archaea have exhibited resilience to water salinity and have been found to promote the growth of certain plants, such as *Phragmites australis*, which is utilized in phytoremediation efforts (Llirós et al. 2014; Wang et al. 2022). *Phragmites australis* is a macrophyte renowned for its remarkable ability to accumulate contaminants in aquatic systems (Ahmad et al. 2014). AI-Homaidan et al. (2020) evaluated the capacity of *Phragmites australis* in wetlands to sequester heavy metals and reported significant accumulation of Zn, Cd, and Pb. A study by A. Bello et al. (2018) demonstrated that *Phragmites australis* achieved removal rates of 93%, 84%, and 95% for Cd, Ni, and Pb, respectively, over 6 weeks. Archaea's adaptation to extreme environments positions them as promising candidates for bioremediation of chemical pollutants. Various archaea species have been effectively utilized in remediating polluted locations, including acid mine drainage, hypersaline environments, and gas and oil production wastewater. For instance, the archaea, Haloarcula EH4 demonstrated significant potential in degrading aromatic hydrocarbons (Bertrand et al. 1990; Tapilatu et al. 2010; Naitam and Kaushik 2021). The utilization of endophytic archaea holds promise as a distinctive approach to minimizing environmental harm caused by contaminated aquatic systems. These archaea have the potential to provide unique solutions toward this goal.

5.4 Application of Endophytes in Biotransformation of Contaminants in Aquatic Systems

Modifying organic and inorganic contaminants from an aqueous environment via potential biotransformation by endophytes is an emerging study area. Limited studies have been conducted on biotransformation as a remedial technique as well as the biotransformation potential of endophytes (Chalifour et al. 2021; Choudhary et al. 2021; Desiante et al. 2021). Biotransformation aims to remove environmental pollutants from aqueous environments via structural and functional modification of their harmfulness, bioavailability, and persistence of chemical compounds in a biological system. It transforms chemical compounds into less toxic products via stereo selection and bio-oxidation (Huang 2022). Using endophytic microbes as biological catalysts is a microbial transformation technique (biotransformation) that aims to mitigate the challenges associated with using conventional catalysts to transform environmental pollutants. Exploration of endophytic entities for their potential to produce nonspecific ligninolytic enzymes capable of cleaving strong molecular bonds is necessary. Different endophytic microbes have been employed in the biotransformation of several contaminants, and they include Serratia, Ochrobactrum, Enterobacter, Pseudomonas, Bacillus, Aspergillus, Sphingomonas, and Cellulosimicrobium (Kuźniar et al. 2019).

An economical and efficient means commonly used in removing heavy metals from the contaminated aqueous environment is biosorption; this involves the adsorption of heavy metal ions onto the active sites of biosorbents. Examples of biosorbents include bacteria, fungi, algae, biopolymers, and plant residues (Beni and Esmaeili 2020). Endophytic fungi like Aspergillus and Penicillium are known for their high biosorbent abilities, as they can adsorb heavy metals from a contaminated aqueous environment (Ahmad et al. 2006; AI-Jawhari 2022). The extensive heavy metals adsorption capacity of the EF, *Portulaca* sp., stems from its high mycelia-producing ability (Kapahi and Sachdeva 2017). The biosorption potential of *Phanerochaete chrysosporium* to remove heavy metals from an aqueous solution was also evaluated by Rudakiya et al. (2018). They deduced that *Phanerochaete chrysosporium* utilized its hyphal cells to absorb/eradicate metals like Cd, Cu, Cr, Pb, Ni, Zn, Co, Se, and Al by binding them to their cell wall component, whereas some fungi also utilize the metals above for their metabolic activities.

Endophytes have been recognized as essential factors that enhance the removal rate of potential aqueous contaminants. Besides their increased stereoselectivity, chemoselectivity, and regioselectivity, the evaluation of EF as a biotransformation agent is due to their ability to utilize less organic solvent during the transformation process and produce little or no toxic by-products (Ekiz et al. 2018). This results in their classification as an eco-friendly, sustainable, renewable, and economical means of removing environmental contaminants from an aqueous environment (Choudhary et al. 2021). The generation of specific products during biotransformation by EF results from their specificity and stereoselectivity characteristics, as EF produces stable enzymes with extensive application. Endophytes utilize organic contaminants for their carbon and energy requirements or co-metabolize them. Contaminants can serve as electron donors due to their ability to be oxidized. Halogen-containing compounds can equally serve as electron acceptors without oxygen to aid oxygen-limited respiration processes. For example, dehalorespiration as reductive dehalogenation or oxidative anaerobic co-metabolism is a biotransformation process (Fester et al. 2014).

Biotransformation of chemical compounds by endophytic microbes has been reported to occur via some mechanisms, which include esterification, dechlorination, dehydrogenation, deoxygenation dechlorination, demethylation, hydrolysis, aldolization, oxidation, hydroxylation, carbonylation, ketonization, deacetylation, and epoxidation (Mishra and Venkateswara Sarma 2018; Tian et al. 2021; Choudhary et al. 2021; Ma et al. 2022). Biotransformation of agrochemicals, aldrin, and dieldrin metabolites into less toxic compounds by the EF *Pleurotus ostreatus* followed the hydroxylation and epoxidation pathways according to Purnomo et al. (2017). The study of Ekiz et al. (2018) exploited the potential of EF as a biocatalyst. They studied the transformation of plant metabolites to potent active by-products by the native EF *Alternaria eureka* 1E1BL1. Biotransformation of naturally occurring toxic surface-active agents in plants, steroidal saponin by endophytes, was mediated by different transformation processes such as sugar metabolism, carbonylation, dehydrogenation, etc. (Huang 2022).

Ginsenoside Rg3 is a scarce and potential anticancer agent. Hu et al. (2020) evaluated the potential of the bacterium *Cellulosimicrobium cellulans* sp. 21 to transform low bioavailable ginsenoside Rb1 to ginsenoside Rg3 through biocatalysis and hydrolysis with insignificant derivatives. Endophytic microbes like *Enterobacter* sp., *Serratia* sp., *Ochrobactrum* sp., *Arthrobacter* sp. have also been implicated in the transformation of ginsenoside Rb1 to ginsenoside Rg3 (Tam et al. 2018). Marín et al. (2018) investigated the potential of different EF to transform petroleum hydrocarbons into less toxic intermediates/products effectively. Verticillium and Xylaria spp. exhibited over 95% hydrocarbon removal rates among the EF evaluated. The potential of endophytic bacteria, *Pseudomonas* sp., to significantly reduce the concentration of an environmental priority pollutant, phenanthrene, has been investigated. A significant reduction of 61% within three days of cultivation was observed as phenanthrene was transformed into less toxic metabolites, thereby reducing the toxicity of the polycyclic aromatic hydrocarbon (Sun et al. 2018).

5.5 Endophyte-Assisted Phytoremediation Strategies for Contaminated Aquatic Systems

Phytoremediation techniques engage the ecological ability of plants and their related microorganisms, such as endophytic microbes, to essentially enhance the removal of contaminants from the environment (Riskuwa-Shehu and Ismail 2018;

He et al. 2020a). Aquatic plants can actively and passively absorb substantial amounts of metals from contaminated water and/or sediment through their roots, stems, and leaves. As a result, these plants are suited for changes brought on by heavy metals in the aquatic environment (Bai et al. 2018). Endophyte-assisted phytoremediation has been noted as a promising approach for in situ remediation of polluted soils (Fig. 5.3) due to the abundance of endophytes that have been proven to be resistant to heavy metals and/or capable of decomposing organic pollutants (Li et al. 2012). Phytoremediation of contaminated aqueous environment follows different physiological and mechanical means of interaction or mechanisms of action, which include phytostabilization, phytovolatilization, phytoextraction, bioaccumulation, chelation, degradation, and translocation (Kafle et al. 2022). Utilizing the abilities mentioned earlier of plants and endophytes is sustainable for detoxifying a contaminated aqueous environment. With the help of metal-tolerant plant species, heavy metals can be immobilized underground and made less bioavailable through phytostabilization. This prevents the metals from migrating into the ecosystem and lowers the risk of getting into the food chain (Marques et al. 2009).

Degradation genes and PGP properties are known to be present in endophytes, which improves the efficacy of phytoremediation and helps their host plant resist

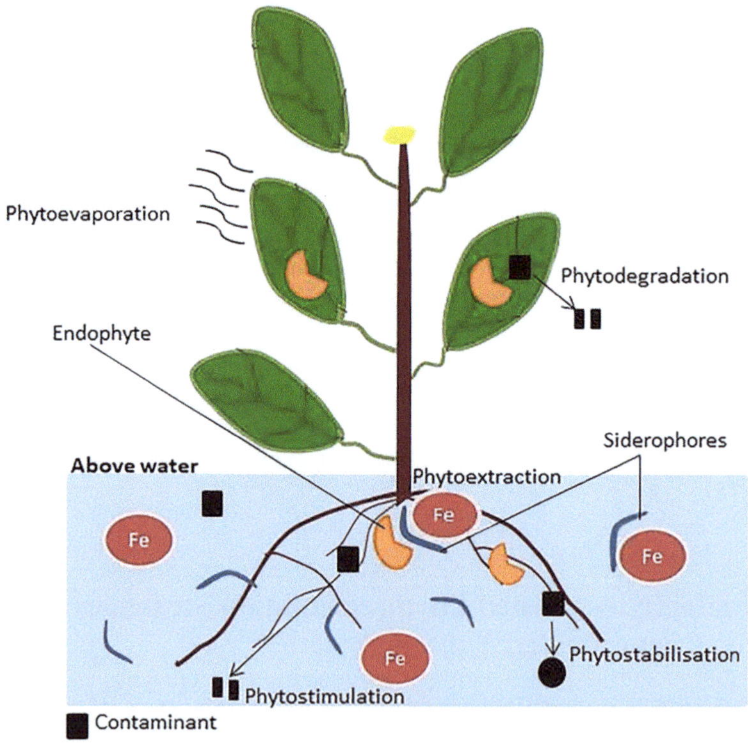

Fig. 5.3 Illustration of some endophyte-assisted phytoremediation strategies for contaminated aquatic systems

environmental challenges. Once inside the plants, endophytes can improve metal translocation and/or contribute to detoxifying organic toxins, perhaps leading to a long-term solution for mixed contaminations (Syranidou et al. 2016). Through the volatilization process, plants absorb specific compounds, such as volatile organic compounds, from the contaminated aquatic systems, convert less volatile compounds into more volatile forms, and release the contaminants into the atmosphere through evapotranspiration.

Phytoextraction involves the removal of contaminants from water or soil using plants. The role of endophytes in the phytoremediation of environmental pollutants is due to their ability to produce siderophores, which are low-molecular-weight compounds with a high affinity for metals. These siderophores aid the phytoextraction of metal ions from a contaminated aqueous environment (He et al. 2020b). Endophytes' metal tolerance mechanisms include precipitation, sequestration, compartmentation, and volatilization (Mishra and Venkateswara Sarma 2018). Endophytes protect plants from the toxic effects of the possible accumulation of excess contaminants. This accumulation process occurs in combination with the root hairs of the plants, and it is more efficient than utilizing each separately (Choppala et al. 2014; Sharma et al. 2018). The EF *Exophiala pisciphila* can accumulate significant Pb and Cd during phytoremediation (Zhang et al. 2008). Immobilization of metalloids and metals in the environment is another characteristic of EF, and plant-endophyte is a helpful process as endophytes support the plant in the secretion of secondary metabolites (Ho et al. 2012). An investigation by Sauvêtre et al. (2018) observed the potential of EF *Rhizobium radiobacter* and *Chryseobacterium nitroreducens* via plant–endophyte interaction to enhance the removal of a prominent aqueous pollutant of pharmaceutical origin, carbamazepine (CBZ). Phytoremediation of metal contaminants with some endophytic fungal isolates showed a high tolerance to heavy metals in the environment, including Cu, Pb, Zn, and Cd.

Textile dyes are water-soluble, including pharmaceuticals and paper wastes; this characteristic enables their penetration and possible contamination of groundwater if and when disposed of indiscriminately (Lellis et al. 2019; Karaś et al. 2021). Removal or detoxifying contaminants in groundwater or wastewater via sustainable and economical means such as phytoremediation is imperative. Endophytes remediate contaminated wastewater through synergistic interactions with its host plants during phytoremediation.

The combination of endophytes and plants for phytoremediation has been reported to be more efficient when compared with endophytes alone (AI-Jawhari 2022). Phytoaugmentation is a variation of phytoremediation that uses plants to stabilize or immobilize contaminants in soil, sediments, or water. Plants are chosen for their ability to tolerate or accumulate pollutants, and their root systems help bind or immobilize contaminants, preventing their movement into the surrounding environment. Phytoaugmentation is an emerging technology that has shown great potential in the remediation of wastewater contaminated with pollutants. This innovative approach harnesses the power of plants and explores their potential applications in treating both domestic and industrial wastewater (Kumar et al. 2021). The inherent ability of plants to uptake and transform specific contaminants can be boosted by incorporating endophytes into the phytoaugmentation process. Endophytic phytoaugmentation involves the introduction of external strains of endophytic bacteria to phytoremediation plants. This process stimulates and enhances the interactions between plants and microbes, improving efficiency in remediation efforts (Redfern and Gunsch 2016). The interaction of plants with associated endophytes for bioremediation of environmental contaminants is a promising technology as the process enhances the degradation capacity of the endophytes (Stępniewska and Kuźniar 2013).

The application of aquatic plants in association with endophytes in the remediation of aquatic systems not only rids the surface water of contaminants but also protects the groundwater from contamination. Decontamination of soils via phytoremediation is a sustainable means of protecting the groundwater. There is no doubt that endophytes are a remarkable collection of microorganisms with the ability to tolerate and eradicate pollutants in aquatic environments in a sustainable manner. Endophytes have an excellent tolerance and eradication potential and could be sampled from various hosts. To benefit their operations, improvements and innovations are also possible for endophytes; this entails researching and comprehending the precise mechanisms that endophytes use for bioremediation. Using updated technology, it is imperative to highlight endophytes' roles in effective environmental remediation and bioresource conservation. This will benefit people and the environment and contribute to preserving the natural world.

5.6 Disinfection of Water for Irrigation Systems

The teeming population of the globe demands abundant food and guaranteed security; however, the effects of global climate change are impeding the production of food crops and other agricultural produce (Bortolini et al. 2018). The agricultural sector is the largest consumer of water, and based on the published report on the FAO-AQUASTAT database (2022), agricultural water withdrawal is the amount of self-supplied water available for irrigation, livestock management and aquacultural practice. The irrigation water withdrawal is the amount of water withdrawn for irrigation purposes, while the irrigation water requirement is the amount needed for optimal crop production. It consists of water (exclusive of precipitation and soil moisture) to meet absolute crop water requirements. It may be expressed in water depth (km^3/year) or water volume (109 m^3/year).

There is an increasing awareness of the efficient use of water resources. One of the ways encouraged to relieve difficulties in water access and ensure water availability for agriculture is by taking full advantage of low-quality water for irrigation needs, such as brackish, high-sand and surface-lake waters as well as reclaimed municipal wastewater, rather than pumping out groundwater from deeper water table (Zhou et al. 2019). The irrigation water quality can be assessed based on its agronomic impact, sanitary risk on public health, ease of field distribution, or management of the water circulation system (Toze 2006). Irrigation water is preponderantly sourced from rivers and dams and conveyed through open channels into the farm water storage before distribution to the crop field (Koech and Langat 2018). Though irrigation water sources with compromised hygiene often do not affect crop yield, but may impose danger on public health due to pathogens transmission, most notably when fresh vegetables are involved (Bortolini et al. 2018).

Low-quality water used for irrigation would unarguably diminish the quality of the soil, increasing amounts of heavy metals in the soil and food crops, encouraging high-level organic contaminants, and thus negates the campaign for food quality and safety toward sustainable development goal (Hass et al. 2010; Allende and Monaghan 2015). Additionally, increasing evidence suggests the occurrence of geogenic contaminants in open water sources. The appearance of trace elements and the increasing use of wastewater have highlighted the vulnerability and complexities of the composition of irrigation water and its long-term effect on food quality and public health. (Malakar et al. 2019) Hence, the rapidly changing quality of irrigation water urgently needs closer attention to understand and predict long-term effects on soils and food crops.

For instance, arsenic, which is notorious for its carcinogenic trait and as a poisonous heavy metal, is widely distributed in natural aquatic environments, constituting an enormous threat to public health (Nordstrom 2002). The plants can easily pick up such geogenic contaminants, among others, through transpiration pull across the irrigation water pathway, and then bio-amplified in the human tissue through consumption of affected crops, where it would wreak havoc on the human metabolic system. Interestingly, recent research dynamics are moving toward adopting a biocontrol approach in cleaning contaminated environments because it guarantees sustainability and cost-effectiveness (Plewniak et al. 2018). Microbial oxidation of arsenic (III) ions is one of the most promising approaches as a precursor step in removing the arsenic ions from contaminated groundwater (Crognale et al. 2019).

The biological cleanup of water resources is dependent absolutely on the application of bacteria strains with no trace of virulent determinants, acting as biocatalysts for biochemical oxidation, degradation of pollutants in the aquatic medium and establishment of the biologically stable water environment to prevent the growth of microorganisms in the water distribution system (Abu Hasan et al. 2020). Most bacterial endophytes are neither phytopathogens nor human pathogens. They invade and live mutually within the tissue of the host plants. Apart from the endophytes enjoying protective shelter from the host, they also benefit from organic nutrients and guaranteed support for life cycle transmission from one generation to the next. In contrast, the host plants become more viable and resistant to stress, herbivores, pests, and pathogens (Mengistu 2020).

Although, only a few studies have been reported on the direct application of endophytes in surface or wastewater treatments. When we consider the effectiveness and inevitable contribution of endophytes in their ecological niche, endophytes may present a reservoir of viable and eco-friendly biological agents for the cleanup of water resources intended for irrigation. More importantly, considering their readily established mutual association with plants, residual endophyte cells pose no threat to the crops or the environment even after the cleanup or treatment of irrigation water. Rather, they confer more crop protection and growth support as a secondary effect.

5.7 Conclusion

Agricultural practice accounts for the highest annual water demand; a substantial amount goes into absolute crop water requirements. The reuse of municipal wastewater is fast becoming an alternative irrigation water source, besides external surface water sources, due to unprecedented water shortage in most of the world driven by the conspicuous changes in the global climatic condition. Unarguably, these water sources are heavily polluted by default, requiring adequate and effective cleanup to avoid the unintending upset of significant biodiversity, carryover effects on public health and gross reduction in food quality. Several studies have established the efficiency of microorganisms, bacteria and fungi in treating wastewater for bioremediation and biotransformation. Endophytes have become handy in this regard, and their potential has not been fully exploited. More attention is required to be paid to this valuable group of microorganisms. Their application in decontaminating and bio-cleaning waste and surface water for irrigation purposes promises to be efficient, eco-friendly and sustainable. It is a cost-effective approach that can be achieved with cheap technology to support global sustainable food security and safety goals.

References

Abu Hasan H, Muhammad MH, Ismail NI (2020) A review of biological drinking water treatment technologies for contaminants removal from polluted water resources. J Water Process Eng 33

Agunbiade FO, Moodley B (2014) Pharmaceuticals as emerging organic contaminants in Umgeni River water system, KwaZulu-Natal, South Africa. Environ Monit Assess 186:7273–7291

Ahmad I, Ansari M I, Aqil F (2006) Biosorption of Ni, Cr and Cd by metal tolerant Aspergillus niger and Penicillium sp. using single and multi-metal solutions

Ahmad SS et al (2014) Phytoremediation potential of *Phragmites australis* in Hokersar wetland-a Ramsar site of Kashmir Himalaya. Int J Phytoremediation 16(12):1183–1191

Al-Jawhari IFH (2022) Recent advancements in mycoremediation. In: Suyal DC, Soni R (eds) Bioremediation of environmental pollutants. Springer, Cham

Al-Jawhari IF, Al-Sead KG (2016) Fate of herbicide Granstar (tribenuron methyl) in a wheat field in Al-Nasiriya governorate. Int Res J Biol Sci 5(8):22–37

Akhtar N, Mannan MAU (2020) Mycoremediation: expunging environmental pollutants. Biotechnol Rep 26:e00452

Al-Homaidan A et al (2020) Accumulation of heavy metals in a macrophyte *Phragmites australis*: implications to phytoremediation in the Arabian Peninsula wadis. Environ Monit Assess 192:1–10

Ali A et al (2017) Phytoextraction of toxic trace elements by *Sorghum bicolor* inoculated with *Streptomyces pactum* (Act12) in contaminated soils. Ecotoxicol Environ Saf 139:202–209
Allende A, Monaghan J (2015) Irrigation water quality for leafy crops: a perspective of risks and potential solutions. Int J Environ Res Public Health 12:7457–7477
Anitha SP (2022) Mycoremediation of heavy metals in marine ecosystems using fungal endophytes. Technoarete Trans Recent Res Appl Microbiol Biotechnol 1:1
Anyasi RO, Atagana HI (2019) Endophyte: understanding the microbes and their applications. Pak J Biol Sci 22(4):154–167
Arfaeinia H et al (2019) The effect of land use configurations on concentration, spatial distribution, and ecological risk of heavy metals in coastal sediments of the northern part along the Persian Gulf. Sci Total Environ 653:783–791
Asemoloye MD et al (2020) Hydrocarbon degradation and enzyme activities of Aspergillus oryzae and Mucor irregularis isolated from Nigerian crude oil-polluted sites. Microorganisms 8(12):1912
Asif MB et al (2017) Degradation of pharmaceuticals and personal care products by white-rot fungi – a critical review. Curr Pollut Rep 3(2):88–103
Bai L et al (2018) Heavy metal accumulation in common aquatic plants in rivers and lakes in the Taihu Basin. Int J Environ Res Public Health 15(12):2857
Barman DN et al (2014) Cloning and expression of ophB gene encoding organophosphorus hydrolase from endophytic Pseudomonas sp. BF1-3 degrades organophosphorus pesticide chlorpyrifos. Ecotoxicol Environ Saf 108:135–141
Bello AO et al (2018) Phytoremediation of cadmium-, lead-and nickel-contaminated water by *Phragmites australis* in hydroponic systems. Ecol Eng 120:126–133
Beni AA, Esmaeili A (2020) Biosorption, an efficient method for removing heavy metals from industrial effluents: a review. Environ Technol Innov 17:100503
Bertrand JC, Almallah M, Acquaviva M, Mille G (1990) Biodegradation of hydrocarbons by an extremely halophilic archaebacterium. Lett Appl Microbiol 11(5):260–263
Bhadouria R et al (2020) Mycoremediation of agrochemicals. In: Agrochemicals detection, treatment and remediation. Butterworth-Heinemann, pp 593–620
Bhadra F, Gupta A, Vasundhara M, Reddy MS (2022) Endophytic fungi: a potential source of industrial enzyme producers. 3 Biotech 12(4):1–17
Bortolini L, Maucieri C, Borin M (2018) A tool for evaluating irrigation water quality in the arid and semi-arid regions. Agronomy 8
Chalifour A et al (2021) Temperature, phytoplankton density and bacteria diversity drive the biotransformation of micropollutants in a lake ecosystem. Water Res 202:117412
Chandra D et al (2022) Impact of microfiltration on water quality. Environ Qual Manag 31(3):253–266
Chlebek D, Hupert-Kocurek K (2019) Endophytic bacteria in the phytodegradation of persistent organic pollutants. Postępy Mikrobiologii-Adv Microbiol 58(1):70–79
Choppala G et al (2014) Cellular mechanisms in higher plants govern tolerance to cadmium toxicity. Crit Rev Plant Sci 33(5):374–391
Choudhary M et al (2021) Endophytic fungi-mediated biocatalysis and biotransformations paving the way toward green chemistry. Front Bioeng Biotechnol 9:419
Chow C et al (2022) An archaic approach to a modern issue: endophytic archaea for sustainable agriculture. Curr Microbiol 79(11):322
Crognale S et al (2019) Biological as(III) oxidation in biofilters by using native groundwater microorganisms. Sci Total Environ 651:93–102
Čvančarová M et al (2015) Biotransformation of fluoroquinolone antibiotics by ligninolytic fungi–metabolites, enzymes and residual antibacterial activity. Chemosphere 136:311–320
Dang C et al (2018) Role of extracellular polymeric substances in biosorption of Pb2+ by a high metal ion tolerant fungal strain Aspergillus niger PTN31. J Environ Chem Eng 6(2):2733–2742
Desiante WL, Minas NS, Fenner K (2021) Micropollutant biotransformation and bioaccumulation in natural stream biofilms. Water Res 193:116846

Ekiz G, Duman S, Bedir E (2018) Biotransformation of cyclocanthogenol by the endophytic fungus Alternaria eureka 1E1BL1. Phytochemistry 151:91–98

Eman A et al (2013) Biodegradation of glyphosate by fungal strains isolated from herbicides polluted soils in Riyadh area. Br J Environ Sci 1:7–29

Esterhuizèn-Londt M, Schwartz K, Pflugmacher S (2016) Using aquatic fungi for pharmaceutical bioremediation: uptake of acetaminophen by Mucor hiemalis does not result in an enzymatic oxidative stress response. Fungal Biol 120(10):1249–1257

Fadiji AE, Babalola OO (2020) Elucidating mechanisms of endophytes used in plant protection and other bioactivities with multifunctional prospects. Front Bioeng Biotechnol 8:467

Fadiji AE, Ayangbenro AS, Babalola OO (2020) Organic farming enhances the diversity and community structure of endophytic archaea and fungi in maize plant: a shotgun approach. J Soil Sci Plant Nutr 20:2587–2599

Fester T et al (2014) Plant–microbe interactions are drivers of ecosystem functions relevant to the biodegradation of organic contaminants. Curr Opin Biotechnol 27:168–175

Food and Agriculture Organization (FAO) (2022) AQUASTAT database. https://www.fao.org/aquastat/statistics/query/index.html;jsessionid=9F2CD7A701908428C61BAE6EA65C014A. Accessed 1 Aug 2022

Gani KM et al (2021) Emerging contaminants in South African water environment – a critical review of their occurrence, sources and ecotoxicological risks. Chemosphere 269:128737. https://doi.org/10.1016/J.CHEMOSPHERE.2020.128737

Gao T et al (2020) Decolourization and detoxification of triphenylmethane dyes by isolated endophytic fungus, *Bjerkandera adusta* SWUSI4, under non-nutritive conditions. Bioresour Bioprocess 7(1):1–12

García-Galán MJ et al (2011) Biodegradation of sulfamethazine by Trametes versicolor: removal from sewage sludge and identification of intermediate products by UPLC–QqTOF-MS. Sci Total Environ 409(24):5505–5512

Govindasamy V, George P, Raina SK, Kumar M, Rane J, Annapurna K (2018) Plant-associated microbial interactions in the soil environment: role of endophytes in imparting abiotic stress tolerance to crops. Adv Crop Environ Interact:245–284

Hass A, Mingelgrin U, Fine P (2010) Heavy metals in soils irrigated with wastewater. In: Treated wastewater in agriculture: use and impacts on the soil environment and crops. Wiley-Blackwell, Hoboken

He G, Wang Z, Cui Z (2020a) Managing irrigation water for sustainable rice production in China. J Clean Prod 245

He W, Megharaj M, Wu CY, Subashchandrabose SR, Dai CC (2020b) Endophyte-assisted phytoremediation: mechanisms and current application strategies for mixed soil pollutants. Crit Rev Biotechnol 40(1):31–45

Ho YN et al (2012) Selection and application of endophytic bacterium Achromobacter xylosoxidans strain F3B for improving phytoremediation of phenolic pollutants. J Hazard Mater 219:43–49

Hu K et al (2020) Exploring the degradation capability of Trametes versicolor on selected hydrophobic pesticides through setting sights simultaneously on culture broth and biological matrix. Chemosphere 250:126293

Huang N (2022) Study of saponin components after biotransformation of Dioscorea nipponica by endophytic fungi C39. J Anal Methods Chem 2022

Hussain H et al (2015) Antimicrobial activity of two mellein derivatives isolated from an endophytic fungus. Med Chem Res 24(5):2111–2114

Hussain A et al (2018) Bioremediation of hexavalent chromium by endophytic fungi; safe and improved Lactuca sativa L. Chemosphere Prod 211:653–663

Ijaz A, Shabir G, Khan QM, Afzal M (2015) Enhanced remediation of sewage effluent by endophyte-assisted floating treatment wetlands. Ecol Eng 84:58–66

Jung J, Kim JS, Taffner J, Berg G, Ryu CM (2020) Archaea, tiny helpers of land plants. Comput Struct Biotechnol J 18:2494–2500

Kafle A et al (2022) Phytoremediation: mechanisms, plant selection and enhancement by natural and synthetic agents. Environmental. Advances 100203
Kapahi M, Sachdeva S (2017) Mycoremediation potential of Pleurotus species for heavy metals: a review. Bioresour Bioprocess 4(1):1–9
Karaś MA, Wdowiak-Wróbel S, Sokołowski W (2021) Selection of endophytic strains for enhanced bacteria-assisted phytoremediation of organic pollutants posing a public health hazard. Int J Mol Sci 22(17):9557
Khan I et al (2019) Mycoremediation of heavy metal (Cd and Cr)–polluted soil through indigenous metallotolerant fungal isolates. Environ Monit Assess 191(9):1–11
Koech R, Langat P (2018) Improving irrigation water use efficiency: a review of advances, challenges and opportunities in the australian context. Water:10
Kulkarni S, Nene S, Joshi K (2017) Production of Hydrophobins from fungi. Process Biochem 61:1–11
Kumar V, Dwivedi SK (2021) Mycoremediation of heavy metals: processes, mechanisms, and affecting factors. Environ Sci Pollut Res 28(9):10375–10412
Kumar V et al (2021) Phytoaugmentation technology for phytoremediation of environmental pollutants: current scenario and future prospects. In: Bioremediation for environmental sustainability. Elsevier, pp 329–381
Kuźniar A, Włodarczyk K, Wolińska A (2019) Agricultural and other biotechnological applications resulting from trophic plant-endophyte interactions. Agronomy 9(12):779
Lellis B et al (2019) Effects of textile dyes on health and the environment and bioremediation potential of living organisms. Biotechnol Res Innov 3(2):275–290
Li HY et al (2012) Endophytes and their role in phytoremediation. Fungal Divers 54:11–18
Liu M et al (2014) Isolation, identification and characteristics of an endophytic quinclorac degrading bacterium Bacillus mega*terium* Q3. PLoS One 9(9):e108012
Llirós M et al (2014) Specific Archaeal communities are selected on the root surfaces of Ruppia spp. and *Phragmites australis*. Wetlands 34:403–411
Ma PY et al (2022) Native endophytes of Tripterygium wilfordii-mediated biotransformation reduces toxicity of celastrol. Front Microbiol 13
Malakar A, Snow DD, Ray C (2019) Irrigation water quality – a contemporary perspective. Water 11
Marco-Urrea E et al (2010) White-rot fungus-mediated degradation of the analgesic ketoprofen and identification of intermediates by HPLC–DAD–MS and NMR. Chemosphere 78(4):474–481
Marín F, Navarrete H, Narvaez-Trujillo A (2018) Total petroleum hydrocarbon degradation by endophytic fungi from the Ecuadorian Amazon. Adv Microbiol 8(12):1029–1053
Marques AP, Rangel AO, Castro PM (2009) Remediation of heavy metal contaminated soils: phytoremediation as a potentially promising clean-up technology. Crit Rev Environ Sci Technol 39(8):622–654
Mengistu A (2020) Endophytes: colonization, behaviour, and their role in defense mechanism. Int J Microbiol:Article: ID 6927219., 8 pages. https://doi.org/10.1155/2020/6927219
Mengistu AG et al (2021) Analysis of the spatio-temporal variability of soil water dynamics in an arid catchment in South Africa. Geoderma Reg 25:e00395. https://doi.org/10.1016/J.GEODRS.2021.E00395
Mesquini JA et al (2015) Detoxification of atrazine by endophytic Streptomyces sp. isolated from sugarcane and detection of nontoxic metabolite. Bull Environ Contam Toxicol 95(6):803–809
Mishra R, Venkateswara Sarma V (2018) Current perspectives of endophytic fungi in sustainable development. In: Fungi and their role in sustainable development: current perspectives, pp 553–584
Mnkandla SM, Otomo PV (2021) Effectiveness of mycofiltration for removal of contaminants from water: a systematic review protocol. Environ Evid 10(1):1–8
Mohammadian E et al (2017) Tolerance to heavy metals in filamentous fungi isolated from contaminated mining soils in the Zanjan Province, Iran. Chemosphere 185:290–296

Müller H et al (2015) Plant genotype-specific archaeal and bacterial endophytes but similar Bacillus antagonists colonize Mediterranean olive trees. Front Microbiol 6:138
Naitam MG, Kaushik R (2021) Archaea: an agro-ecological perspective. Curr Microbiol 78(7):2510–2521
Ngieng NS et al (2013) Decolourisation of synthetic dyes by endophytic fungal flora isolated from senduduk plant (*Melastoma malabathricum*). Int Sch Res Not 2013
Nordstrom DK (2002) Worldwide occurrences of arsenic in ground water. Science 296(5576):2143–2145
Patience MT, Elumalai V, Rajmohan N, Li P (2021) Occurrence and distribution of nutrients and trace metals in groundwater in an intensively irrigated region, Luvuvhu catchment, South Africa. Environ Earth Sci 80:1–15
Phakathi S et al (2021) Farmer-led institutional innovations in managing smallholder irrigation schemes in KwaZulu-Natal and Eastern Cape Provinces, South Africa. Agric Water Manag 248:106780
Pini AK, Geddes P (2020) Fungi are capable of mycoremediation of river water contaminated by E. coli. Water Air Soil Pollut 231(2):1–10
Plewniak F et al (2018) A genomic outlook on bioremediation: the case of arsenic removal. Front Microbiol 9. https://doi.org/10.3389/fmicb.2018.00820
Prieto A et al (2011) Degradation of the antibiotics norfloxacin and ciprofloxacin by a white-rot fungus and identification of degradation products. Bioresour Technol 102(23):10987–10995
Purnomo AS et al (2017) Biodegradation of aldrin and dieldrin by the white-rot fungus *Pleurotus ostreatus*. Curr Microbiol 74(3):320–324
Raimi A, Adeleke R (2021) Bioprospecting of endophytic microorganisms for bioactive compounds of therapeutic importance. Arch Microbiol 2035(203):1917–1942
Rana KL (2019) Endophytic fungi: biodiversity, ecological significance, and potential industrial applications. In: Recent advancement in white biotechnology through fungi. Springer, Cham, pp 1–62
Redfern LK, Gunsch CK (2016) Endophytic phytoaugmentation: treating wastewater and runoff through augmented phytoremediation. Ind Biotechnol
Rimayi C et al (2018) Contaminants of emerging concern in the Hartbeespoort Dam catchment and the uMngeni River estuary 2016 pollution incident, South Africa. Sci Total Environ 627:1008–1017. https://doi.org/10.1016/J.SCITOTENV.2018.01.263
Riskuwa-Shehu ML, Ismail HY (2018) Isolation of endophytic bacteria and phytoremediation of soil contaminated with polycyclic aromatic hydrocarbons using *Cajanus cajan* and Lablab purpere*us*. Bioremediat Sci Technol Res 6(1):26–30
Rodriguez RJ et al (2009) Fungal endophytes: diversity and functional roles. New Phytol 182(2):314–330
Rudakiya DM et al (2018) Biosorption potential of *Phanerochaete chrysosporium* for arsenic, cadmium, and chromium removal from aqueous solutions. Global Chall 2(12):1800064
Sandesh K et al (2019) Optimization of Direct Blue-14 dye degradation by Bacillus fermus (Kx898362) an alkaliphilic plant endophyte and assessment of degraded metabolite toxicity. J Hazard Mater 364:742–751
Saravanan A (2021) Effective water/wastewater treatment methodologies for toxic pollutants removal: processes and applications towards sustainable development. Chemosphere 280:130595
Sauvêtre A et al (2018) Metabolism of carbamazepine in plant roots and endophytic rhizobacteria isolated from *Phragmites australis*. J Hazard Mater 342:85–95
Shang N et al (2019) Biodegradation of malachite green by an endophytic bacterium Klebsiella aerogenes S27 involving a novel oxidoreductase. Appl Microbiol Biotechnol 103(5):2141–2153
Sharma S, Roy S (2015) Biodegradation of dye reactive black-5 by a novel bacterial endophyte. Int Res J Environ Sci 4:44–53
Sharma S, Tiwari KL, Jadhav SK (2018) Diversity of fungal endophytes in Typha latifolia (L.) and their lead biosorption activity. Euro-Mediterr J Environ Integr 3(1):1–8

Siebert S (2010) Groundwater use for irrigation – a global inventory. Hydrol Earth Syst Sci 14:1863–1880
Sim CSF, Chen SH, Ting ASY (2019) Endophytes: emerging tools for the bioremediation of pollutants. In: Emerging and eco-friendly approaches for waste management, pp 189–217
Singh P et al (2020) Bioremediation: a sustainable approach for management of environmental contaminants. In: Abatement of environmental pollutants. Elsevier, pp 1–23
Stępniewska Z, Kuźniar A (2013) Endophytic microorganisms—promising applications in bioremediation of greenhouse gases. Appl Microbiol Biotechnol 97(22):9589–9596
Sun K et al (2018) Subcellular distribution and biotransformation of phenanthrene in pakchoi after inoculation with endophytic Pseudomonas sp. as probed using HRMS coupled with isotope-labeling. Environ Pollut 237:858–867
Syranidou E et al (2016) Exploitation of endophytic bacteria to enhance the phytoremediation potential of the wetland helophyte Juncus acutus. Front Microbiol 7:1016
Tam NT et al (2018) Isolation and screening of endophytic bacteria from Ngoc Linh ginseng (Panax vietnamensis Ha et Grushv) for biosynthesis β-glucosidase. Acad J Biol 40(2):153–161
Tapilatu YH, Grossi V, Acquaviva M, Militon C, Bertrand JC, Cuny P (2010) Isolation of hydrocarbon-degrading extremely halophilic archaea from an uncontaminated hypersaline pond (Camargue, France). Extremophiles 14:225–231
Tian H et al (2021) Biotransformation of artemisinic acid to bioactive derivatives by endophytic Penicillium oxalicum B4 from Artemisia annua L. Phytochemistry 185:112682
Toze S (2006) Reuse of effluent water—benefits and risks. Agric Water Manag 80:147–159
Vasseghian Y et al (2021) The concentration of persistent organic pollutants in water resources: a global systematic review, meta-analysis and probabilistic risk assessment. Sci Total Environ 796:149000. https://doi.org/10.1016/J.SCITOTENV.2021.149000
Verlicchi P, Grillini V (2020) Surface water and groundwater quality in South Africa and Mozambique – analysis of the most critical pollutants for drinking purposes and challenges in water treatment selection. Water 12:305
Wang Y, Dai CC (2011) Endophytes: a potential resource for biosynthesis, biotransformation, and biodegradation. Ann Microbiol 61(2):207–215
Wang D, Bai Y, Qu J (2022) The Phragmites root-inhabiting microbiome: a critical review on its composition and environmental application. Engineering 9:42–50
Yang SO et al (2017) Decolorization of acid, disperse and reactive dyes by Trametes versicolor CBR43. J Environ Sci Health A 52(9):862–872
Zhang Y et al (2008) Dark septate endophyte (DSE) fungi isolated from metal polluted soils: their taxonomic position, tolerance, and accumulation of heavy metals in vitro. J Microbiol 46(6):624–632
Zhou B et al (2019) Variation of microorganisms in drip irrigation systems using high-sand surface water. Agric Water Manag 218:37

Chapter 6
Estimating Crop Water Productivity Using Remote Sensing Data at Plot Scale in an Irrigation System: The Case of Chisumbanje and Ratelshoek Estate

B. Mukandiwa, W. Gumindoga, D. T. Rwasoka, and L. Chikwiramakomo

Abstract As a key water user, irrigation is critical to food production and security. The study's main objective is to estimate crop water productivity using remotely sensed data at a plot scale in an irrigation system. To achieve this, the study assessed the biophysical site-specific factors affecting crop water productivity and irrigation performance in Chisumbanje sugarcane and Ratelshoek wheat farms of the Chipinge district of Manicaland province, Zimbabwe. This study estimated and compared the spatial distribution of seasonal actual evapotranspiration (ET_a) of sugarcane and wheat for two contrasting irrigation schemes using the Surface Energy Balance Systems algorithms (SEBS) and WaPOR-derived products from 2012 to 2020. The results show substantial seasonal variation in actual evapotranspiration, with the maximum ET_a in the summer season of 9 mm/day, a minimum of 3.98 mm/day in the winter season, and a mean ET_a of 5.85 mm/day and a standard deviation of 2.02 mm/day. The actual evapotranspiration is high (>7.5 mm/day) in September, October, and December. The spatial–temporal variability of ET_a maps in the Chisumbanje sugarcane estate and Ratelshoek wheat estate reveals that the sugarcane estate has higher ET values than the wheat estate. The findings from SEBS and WaPOR were used to assess the crop water productivity in both estates. Crop water productivity (CWP) varies from 2.4–3.0 kg m^{-3} (for wheat) to 1.2–1.6 kg m^{-3} (sugarcane). The findings from this research demonstrate the potential for irrigation managers to use remote sensing-based models to monitor irrigation water usage for efficient and sustainable use of water resources.

Keywords Crop water · Evapotranspiration · Remote sensing · Surface energy balance systems

---

B. Mukandiwa · W. Gumindoga (✉) · L. Chikwiramakomo · D. T. Rwasoka
Department of Construction and Civil Engineering, University of Zimbabwe, Harare, Zimbabwe

T. Mabhaudhi et al. (eds.), *Enhancing Water and Food Security Through Improved Agricultural Water Productivity*,
https://doi.org/10.1007/978-981-96-1848-4_6

6.1 Introduction

Irrigated agriculture is the leading water consumer worldwide, accounting for about 70% of all available freshwater (Kharrou et al. 2021). However, future water supply for irrigated agriculture will be impacted by growing strain on existing water resources, especially in semi-arid regions, due to population expansion, climate change, and competition from other economic sectors (Ashktorab and Zibaei 2021). For the optimal and sustainable use of water resources in this setting, evaluating irrigation performance through precise assessment of crop water yield and enhancing irrigation water management by utilizing creative and economical technologies are essential (Montazar 2021). Monitoring crop water usage is essential to managing water resources well. This helps boost water use efficiency, a target of the UN Sustainable Development Goals (SDG 6.4) that should rise significantly by 2030 (UN 2016).

Crop performance, water productivity, and the financial viability of crop cultivation must all be evaluated globally using water-saving crop production techniques and appropriate integrated plant nutrition management plans that make use of inexpensive, locally accessible fertilizer sources (Midya et al. 2021). By 2030, the world's crop output must increase by 75% to feed 8 billion people (FAO 2021). This can be accomplished by raising crop yields by 62%, intensifying cropping by 12%, and increasing arable land by 25%. (Gilland 2002). Irrigated agriculture is, therefore, essential to increasing food security and availability. Globally, crop yields from irrigation are predicted to increase from 2000 to triple or perhaps quadruple by 2030 due to more crops being produced per unit of land or water supplied. The area of irrigated land in emerging nations is projected to rise by 27% between 1996 and 2030 (Gilland 2002; FAO 2021). The sub-Saharan African (SSA) region continues to face issues that present chances for further advances in water use efficiency through the adoption of efficient technologies that are suited for the given context (Poudel et al. 2021). The technologies include high unrealized potential for irrigation, rainwater harvesting to increase water availability, low-cost irrigation technologies that can be adjusted to local conditions, and the restoration of traditional schemes. Other management practices include a strong commitment from donors, non-governmental organizations (NGOs), and national governments to expand smallholder irrigation.

A large portion of the populace in most SSA countries lacks consistent access to food sources and meets minimal nutritional needs, resulting in food shortages and undernourishment. Africa's future is inextricably connected to its demographic shifts, necessitating careful consideration of measures to curtail excessive reproduction and adjust to the fast growth of middle-aged population segments as life expectancy rises. Only 331,000 ha of Zimbabwe's more than 550,000 ha are irrigated because of the country's inadequate water supply. Additionally, due to the nation's droughts, the irrigated area decreased from 165,000 ha to 136,000 ha between 1982 and 1985 (Scoones et al. 2019). The availability of water resources is a key factor in determining the performance of irrigation projects as assessed by water delivery, productivity, gross margins, and return on investment (Benavides et al. 2021). Water delivery performance measures include adequacy, equity, reliability, and timeliness of water supply.

Water is an essential part of our life, be it for plants or humans; it is an integral part of the ecosystem (Vasilachi et al. 2021). However, a variety of circumstances may impact the quantity, quality, or accessibility of this resource. Even if the existence of climate change is up for debate, there is no denying that extreme weather events have increased, making it critical to use water resources responsibly. Water is limited, but its availability is a major concern, particularly in Southern African nations where there is a growing expectation that resources will be depleted and demand from various water consumers will rise. The measures to meet the demand for power, food, industry, and agricultural water are more important now that supply unpredictability has increased due to climate change (Bhavsar and Gohil 2015).

The Sustainable Development Goals (SDGs) of the United Nations (UN) highlight the need to enhance vulnerable people's food security, particularly in developing nations where several interrelated issues contribute to food insecurity. Most of Southern Africa's economies depend heavily on agriculture, a major factor in the region's food security. The Zimbabwe Vulnerability Assessment Committee (ZIMVAC 2019 report) estimated that 3.8 million people in Zimbabwe's rural areas needed food assistance, and 5.5 million people were expected to be food insecure. Zimbabwe has pushed for laws and measures that increase the overall food security of its people in recent years. For example, the Zimbabwe Agenda for Sustainable Socio-Economic Transformation (ZIM ASSET), Transitional Stabilisation Program (TSP), and the most recent National Development Strategy 1 (NDS1), among other economic and development strategies, also seek to improve the food situation in the country. Some policies, such as Command Agriculture introduced in 2016 and Pfumvudza 2020/21, have also improved crop productivity and self-sufficiency at the household level. In 2018, the nation depleted its strategic grain store despite policies and tactics put in place. Commercial imports could not meet 70% of the nation's annual corn demand because of the lack of foreign money. Global food aid reserves emerged as a key means of preventing catastrophes related to national food security. To reduce food insecurity, the agriculture industry must evaluate crop water productivity.

A quantitative term that describes the link between crop production and water input is called crop water productivity (CWP). The status of CWP was impacted by the amount of irrigation water provided at different phases of crop growth and the crop's response to moisture stress during growth (Chai et al. 2016). Even though CWP measures the benefit of using a unit of water for agricultural production, it is important to remember that it also addresses the following SDGs more comprehensively: No Poverty, SDG 1. To lessen their exposure to and vulnerability to extreme events connected to climate change, it is necessary to increase the resilience of those with low incomes and those in vulnerable situations. Goal 2: Zero Ending hunger and ensuring everyone has year-round access to enough and safe food are urgent priorities. Small-scale farmers' earnings and agricultural production must also be doubled. SDG 6: Sanitation and clean water. Water resource management that is integrated across all levels. Jiang et al. (2022) claim that whereas irrigated crops in underdeveloped nations have higher water productivity than rain-fed crops, in industrialized nations, agricultural output is lower. Consequently, the essential crop water productivity component in Zimbabwe's Chisumbanje and Ratelshoek Estate is estimated in this study.

Remote sensing has been used to create several models that estimate crop water requirements or actual evapotranspiration (ET_a). The first is the Internalized Calibration (METRIC) model for Mapping Evapotranspiration at High Resolution. This model has been compared to other evapotranspiration estimating techniques such as pan evaporation, weighted lysimeters, Eddy Covariance (EC), Bowen Ratio Energy Balance System (BREBS), and sap flow. Any study, including remote sensing, must validate estimates of surface characteristics obtained through remote sensing. Given this, the study aims to calculate crop water productivity in an irrigation system utilizing remotely sensed data at the plot scale.

6.2 Materials and Methods

6.2.1 Description of the Study Area

Ratelshoek Estate and Chisumbanje Estate served as the study's locations. Chisumbanje is 403 meters above sea level, while the Ratelshoek Estate is a farm in Zimbabwe's Manicaland. Ratelshoek Estate's landscape is thought to be 845 m above sea level. Chisumbanje receives 488 mm of rain annually, most of which falls during the scorching summer months of October through March (Mushay et al. 2023). Summer temperatures average 25.6 °C, while winter temperatures average 16.7 °C.

The area of the Chisumbanje estate is roughly 50 km^2. The dark grey and black expanding montmorillonite clay soils in the Chisumbanje Sugarcane Estate are distinguished by loose, granular surface layers that range in thickness. The soil's bottom is highly calcareous, gravelly, weathered basalt, typically more permeable than the soils above it. The average depth of the soil was between one and two meters (Chikodzi et al. 2013). On the other hand, Ratelshoek Estate has an average rainfall of 800 mm/annum with maximum and minimum temperatures of 25.9 and 11.9 °C, respectively, based on the Meteorological Service Department (MSD) statistics. The dominant rock types are dolerite, quartzite, siltstone, and sandstone. Sandy loam soils on higher slopes and black clay soils on the lower valleys characterize the area. The soils are acidic, with pH ranging from 4 to 4.9. The study area Chisumbanje Estate is in ecological region 1, so an irrigation system is being used to supplement the water required by sugarcane to attain its maturity as it requires between 1200 and 1500 mm/year of rainfall in each growing season. The project area is mainly a sugarcane plantation area responsible for providing ethanol products to the country.

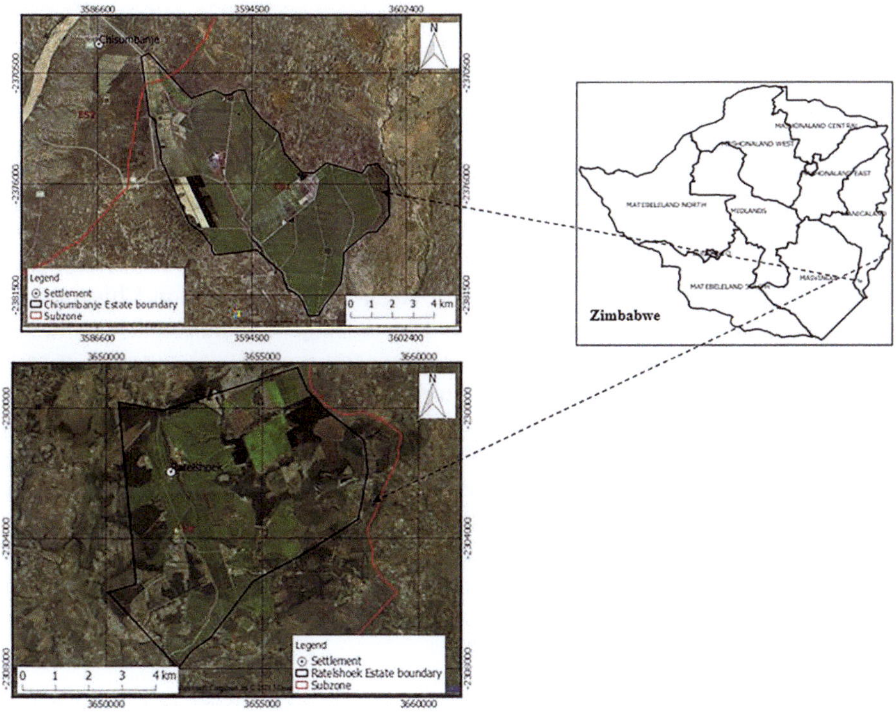

Fig. 6.1 (**a**) The project study area showing the Chisumbanje Estate and (**b**) Ratelshoek Estate overlaying settlements and subzones on the Landsat 8 OLI imageries. (Source: https://earthexplorer.usgs.gov/ 2021 images)

Chisumbanje has 8.5 km^2 of sugarcane, while Middle Sabi has 3.5 km^2. Ratelshoek estate is in Region 5 in Chipinge district, Manicaland Province. The Estate is located at the 35 km peg along the Chipinge to Tanganda road, which runs east of Chipinge town into neighbouring Mozambique. The Estate is quite extensive and employs about 900 workers, some seasonal. Figure 6.1a, b shows the study area project of Chisumbanje and Ratelshoek Estate, respectively, in Chipinge district, Manicaland province of Zimbabwe.

Hydrology

The major river in the catchment is the Save River, which has a catchment area of approximately 48,448 km^2. Subzone ES1 (2763 km^2) and ES2 (4072 km^2) are in the Lower Save sub-catchment, major tributaries of the catchment. The mean annual precipitation ranges from 621 mm/year to 647 mm/year, while the evaporation is 2000 mm/year. The major river that supplies water to the Ratelshoek estate is the Budzi River in the Budzi sub-catchment. The mean annual rainfall in the area is around 1200 mm per year, while the mean annual evaporation is approximately 1600 mm/year.

Socio-economic Activities

The project areas are largely sugarcane for Chisumbanje and wheat for the Ratelshoek farming community. Their economy is mainly agricultural, with the most common crops being maize, millet, rapoko, cotton, wheat, and sugarcane. Interestingly, it is taboo to grow millet in Chief Musikavanhu's area. The local varieties of crops contrast sharply with the commercial estates where coffee, tea, and bananas are grown. The Chisumbanje Ethanol Project is a national project of great strategic importance where ethanol is produced from sugarcane. It is seen as one of Africa's most significant ethanol projects. The Ratelshoek Estate is where wheat and tea are grown. The project consists of sugarcane plantations in Chisumbanje and Middle Sabi, with the ethanol plant in Chisumbanje and the Ratelshoek Estate for wheat in the Chipinge district. It is also a consortium of local investors in partnership with the government's Agriculture and Rural Development Authority, ARDA. At its peak, the Chisumbanje ethanol project and ARDA's cane-growing adjacent farms operated by the private investors trading as rating (at Middle Sabi) and Macdom (at Chisumbanje) have been projected to create employment for more than 8000 people, becoming one of the single largest job creation ventures in Zimbabwe in recent years. World-class irrigation infrastructure has been established, and an outstanding ethanol-producing plant has been constructed. In terms of population distribution, Chipinge is rural, whereas Chisumbanje has a population of about 300,000 people (Zimstats 2012).

Geology and Soils

There is a wide range of soils in the Save catchment area, which is in the lowlands as well as the highlands, which are generally alluvial type: loose unconsolidated sediments made up of a variety of materials (clay silt, sand, and gravel) deposited relatively recently. The catchment generally consists of moderately well-drained soils, except around the lake, where the soils are less well drained. The topsoil is mostly deep (over 1.5 m). In terms of geology, a large proportion of the Save Catchment is made up of igneous and metamorphic rocks of the Basement Complex of Precambrian, both Archean and Proterozoic, age comprising Archean charnockitic gneiss, granulites, meta-sediments and ultrabasic, Proterozoic sequences are dominated by meta-sediments and older granites. The Save catchment region is made up of both highland and lowland soils, most of which are of the alluvial type, which are loose, unconsolidated sediments composed of a variety of elements (clay silt, sand, and gravel) that were produced relatively recently. Except for the area surrounding the lake, where the soils are less well drained, the watershed is typically made up of reasonably well-drained soils. Most of the topsoil is deep (over 1.5 m). Regarding geology, the Save Catchment is mostly composed of igneous and metamorphic rocks from the Basement Complex of the Precambrian, spanning the Archean and Proterozoic ages. These rocks include ultrabasic, meta-sediments, granulites, and Archean charnockitic gneiss. Older granites and meta-sediments dominate Proterozoic sequences.

Climate

Chipinge District's climate is classified as subtropical. The inter-tropical convergence zone (ITCZ), a region of low pressure with calm prevailing winds around the equator, significantly impacts it. The ITCZ brings rainy weather, which swings between the continent's north and south. Thus, November through April is the wet season, and May through October is the dry season. The cold season, which lasts from May to August, and the hot season, which lasts from September to October, are the two divisions of the dry season. Due to the obvious effects of its vicinity to the Indian Ocean and the highlands, the area is also vulnerable to several cyclones. The area is arid because it lies in a valley, and crop yields are generally poor; thus, it is prone to food insecurities. It differs from the other northern part of Chipinge district in Zimbabwe's region.

6.2.2 Methodology for the Review of Factors that Affect Crop Water Productivity

The factors that affect water productivity include climate, crop type, soil type, evapotranspiration, cropping intensity, crop coefficient, and fertilization. This study derived these factors from different platforms and spatial techniques used to develop spatial variation maps, such as Evapotranspiration (ET_a), soils, and geological maps. Chisumbanje Estate is mainly for sugarcane, whereas Ratelshoek is for wheat. Evapotranspiration is a factor to be considered for crop water productivity in this study. The ET_a maps were derived from Moderate Resolution Imaging Spectroradiometer (MODIS) daily ET_a images using the Surface Energy Balance System (SEBS) techniques to estimate the daily evapotranspiration rate for both Chisumbanje and Ratelshoek estates. This was done in a GIS environment, computing the SEBS algorithm to develop ET_a spatial variation maps. The soil type is also a fundamental factor in considering crop water productivity. The soil type was derived from the soil maps of Zimbabwe; this was done in a GIS environment by clipping the study area sites from the Zimbabwe soil maps. The soils for Chisumbanje and Ratelshoek were categorized based on the soil type according to the World Relational Base (WRB). The soil retention capacity was also measured to determine the rate of soil retention in the study area.

6.2.3 Methodology for the Estimation and Comparison of the Spatial Distribution of Seasonal Actual Evapotranspiration

The daily evapotranspiration was calculated using the surface Energy Balance System (SEBS) algorithm. It is suggested that SEBS be used to estimate air turbulent fluxes and evaporative fractions at appropriate scales by utilizing data from satellite earth observation and ground-based meteorological information (Su 2002).

Data Acquisition

The actual evapotranspiration was estimated using ground-based meteorological data and data from remote sensing. Ten cloud-free TERRA MODIS (Moderate-resolution Spectroradiometer sensor) Level 1b images were downloaded from http://ladsweb.nascom.nasa.gov/browse images, covering the Chisumbanje and Ratelshoek estate for the month of August in the year 2020. The ASCII files that describe the sensor calibration curve and responses for a specific set of ground-based meteorological data from the Ratelshoek Estate Meteorological Station and the Chisumbanje research station are the MODIS coefficients for each band. The study considered decadal seasons due to the challenge of analysing a large number of satellite images. WaPOR was used to extract the necessary parameters from the time series data from WaPOR as CSV to determine the degree of water productivity.

Projecting and Converting MODIS Level-1B Data with ModisSwathTool

The MODIS L1B data are in swath (orbit-based) format. Therefore, the data was reprojected to standard projection and format compatible with GIS software. The MODIS Level-1B and geolocation data products were converted to GeoTIFF format and reprojected to Geographic projection ready for ILWIS importing. In the Modis Swath Tool, desired bands and coordinates for each image were selected. Table 6.1 shows bands that were selected, converted to TIFF file format and used for this research:

The same steps as above were applied to the Geolocation file; in this case, Solar Zenith, Solar Azimuth, Sensor Azimuth, Sensor Zenith and Height bands were selected and converted to GeoTIFF files (Sobrino 2003).

MODIS Image Pre-processing for SEBS

All the selected and reprojected bands were imported to ILWIS for further processing. The further processing of the bands includes the computation of the SEBS

Table 6.1 MODIS bands

Band	Band	Bandwidth	Primary use
EV_250_Aggr1km_RefSB_b0	Channel 1	620–670 nm	Land boundaries
EV_250_Aggr1km_RefSB_b1	Channel 2	841–876 nm	Land boundaries
EV_500_Aggr1km_RefSB_b0	Channel 3	459–479 nm	Land properties
EV_500_Aggr1km_RefSB_b1	Channel 4	545–565 nm	Land properties
EV_500_Aggr1km_RefSB_b2	Channel 5	1230–1250 nm	Land properties
EV_500_Aggr1km_RefSB_b3	Channel 6	1628–1652 nm	Land properties
EV_500_Aggr1km_RefSB_b4	Channel 7	2105–2155 nm	Land properties
EV_1KM_Emissive_b10	Channel 31	10.780–11.280 μm	Surface temperature
EV_1KM_Emissive_b11	Channel 32	11.770–12.270 μm	Surface temperature

algorithm to produce the ET spatial variation maps. SEBS for ILWIS helps process the satellite images; the results are maps.

Converting Raw to Radiances/Reflectance (MODIS)
The MODIS Level 1b data are given in SI (simplified integer number); therefore, the data were converted to reflectance and radiances. MODIS channels 1 to 7 were converted to reflectance, and channels 31 and 32 were converted to radiances.

The SEBS in ILWIS provided the tools to convert the imported MODIS channels in digital number/simplified integers into radiances or reflectance, which is done by applying the proper calibration coefficients. The calibration coefficients consist of a scale and offset and are provided in the HDF header file. The calibration coefficients were extracted using the HDF view. The conversion from SI (simplified integer)/ raw data to reflectance was conducted using Eq. 19:

$$\text{Reflectance} = \text{reflectance\_scale}\ \left(\text{SI} - \text{reflectance\_offset}\right) \tag{6.1}$$

The conversion from SI/raw data into radiance was conducted using this equation:

$$\text{Radiance} = \text{radiance\_scale}\ \left(\text{SI} - \text{radiance\_offset}\right) \tag{6.2}$$

Solar and satellite zenith and azimuth angles needed were corrected by scale factor 0.01; this was done using the ILWIS map calculator, applying the following formulas in the command line:

$$\text{Solarzenith angle} = \text{solarzenith\_dn}^{*}\ 0.01. \tag{6.3}$$

$$\text{Solaraazimuth angle} = \text{solaazimuth\_dn}^{*}\ 0.01 \tag{6.4}$$

$$\text{Sensorzenith angle} = \text{sensor zenith\_dn}^{*}\ 0.01 \tag{6.5}$$

$$\text{Sensorazimuth angle} = \text{sennsorazimuth\_dn}^{*}\ 0.01 \tag{6.6}$$

6.3 Methodology for the Estimation of the Water Productivity of Sugarcane and Wheat Under Irrigation

WaPOR Datasets
FAO WaPOR system provides open access to the water productivity database and underlying map layers. It allows for direct data queries, time series analyses, area statistics, and data download of key water and land productivity assessment variables. The portal's services are directly accessible through dedicated FAO WaPOR

APIs, which are gradually published and documented through the FAO API site. Google Earth Engine powers water productivity assessments and other computation-intensive calculations. Datasets from FAO's portal to monitor water productivity through Open Access Remotely sensed derived data (WaPOR; URL: https://wapor.apps.fao.org/home/WAPOR_2/1) are used for the analyses as it provides the required layers to estimate both land and water productivity. The database covers Africa and the Near East regions in near real time for the period between 2009 and to date (2021) (FAO 2020a). WaPOR datasets are available at the continental scale (level 1 at 250 m), country (level 2 at 100 m), and project level (level 3 at 30 m). The latest WaPOR version (WaPOR v2.1) is an improvement from WaPOR v1.0 following the quality assessments by IHE Delft and ITC (Mul and Bastiaanssen 2019; FAO 2020c).

The methodology used for compiling the actual evapotranspiration of WaPOR is based on the ET Look up method (Bastiaanssen et al. 1998) and further developed by the FRAME consortium. A full description of the methodology is provided by the FAO (2020b). WaPOR v2.1 was suitable for inter-plot comparison of irrigation performance indicators for plots larger than 2 ha (Blatchford et al. 2020). All data collected from FAO WaPOR databases, including land use/land cover, actual evapotranspiration, and crop water productivity, were then further analysed to give results on the irrigation systems being employed in the Chisumbanje and Ratelshoek estates. Water productivity is monitored for Chisumbanje and Ratelshoek irrigation schemes to assess the irrigation system's functioning and propose improvements to these systems.

6.3.1 A Framework for Assessing Irrigation Performance Using WaPOR Data

Irrigation performance indicators are derived from WaPOR and field data in three main steps. First, actual evapotranspiration ($ET_a = E + T$), reference evapotranspiration (ETref) and net primary production (NPP) layers of FAO WaPOR are pre-processed to match the spatial resolution, remove non-crop pixels and undergo a quality check. Second, the seasonal ET_a seasonal potential evapotranspiration (ETp, s) and seasonal NPP (NPPs) are calculated from their respective WaPOR layers between the start of the season (SOS) and end of the season (EOS) for each plot. ETp is derived from ETref and crop coefficient (Kc). Finally, the irrigation performance indicators are analysed. At this stage, NPPs are translated to above-ground biomass (hereafter referred to as biomass (B)) using crop-specific information (above over total biomass (AOT), light use efficiency correction factor (fc) and moisture content of fresh biomass (mc)). The biomass is 166 multiplied by the harvest index (HI) to derive the crop yield (Chukalla et al. 2021).

6.4 Land Use and Land Cover Classification

The worldwide 100 m resolution prototype legend maps were utilized in this investigation to classify land cover and land use. The primary data sources are satellite observations from PROBA-V, arranged into millions of 110 × 110 km Sentinel-2 equivalent tiles. Sentinel-2 observations are processed with good quality and continuity thanks to the tiling grid and UTM projection. This novel algorithm takes these satellite observations and applies geometric and atmospheric corrections to pre-process the satellite data; uses sensor-specific status masks and (temporal) outlier detection techniques to clean the data; calculates the input data density indicator data fusion between five-daily 100 m resolution and daily 300 m resolution data; extracts 183 metrics, such as base reflectances, vegetation indicators, time series harmonics, and descriptive statistics. The use of 168 K training points, collected at 10 m resolution, from GeoWIKI's crowd-sourcing for the year 2015, the use of well-established, external datasets for the shoreline masking, ecological regionalization, built-up (urban) cover, permanent and seasonal water cover, arctic vegetation, weather and topography; supervised classification and regression.

The classified metrics are calculated over three years (epoch) in three processing modes: base maps for epoch 2015, which serve as a reference for the classifier and regression models; consolidated maps (epochs 2016, 2017, 2018, 2019, and 2020) data of the same season. Time series break maps are computed using a BFAST break detection algorithm on a time series of MODIS Near-Infrared Reflectance of Vegetation (NIRv) input data and a Hidden Markov Model. These break maps show the areas where changes occurred between years and are used in temporal post-processing rules to improve the cover fraction time series. The changes in land cover maps also align with these break predictions. The final maps are validated using an independent set of 21.7 K validation points sourced from Geo-WIKI. The processing continued to use innovative Big Data techniques on the PROBA-V Mission Exploitation Platform. The land is mostly cropland, with large swathes of sugarcane in Chisumbanje, while tea and wheat dominate Ratelshoek. Ratelshoek, in ecological region 1, has a dense population of savannah trees compared to Chisumbanje, which is in natural region 5.

6.5 Validation of Results

The estimated wheat and sugarcane yield was validated by comparing the simulated and observed yields. The estimated average daily ET_a from SEBS was compared to the daily ET_a calculated from meteorological data collected from 2012 to 2021. Even though the Eta depends on the climate, availability of water, the crop's growth cycle (crop age), the type of crop and other factors, the assessment result of the ET_a, wheat and sugarcane yield found in this study with the results of other studies is helpful.

6.6 Results and Discussion

Crop water productivity is an important indicator of agricultural water use. Understanding the impacts of climatic and agricultural factors variation on crop water productivity will provide a theoretical basis for regional crop productivity improvement. Climate, crop type, soil type, geology, evapotranspiration, cropping intensity, crop coefficient, fertilization, and land use/land cover affect the crop water productivity. These factors were measured and mapped for both Chisumbanje and Ratelshoek Estate. Figure 6.2a, b shows the mean annual rainfall for Chisumbanje and Ratelshoek Estate, respectively. Figure 6.3a, b demonstrates the average annual ET for Chisumbanje and Ratelshoek Estate, respectively. These climatic thematic layers determine the climate characteristics of the study region.

The Topographic Wetness Index shows the degree of wetness in an area. The wetness index sets the catchment area with the slope gradient (Beven and Kirkby 1993). The WI has been used to study spatial scale effects on hydrological processes. Figure 6.4a, b shows the spatial variation of the wetness index in Chisumbanje and Ratelshoek Estate, respectively.

Soil Type and Water Retention

The soil type determines the soil water retention capacity, hence the amount of water available to the plant (Fig. 6.5). High water retaining soils give a chance to the plant to realize the maximum possible water use hence a high level of crop water productivity. Clay has a high crop water productivity index because of soil depth retention compared to sand. By so doing, it elaborates that the soil type can significantly affect crop water productivity. Based on the measurements done using A 2-foot-deep hole offers to measure water retention. The holes were filled with water and allowed to drain. It is filled a second time, and the time it takes to drain again is measured.

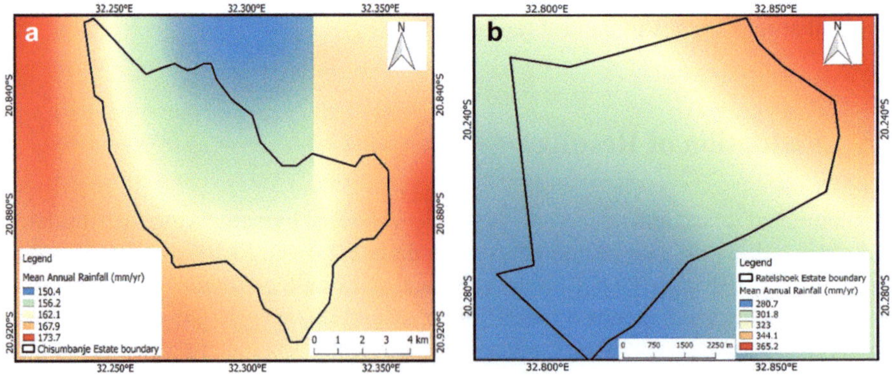

Fig. 6.2 (**a**) The mean annual rainfall for Chisumbanje Estate and (**b**) for Ratelshoek Estate

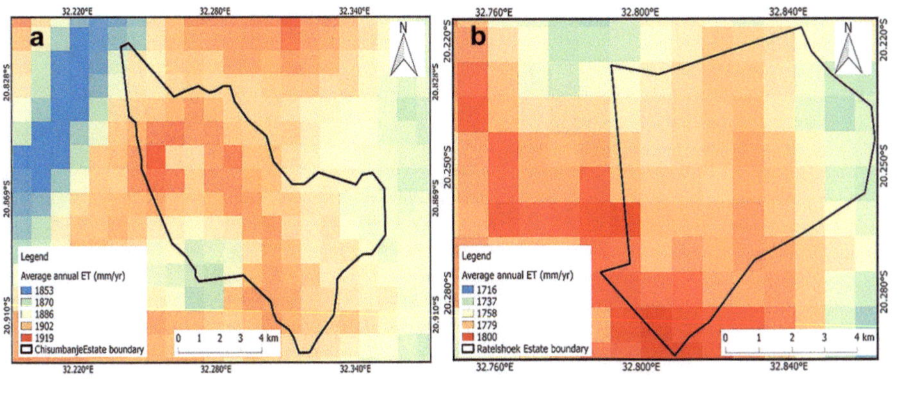

Fig. 6.3 (**a**) The average annual ET for Chisumbanje Estate and (**b**) Ratelshoek Estate

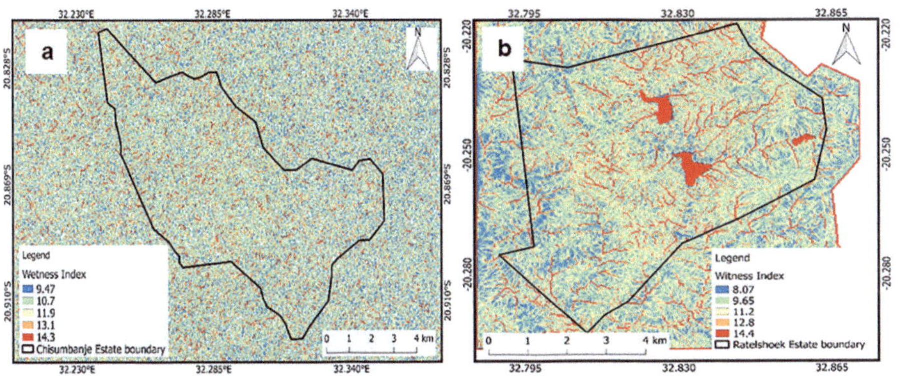

Fig. 6.4 (**a**) The wetness index for Chisumbanje Estate and (**b**) Ratelshoek Estate

The water drains away within 2–3 h, and the drainage is considered good; hence, the study area has good water retention capacity. Figure 6.6a–d shows the spatial variation percentage of sand, silt, clay and organic carbon in the Chisumbanje Estate. The sand soil varies from 19% up to 54%, silt from 16% to 65%, clay content from 17% to 69%, and organic content from 16% to 59%. Comparatively, Fig. 6.7a–d illustrates the spatial variation percentage of sand, silt, clay and organic carbon in the Ratelshoek Estate. The sand soil changes from around 24% up to around 66%, silt from 15% to 72%, clay content from 13.3% to 73%, and the organic content from 24.2% to 73.2%.

Soil Type and Geology

Figure 6.8 shows the soil type and geology in the Chisumbanje area. Eutric vertsoils dominate this study area, and the basalt rock on geology dominates the project area. Leptosols and haplic ferralsols dominate Fig. 6.9 Ratelshoek Estate, and the underlain geology is limestone for the whole area.

Fig. 6.5 Soil type water retention capacity

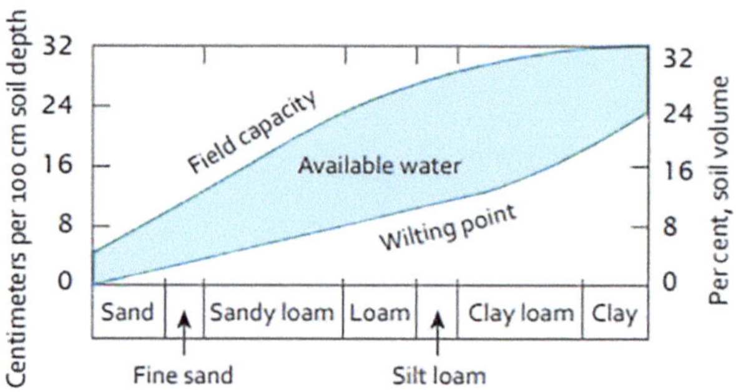

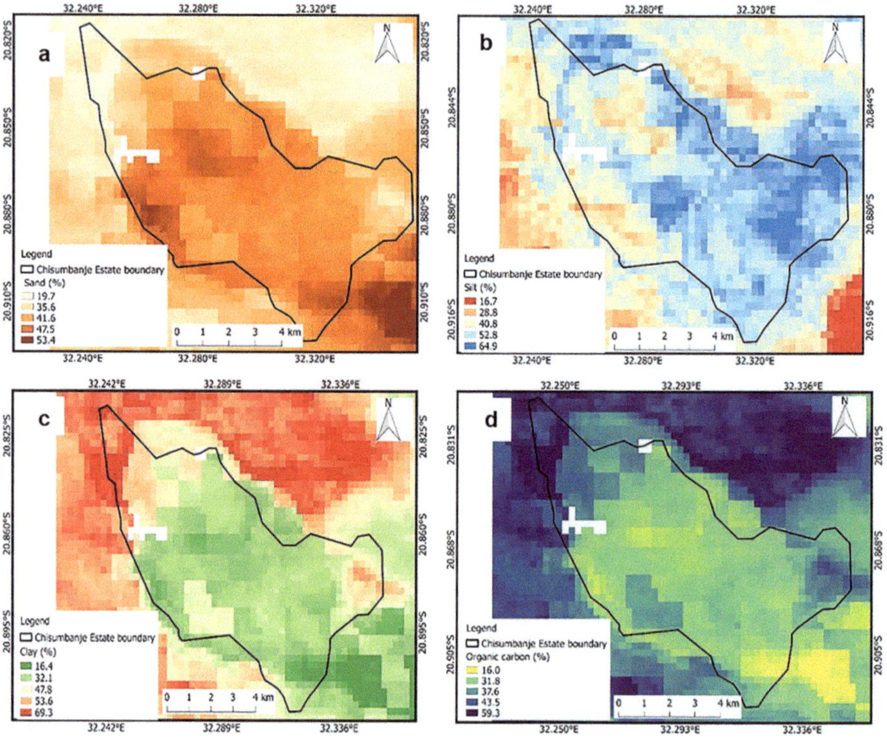

Fig. 6.6 (**a**) The sand soil, (**b**) silt, (**c**) clay content, and (**d**) organic carbon for Chisumbanje Estate

The SEBS algorithm was employed to estimate the actual evaporation of the Chisumbanje and Ratelshoek estates. SEBS algorithm successfully retrieved actual evapotranspiration. Figure 6.10 shows the actual daily evapotranspiration maps for Chisumbanje and Ratelshoek estates. SEBS uses the MODIS Level 1b data to estimate daily actual evaporation. A total of five images were used in this study. The images were obtained between 2012 and 2020 for August for all images with no cloud cover. ET_a can provide an effective tool for fast hydrological monitoring, agricultural management, and climate change studies. Similar to the unirrigated area,

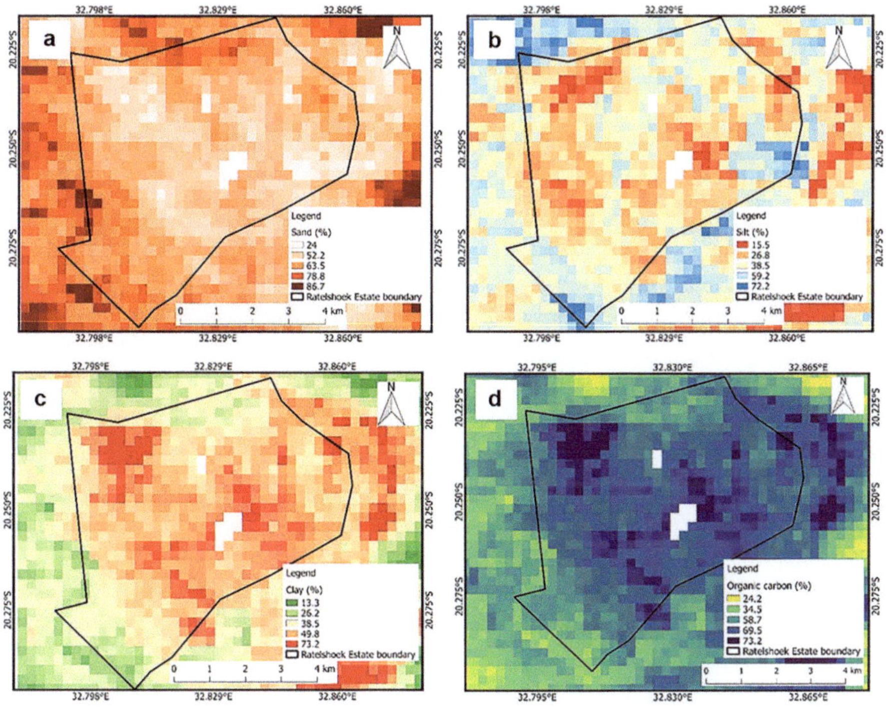

Fig. 6.7 (**a**) The sand soil, (**b**) silt, (**c**) clay content, and (**d**) organic carbon for Ratelshoek Estate

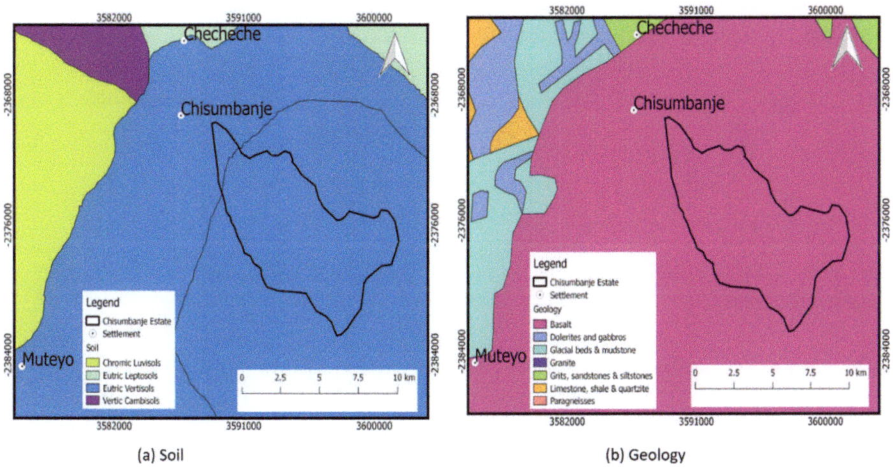

Fig. 6.8 (**a**) soil type and (**b**) geology for Chisumbanje Estate

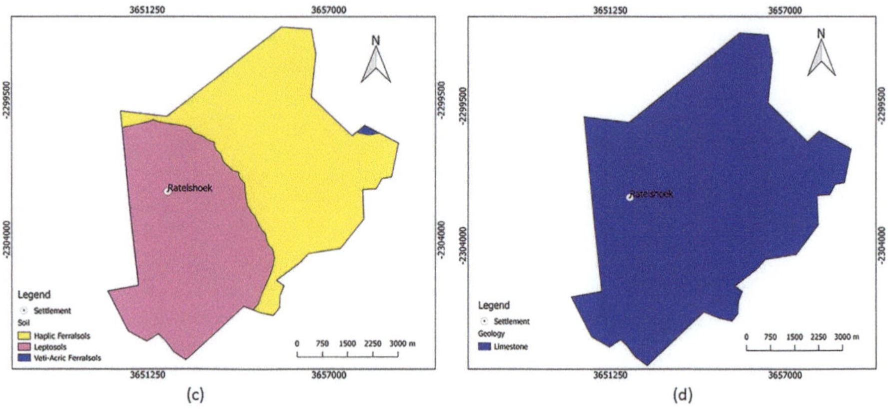

Fig. 6.9 (c) soil type and (d) geology for Ratelshoek Estate

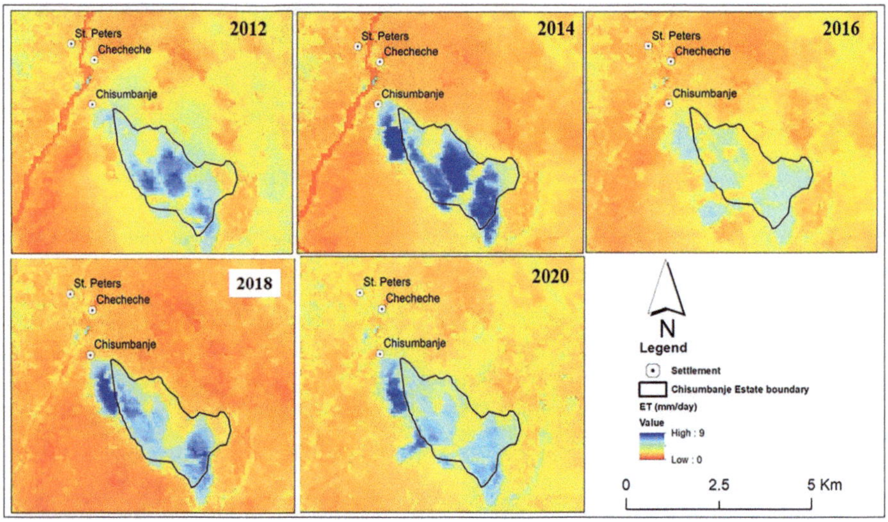

Fig. 6.10 Actual evapotranspiration estimated by SEBS algorithm for 2012–2020 in Chisumbanje Estate

both estates' areas observed high values of actual evapotranspiration. This may be due to the availability of soil moisture. This, combined with a higher difference in the surface air temperature, leads to high values of the instantaneous heat flux only by the wet sensible moisture limit (Rwasoka et al. 2011). This leads to high actual evapotranspiration values and daily evaporation and evaporative fraction values. The spatial-temporal variation of the ET maps for Chisumbanje shows a high daily evapotranspiration rate in the Estate, especially considering 2012, 2014, 2018 and 2020.

The Chisumbanje sugarcane Estate has got high evapotranspiration rate at daily time-step. This is evidenced by the maps in Fig. 6.10, where there is a low evapotranspiration rate outside the Chisumbanje sugarcane Estate. In Fig. 6.11, the Ratelshoek Estate for wheat shows relatively high evapotranspiration rates in the entire Estate. These results can be compared on the regional and national levels where SEBS is widely used. The daily ET_a varied between 0.2 to 9 mm in 2012, 0.0 to 8.7 mm in 2014, 0.10 to 10.2 mm in 2016, 0.1 to 10.1 mm in 2018, and 0.4 to 7.8 in 2020 (Figs. 6.10 and 6.11). As much as is practical, crop producers should use crop actual evapotranspiration in place of the reference evapotranspiration for irrigation scheduling in arid and semi-arid regions similar to the New Mexico environment (Reyes-González et al. 2019). Several studies showed strong relationships between crop yield and seasonal ET_a (Tadesse et al. 2015; Djaman et al. 2018; Poudel et al. 2021).

Table 6.2 shows the mean actual evapotranspiration of the Chisumbanje and Ratelshoek estates over the study period. The average daily ET_a was obtained using a point sampling tool to extract the ET_a daily values from the spatial maps in the GIS environment.

The results show that the seasonal actual evapotranspiration varies in a season. The maximum ET_a in the season is 9 mm/day, and the minimum is 3.98 mm/day. The mean ET_a is 4.75 mm/day with a standard deviation of 2.02. The actual evapotranspiration is high in September, October and December. These months fall in the summer season and drop from January to June (rainy season and winter). The rising

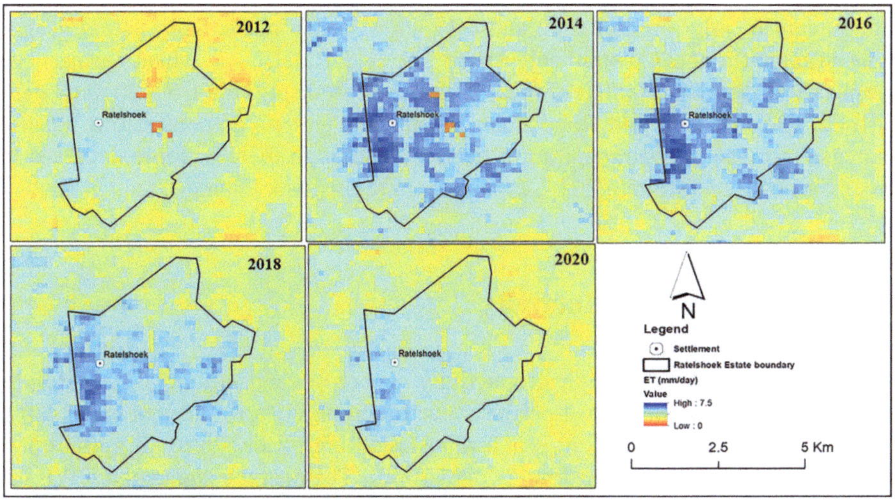

Fig. 6.11 Actual evapotranspiration estimated by SEBS algorithm for 2012–2020 in Ratelshoek Estate

Table 6.2 Mean actual evapotranspiration for Chisumbanje and Ratelshoek estates

Year	2010	2012	2014	2016	2018	2020	Mean	Max	Min	STDEV
Chisumbanje Eta (mm/day)	4.34	3.98	4.32	5.11	7.25	9	4.75	9	3.98	2.02
Ratelshoek Eta (mm/day	5.64	5.28	3.02	6.41	5.95	10.3	7.15	8.65	5.28	2.53

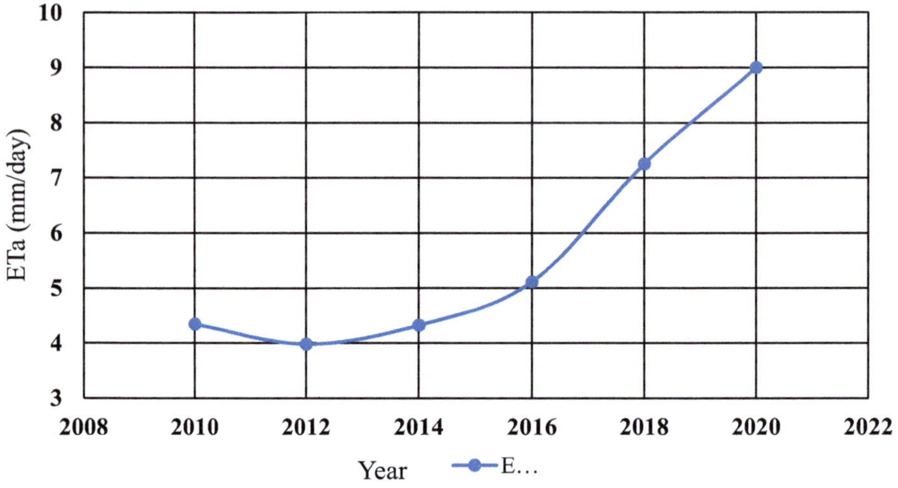

Fig. 6.12 Trend analysis of the mean actual evapotranspiration for Chisumbanje Estate

starts again in July. Figures 6.12 and 6.13 show trends in the daily actual evapotranspiration for the Chisumbanje sugarcane and Ratelshoek wheat farms. The graphs show the general trend of mean actual evapotranspiration from 2010 to 2020 on both Chisumbanje and Ratelshoek Estate. The graphs demonstrate a general increase in mean evapotranspiration from 2010 though there is some decrease in mean evapotranspiration in 2014, especially in Ratelshoek Estate.

6.7 Crop Water Productivity

This study computed the water productivity as a crop yield ratio to ET_a. WP indicates plot, block, or farm-level performance based on choices farmers make using available factors of production. It is a robust measure of the ability of agricultural systems to convert water into food (Zoebl 2006). The average sugarcane CWP for the Valley is 1.328 kg/m^3, with a CV of 32.92%. Although the sugarcane CWP ranges from 0.4 to 2.062 kg/m^3, approximately 99% of the fields had CWPs of less than 0.8 kg/m^3. The remaining fields were likely associated with mixed pixels with other crop types during the classification. The wheat CWP resulted in an average of 1.387 kg/m^3, with a slightly lower CV (25.4%) than sugarcane. Like sugarcane,

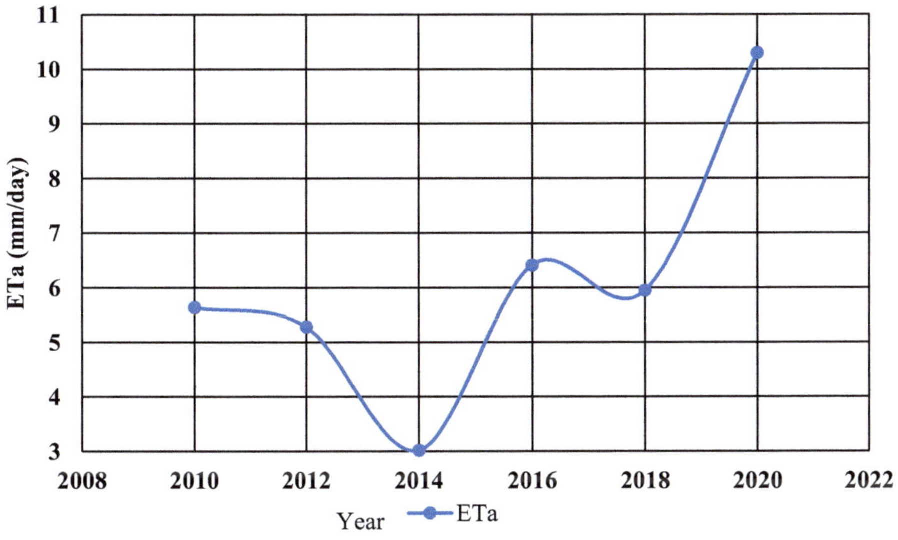

Fig. 6.13 Trend analysis of the mean actual evapotranspiration for Ratelshoek Estate

though the CWP exhibited a higher range for wheat, about 99% of the fields with CWPs of less than 2.06 kg/m^3 were observed, with the attributed reason being similar to that of the sugarcane.

The average CWP of sugarcane obtained in the study was close to the range of 1.2–1.6 kg/m^3, recommended by FAO for Zimbabwe; however, it was slightly lower than that reported by the Department of Research and Specialist Services, which was 0.55 kg/m^3. The variation in the climate and sugarcane productivity may have caused the differences. Figures 6.14 and 6.15 show the spatial variation of crop water productivity for sugarcane in the Chisumbanje estate and wheat in the Ratelshoek estate. Bandyopadhyay et al. (2010) also found a linear relationship between wheat productivity and water supply in central India. Any increase in water supply resulted in a large improvement in crop WP.

To improve wheat production in Iran, Faramarzi et al. (2010) recommended increasing the quantity of cereal production through more efficient use of land and water resources, improving activities related to soil moisture conservation and retention, and optimizing fertilizer application. This can also be done in Ratelshoek Estate to increase Wheat production. Using agricultural WP is useful to benchmark the performance of irrigated agriculture but has sometimes been questionable because of the overemphasis on the performance of just one production factor. It depends on many inputs, and each farmer will strive to use the proper mix of these inputs to obtain. It is defined as crop yield produced per unit of irrigation water used. Matching crop water requirements and local water supply was problematic since many factors made it difficult to arrive at values that were 100% correct. Due to the shortage of resources, the effects of other confounding factors like field water

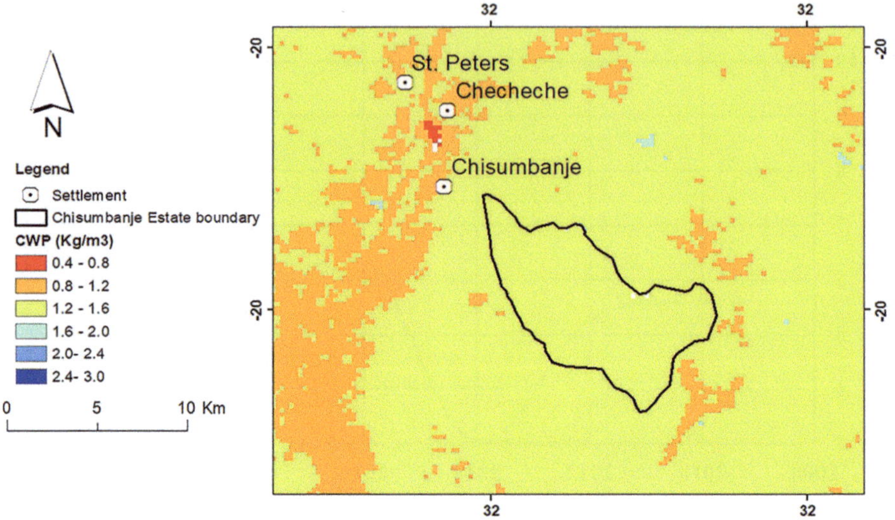

Fig. 6.14 The crop water productivity spatial variation map for sugarcane in Chisumbanje Estate

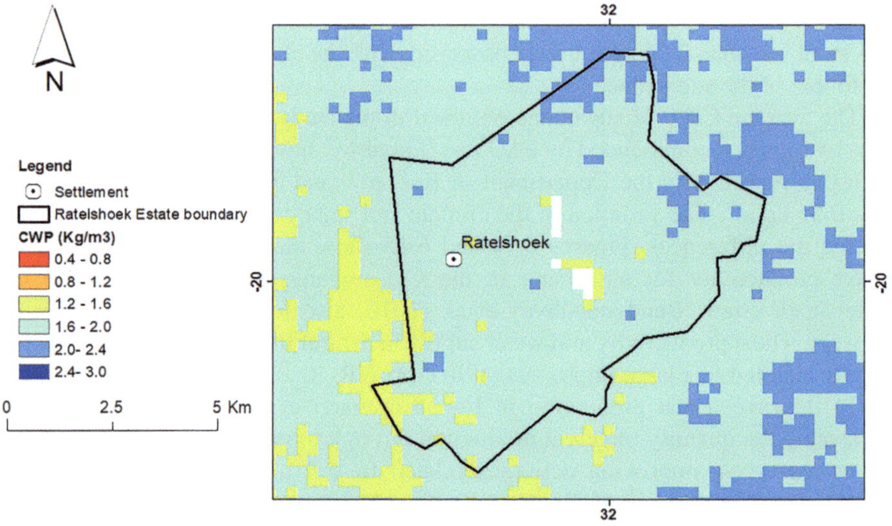

Fig. 6.15 The crop water productivity spatial variation map for wheat in Ratelshoek Estate

utilization and fertilizer input levels were not considered. Research on such factors can help to explain the variability of crop water productivity.

The CWP values obtained in this study are consistent with those obtained by Schwabe and Connor (2012), who discovered that wheat CWP in semi-arid parts of the Southwest United States under sprinkler irrigation was 1.24, 1.75, and 2.03 kg/m^3. Irmak and Djaman (2018) published similar results with subsurface drip-irrigated sugarcane, finding that the sugarcane CWP in Nebraska varied with plant density and ranged from 1.20 kg/m^3 to 5.22 kg/m^3. According to reports, the CWP range for hybrid wheat was 1.65–1.70 kg/m^3. In south-central Nebraska, irrigated sugarcane CWP ranged from 1.89 to 2.50 kg/m^3, according to Djaman et al. (2013). Wheat CWP values were reported by Jancic Tovjanin et al. (2019) to range from 0.67 to 2.34 kg/m^3, and range from 1.79 to 2.38 kg/m^3. According to Mishra et al. (2017), wheat CWP in India was 1.58 kg/m^3, which is lower than the yield in the United States. In south-central Nebraska, irrigated sugarcane CWP ranged from 1.89 to 2.50 kg/m^3, according to Djaman et al. (2013). Wheat CWP values were reported by Jancic Tovjanin et al. (2019) to range from 0.67 to 2.34 kg m/^{3} and range from 1.79 to 2.38 kg/m^3. According to Mishra et al. (2017), wheat CWP in India was 1.58 kg/m^3, which is lower than the yield in the United States.

Reducing different non-productive fluxes can be indirectly aided by good crop management. A robust and healthy crop generates a lot of biomass, which makes transpiration a potent "competitor" for water in comparison to other outflows. For instance, studies by Breman et al. (2001) and Oweis and Hachum (2003) demonstrated that higher nutrient treatments enhanced biomass growth, crop water intake, and crop water productivity with regard to evapotranspiration in semi-arid climates (WPET). Falkenmark and Rockström (2004) found that a change from (less) soil evaporation to (more) crop transpiration is the primary way that improved crop management increases CWP for the same conditions. A robust and healthy canopy is facilitated by managing pests and diseases in addition to water, nutrients, and weeds. According to Passioura's (2006) findings, Australia's water productivity was indirectly enhanced by breeding for tolerance to root and leaf diseases. Any action that boosts crop transpiration early in the growing season should, however, be weighed against the possibility that it would cause soil water stores to run out too quickly, leading to water shortages later in the season (Debaeke and Aboudrare 2004).

Crop water production estimates for the wheat and sugarcane irrigation schemes in Zimbabwe's Manicaland province (Ratelshoek and Chisumbanje Estate). These figures illustrate how difficult it is to draw conclusions from estimates of crop water productivity that differ in terms of time, place, and crop type. The data in Table 6.3 were compiled originally by the author, showing the observed wheat and sugarcane yields, water productivity for wheat and sugarcane and the actual evapotranspiration (ET_a) in the study area. In the Chisumbanje estate, the average daily ET_a measured by SEBS varied from 0.26 to 9.01 mm/day, with an overall average of approximately 6.21 mm/day; in the Ratelshoek estate, the average varied from 1.48 to 7.04 mm/day, with an overall average of 4.06 mm/day. The model's estimated result is similar to the 3.5–5.5 kg/m^3 value that Steduto et al. reported (2012). According to research done on 23 Brazilian sugarcane varieties, Leal et al. (2017) discovered that crop water productivity in stem fresh biomass ranged from 11.45 to 18.45 kg/m3. In the middle Awash basin, Tilahun et al. (2011) calculated crop water productivity in yield for the three main crops: cotton, wheat, and sugarcane. The study revealed that the crop water productivity for sugarcane varied from 5.3 to 13.8 kg/m^3, which agrees with (Tilahun et al. 2011).

Table 6.3 Wheat and sugarcane yields, CWP, and ET_a for Ratelshoek and Chisumbanje Estate

Estate	Year	Crop type	Yield (tonnes/ha)	CWP (kg/m^3)	ET_a (mm/day)
Ratelshoek	2012	Wheat	6.3	0.4	12.5
	2014	Wheat	7.1	3.7	5.3
	2016	Wheat	6.8	0.9	10.3
	2018	Wheat	9.2	3.2	9
	2020	Wheat	10.3	4.0	5.4
	2021	Wheat	9.1	2.3	3.2
Chisumbanje	2012	Sugarcane	88.4	1.2	7.7
	2014	Sugarcane	94.5	0.4	8.5
	2016	Sugarcane	97.2	0.7	3.2
	2018	Sugarcane	78.8	3.1	4.7
	2020	Sugarcane	118.1	3.3	9.4
	2021	Sugarcane	98.3	1.8	5.1

6.8 Conclusion

Water scarcity has been intensifying and posing a threat to the sustainability of agricultural production in arid and semi-arid regions. Hence, understanding crop yield–water relations is essential for sustainable production. This work assesses crop water productivity of the Chisumbanje and Ratelshoek irrigation schemes in Zimbabwe.

Climate, crop type, soil type, geology, evapotranspiration, cropping intensity, crop coefficient, fertilization, and land use/land cover affect crop water productivity. The spatial and temporal variation of these factors in the study area influences the crop water productivity in both Chisumbanje and Ratelshoek Estate.

The study identified spatial variation in the actual evapotranspiration (ET_a) for Chisumbanje and Ratelshoek estates. The minimum observed ET_a for the Ratelshoek wheat farm was 434 mm/year, while the maximum observed actual evapotranspiration was 1022 mm/year, and the minimum observed actual evapotranspiration for Chisumbanje sugarcane farm was 585 mm/year while the maximum observed is 1347 mm/year. The differences in the values are mainly due to the difference in the ecological regions the two estates find themselves: Chisumbanje Estate is in natural region one, and Ratelshoek is in natural region two. Seasonal variation of ET_a is also shown for both schemes.

Crop water productivity (gross biomass water productivity) varies from 1402 kg DM/m^3/year to 1882 kg DM/m^3/year for the Chisumbanje estate compared to a range of 1987 kg DM/m^3/year–2921 kg DM/m^3/year for the 2010 and 2020, respectively. The difference in the maximum gross biomass water productivity is because the weather conditions in natural region one are conducive for any crop to thrive and to use its water capacity to the maximum. In contrast, the heat units in natural region five promote growth and sustainability of the crop water productivity results say otherwise. The findings of this study also show that sugarcane crop water productivity was decreasing for the period, whereas for wheat, it was generally increasing.

Based on the research findings, there is a need to introduce water harvesting and water conservation systems in the scheme to ensure the sustainability of the water supply.

References

Ashktorab N, Zibaei M (2021) Future virtual water flows under climate and population change scenarios: focusing on its determinants. J Water Clim Change 00(0):1–17. https://doi.org/10.2166/wcc.2021.190

Bandyopadhyay KK, Misra AK, Ghosh PK, Hati KM, Mandal KG, Moahnty M (2010) Effect of irrigation and nitrogen application methods on input use efficiency of wheat under limited water supply in a Vertisol of Central India. Irrig Sci 28:285–299

Bastiaanssen W et al (1998) A remote sensing surface energy balance algorithm for land (SEBAL). 1. Formulation. J Hydrol 212:198–212

Benavides J et al (2021) A global analysis of irrigation scheme water supplies concerning requirements. Agric Water Manag 243:106–457

Bhavsar MM, Gohil KB (2015) Review on study of reservoir sedimentation by remote sensing technique. Int J Innov Res Sci Technol 1(12):251–254

Blatchford ML, Mannaerts CM, Njuki SM, Nouri H, Zeng Y, Pelgrum H et al (2020) Evaluation of WaPOR V2 evapotranspiration products across Africa. Hydrol Process 34(15):3200–3221

Breman H, Groot JR, van Keulen H (2001) Resource limitations in Sahelian agriculture. Glob Environ Chang 11(1):59–68

Chai Q, Gan Y, Zhao C, Xu HL, Waskom RM, Niu Y, Siddique KH (2016) Regulated deficit irrigation for crop production under drought stress. A review. Agron Sustain Dev 36:1–21

Chikodzi D, Murwendo T, Simba FM (2013) Climate change and variability in Southeast Zimbabwe: scenarios and societal opportunities. Department of Physics, Geography and Environmental Sciences, Great Zimbabwe University, Masvingo. American Journal of Climate Change: 38–42

Chukalla AD et al (2021) A framework for irrigation performance assessment using WaPOR data: the case of a sugarcane estate in Mozambique. September 1–26

Debaeke P, Aboudrare A (2004) Adaptation of crop management to water-limited environments. Eur J Agron 21(4):433–446

Djaman K, Irmak S, Rathje WR, Martin DL, Eisenhauer DE (2013) Maize evapotranspiration, yield production functions, biomass, grain yield, harvest index, and yield response factors under full and limited irrigation. Trans ASABE 56(2):373–393

Djaman K, O'neill M, Owen CK, Smeal D, Koudahe K, West M et al (2018) Crop evapotranspiration, irrigation water requirement and water productivity of maize from meteorological data under semi-arid climate. Water 10(4):405

Falkenmark M, Rockström J (2004) Balancing water for humans and nature: the new approach in ecohydrology. Earthscan

FAO (2021) Unlocking the water potential of agriculture. FAO, Rome

Faramarzi M, Yang H, Schulin R, Abbaspour KC (2010) Modeling wheat yield and crop water productivity in Iran: implications of agricultural water management for wheat production. Agric Water Manag 97(11):1861–1875

Food and Agriculture Organization (2020a) Perspectivas a mediano plazo: perspectivas para la producción y el comercio mundial de bananos y frutas tropicales 2019–2028. http://www.fao.org/economic/est/est-commodities/banano/es/

Food and Agriculture Organization (2020b) Base datos: Superficie, producción y exportación de banano y plátano. FAOSTAT. http://www.fao.org/faostat/es/#home (consultado 05 mar. 2020).

Food and Agriculture Organization (2020c) Análisis del mercado del banano: resultados preliminares 2019. http://www.fao.org/economic/est/est-commodities/banano/es/
Gilland B (2002) World population and food supply: can food production keep pace with population growth in the next half-century? Food Policy 27(1):47–63
Impacts, in Office, H.D.R. (ed) UNDP. In Africa, Forestry Research Network of Sub-Saharan Africa, Vienna
Jancic Tovjanin M, Djurdjevic V, Pejic B, Novkovic N, Mutavdzic B, Markovic M, Mackic K (2019) Modeling the impact of climate change on yield, water requirements, and water use efficiency of maize and soybean grown under moderate continental climate in the Pannonian lowland. Időjárás/Q J Hungarian Meteorol Serv 123(4):469–486
Jiang T, Sun S, Li Z, Li Q, Lu Y, Li C et al (2022) Vulnerability of crop water footprint in rainfed and irrigation agricultural production system under future climate scenarios. Agric For Meteorol 326:109164
Kharrou MH et al (2021) Assessing irrigation water use with remote sensing-based soil water balance at an irrigation scheme level in a semi-arid region of Morocco. Remote Sens 13(6):1–22. https://doi.org/10.3390/rs13061133
Kirkby MJ, 1993 'Longterm interactions between networks and hillslopes', in Beven KJ, and Kirkby MJ (Eds), Channel Network Hydrology. John Wiley, Chichester, pp 254–293
Leal DP, Coelho RD, Barbosa FDS, Fraga EF, Mauri R, Santos LDC (2017) Water productivity for sugar and biomass of sugarcane varieties. Revista Brasileira de Engenharia Agrícola e Ambiental 21:618–622
Midya A et al (2021) Crop establishment methods and integrated nutrient management improve. Part I. crop performance, water productivity and profitability of rice (Oryza sativa L.) in the lower indo-Gangetic plain, India. Agronomy 11(9). https://doi.org/10.3390/agronomy11091860
Mishra J, Singh R, Arora NK (2017) Alleviation of heavy metal stress in plants and remediation of soil by rhizosphere microorganisms. Front Microbiol 8:1706
Montazar A (2021) Irrigation tools and strategies to conserve water and ensure a balance of sustainability and profitability. Agronomy 11(10):2037. https://doi.org/10.3390/agronomy11102037
Mul M, Bastiaanssen W (2019) WaPOR quality assessment – technical report on the data quality of the WaPOR database version 1.0. FAO Report, Rome
Mushay MM, Kusangaya S, Mujere N (2023) Use of remote sensing to determine rainwater harvesting sites for piped micro-irrigation schemes in Chimanimani District, Zimbabwe. Water SA 49(1):56–63
Oweis TY, Hachum AY (2003) Improving water productivity in the dry areas of West Asia and North Africa. In: Water productivity in agriculture: limits and opportunities for improvement. CABI Publishing, Wallingford, pp 179–198
Passioura J (2006) Increasing crop productivity when water is scarce – from breeding to field management. Agric Water Manag 80(1–3):176–196
Poudel U, Stephen H, Ahmad S (2021) Evaluating irrigation performance and water productivity using EEFlux ET and NDVI. Sustain For 13(14):7967. https://doi.org/10.3390/su13147967
Reyes-González A, Kjaersgaard J, Trooien T, Reta-Sánchez DG, Sánchez-Duarte JI, Preciado-Rangel P, Fortis-Hernández M (2019) Comparison of leaf area index, surface temperature, and actual evapotranspiration estimated using the METRIC model and in situ measurements. Sensors 19(8):1857
Rwasoka D, Gumindoga W, Gwenzi J (2011) Estimation of actual evapotranspiration using the Surface Energy Balance System (SEBS) algorithm in the Upper Manyame catchment in Zimbabwe. Phys Chem Earth Parts A/B/C 36(14):736–746
Schwabe KA, Connor JD (2012) Drought issues in semi-arid and arid environments. Choices 27(3):1–5
Scoones I, Murimbarimba F, Mahenehene J (2019) Irrigating Zimbabwe after land reform: the potential of farmer-led systems. Water Altern 12(1):88–106
Sobrino JA, El Kharraz J, Li ZL (2003) Surface temperature and water vapour retrieval from MODIS data. Int J Remote Sens 24(24):5161–5182.
Steduto P (2012) Crop yield response to water. FAO Irrigation and Drainage Paper No, 66, pp 124–131

Steduto P, Hsiao TC, Fereres E, Raes D (2012) Crop yield response to water, vol 1028. FAO, Rome, p 99

Su Z (2002) The surface energy balance system (SEBS) for estimation of turbulent heat fluxes. Hydrol Earth Syst Sci 6(1):85–100

Tadesse T, Senay GB, Berhan G, Regassa T, Beyene S (2015) Evaluating a satellite-based seasonal evapotranspiration product and identifying its relationship with other satellite-derived products and crop yield: a case study for Ethiopia. Int J Appl Earth Obs Geoinf 40:39–54

Tilahun H, Teklu E, Michael M, Fitsum H, Awulachew SB (2011) Comparative performance of irrigated and rainfed agriculture in Ethiopia. World Appl Sci J 14(2):235–244

UN-Water (2016) Integrated monitoring guide for SDG 6: targets and global indicators. UN-Water, Geneva

Vasilachi IC, Asiminicesei DM, Fertu DI, Gavrilescu M (2021) Occurrence and fate of emerging pollutants in water environment and options for their removal. Water 13(2):181

Zimstats (2012) The Zimbabwe 2012 population census national report. https://www.zimstat.co.zw/wpcontent/uploads/publications/Population/population/census-2012-national-report.pdf

Zoebl D (2006) Is water productivity a useful concept in agricultural water management? Agric Water Manag 84(3):265–273

Chapter 7
Sustainable Intensification of Mixed Farming System in West Africa: Concepts, Practices, and Challenges

H. E. Igbadun, O. A. Ojeleye, Tafadzwanashe Mabhaudhi, and O. Cofie

Abstract Sustainable intensification (SI) is a system of production that increases output without causing significant environmental damage. It focusses on enhancing agricultural land production while managing its environmental impact. Evaluation frameworks have evolved to include non-environmental aspects, such as social concerns, economics, and the human condition. Agricultural sustainability assessment now uses indicator frameworks, which are structured into five domains: productivity, economic, environment, the human condition, and social domains. Mixed farming systems (MFS) is an approach to sustainable agriculture where farmers produce crops and animals in the same location under the same ownership. MFS provides enough food for consumers and income for farmers while ensuring soil fertility, biodiversity, and pest control. Several characterizations and typologies of MFS in West Africa have been identified, but the level of development varies due to farmers' preferences. Despite the benefits of MFS, socio-economic factors, such as skills and competencies, the role of the agricultural knowledge and innovation system, the economy, and the policy environment, pose major challenges and obstacles to its growth in West Africa.

Keywords Rainfed agriculture · Irrigation · Water productivity · Sustainable development · Planetary health

---

H. E. Igbadun (✉)
Department of Agricultural and Bioresources Engineering, Ahmadu Bello University, Zaria, Nigeria

O. A. Ojeleye
Department of Agricultural Economics, Ahmadu Bello University, Zaria, Nigeria

T. Mabhaudhi
Centre on Climate Change and Planetary Health, London School of Hygiene and Tropical Medicine, London, UK

Centre for Transformative Agricultural and Food Systems, University of KwaZulu-Natal, Carbis Road, Scottsville, Pietermaritzburg, KwaZulu-Natal, South Africa

O. Cofie
Internaional Water Management Institute (IWMI), West Africa Regional Office, Accra, Ghana

T. Mabhaudhi et al. (eds.), *Enhancing Water and Food Security Through Improved Agricultural Water Productivity*,
https://doi.org/10.1007/978-981-96-1848-4_7

7.1 Introduction

Sustainable intensification (SI) is a term used when exploring production resources to increase output with a consciousness of minimizing adverse effects on the production environment. In an agricultural context, SI is a process or system of increasing agricultural output while minimizing environmental effects. It is about techniques and practices that maximize production output without devastating production resources such as land and water, without expanding non-agricultural land. It is about intensive use of the production environment while deploying conservation techniques to replenish the resources used.

Early agricultural revolutions focussed on extending agricultural areas to boost aggregate food output. To feed increasing populations and change consumption habits, it became evident that agriculture would need to extend into non-agricultural territories. This puts competition for land and water from other human activities, approaching a more expensive alternative, especially if biodiversity and the public benefits of natural ecosystems (e.g. carbon storage in forests) are prioritized (MEA 2005). Extensification has to give way to a new concept of intensified utilization of resources in the same area.

With SI in agriculture being about producing more food on the same land without or with minimal negative environmental impacts, one of many agricultural practices with high potential for achieving SI is the mixed farming system (MFS). MFS is an agricultural practice where farmers cultivate crops and keep livestock on the same farmland. It allows the farmer to integrate and diversify his resource base, increase profitability and improve livelihood. The system is operated symbiotically; for example, the livestock can provide draught power for crop cultivation, the manure from the livestock is used to fertilize the soil, and the crop residues are used as livestock feed. Therefore, SI-MFS is an attempt to sustainably intensify crop and livestock production integration without a negative environmental impact. This chapter reviews the concept of IS in agriculture, vis-à-vis MFS, the typology and practices of MFS in West Africa and the challenges of MFS in the premise of sustainable intensification.

7.2 Concept of Sustainable Intensification in Agriculture

Although SI has been practised for decades, the concept first appeared in the publications of eminent rural development scholars in the 1990s (see, for example, Conway 1997; Pretty 1997); its regularity of use in scientific works and policy reports has significantly increased over the last three decades, reaching unprecedented recognition in recent years. SI is essential to UN Sustainable Development Goal 2, implying that it will stay on development policies for some time (Gunton 2016). SI in agriculture is concerned with maximizing production output per unit of land while minimizing environmental impact. The philosophy emerged in response to a perceived lack of social and environmental sustainability due to the usage and prevalence of pesticides, inorganic fertilizers, machinery, land degradation, expansion of agricultural areas, etc. (Conway 1997; Pretty 1997). Pretty (1997), for instance, saw SI as a means to achieve substantial production in currently unimproved or degraded areas while protecting or even regenerating natural resources. It is a process or system where yields are increased without adverse environmental impacts or the cultivation of more land. The concept does not promote any particular vision or method of agricultural production; instead, it emphasizes the end rather than means and does not pre-determine technologies, species mix, or particular design components' (Pretty and Bharucha 2014). The SI approach recognizes the interdependence of cultivated and uncultivated systems and the necessity of agriculture's direct contribution to mitigating significant global concerns. The concept is that it can enhance agricultural yields per unit of land without endangering the environment. This signified a paradigm change in environmentalist ideals and high external input agriculture.

Before the late 1990s, agriculture intensification was associated with "inevitably harming while producing food" (Pretty and Bharucha 2014). Early in the 2000s, as the climate change agenda gained traction, the idea gradually came to be associated with interlinkages (water, energy, climate, etc.) outlooks), and the advancement in food production and the decrease in greenhouse gas emissions from agriculture became the primary justifications for SI. As a result, IS is "Producing more from the same area of land while conserving resources, reducing negative impacts on the environment, and enhancing natural capital and the flow of ecosystem services" (FAO 2011). It provides a realistic route to increasing food production while minimizing environmental damage, guaranteeing that the natural resource base on which the agriculture system depends is preserved and even strengthened for future generations (The Montpellier Panel 2013). SI involves agricultural land production enhancement while also managing its environmental impact. The phrases "sustainable" and "intensification" are used to suggest that various strategies might be used to provide the desired results in terms of both more food and better environmental products and services (Pretty and Bharucha 2014).

7.3 Sustainable Intensification Indicators

According to Cook et al. (2015), sustainable intensification focusses on increasing yields over the long run in Africa's vulnerable settings. Intensification is thought to have the ability to lessen demand for the conversion of natural areas to agriculture (Pretty 1997; Reardon et al. 1995). Hence, a framework for evaluating SI is necessary since the scope of SI has expanded to encompass non-environmental aspects, including social concerns, economics, and the human condition (Loos et al. 2014). Developing an index system has been a popular way to determine if a system achieves sustainable agricultural intensification goals due to the term's diversity. As a result, the agricultural sustainability assessment moved closer to utilizing indicator frameworks, which serve as a foundation for choosing a core set of indicators from a large list. Several indicator frameworks have been created to evaluate the development of sustainable agriculture. Many of these frameworks include details on only one system component, including soil health, nutrition, or eradicating poverty (Bockstaller and Girardin 2003; Gustafson 2016). Other system-based frameworks assess various system characteristics, including dependability, equality, self-reliance, stability, and resilience (López-Ridaura et al. 2005; Conway 1994).

The goal-oriented approach by Olsson et al. (2009) gave rise to the indicator framework assessment of SI. Olsson et al. (2009) suggested a framework in which indicators are linked to the objectives once the innovation goals are determined to evaluate performance balanced across domains. Musumba et al. (2017) presented the objective-oriented indicators forming the current SI evaluation indicator framework. These indicators structured into five domains crucial to sustainability are: productivity, economic, environment, the human condition, and social domains.

7.4 Five Domains of Sustainable Intensification

The five (5) domains of sustainable intensification presented by Musumba et al. (2017) for evaluating SI include productivity, economics, environment, the state of humanity, and society. These indicators were identified during a stakeholder engagement at a summit in Ghana in 2013. In contrast to the three domains employed by past sustainability evaluations, namely, economic, environmental, and social domains (López-Ridaura et al. 2002; Van Cauwenbergh et al. 2007), Musumba et al. (2017) identified other key features of sustainable intensification which makes sure that important issues like equality (gender, age, and class), nutrition, and community elements like social cohesiveness and cooperative action are included in the process of choosing indicators (Musumba et al. 2017). The domains, as characterized by Musumba et al. (2017), include the following.

Productivity: This measures the production of both crop and livestock systems. Land and water are regarded as crucial elements of this indicator. The purpose of intensification is to increase output per unit of input over a specific period, and improving seasonal or yearly productivity is its key attribute. In contrast to cropping systems, where intensification focusses on yields, livestock systems may employ stocking rates or offtake as a metric of intensification (Mahon et al. 2017).

Economic: The domain focusses on the economic viability of the agricultural system. It includes factors of production and economic returns. It considers indicators for productivity inputs other than land. Still, water, nutrients, labour, and capital are also the farmers' markets, the diversification of sources of income, and the level and progression of high-value crop production.

Environment: This examines the natural resources that can be used for agriculture, such as soil, water, and air, as well as the environmental services that are directly impacted by agricultural practices, such as habitat, soil water holding capacity, and biodiversity, as well as the amount of pollution that agriculture produces (pesticides, eutrophication, greenhouse gases). The economic domain describes improved efficiency measurements, which are also essential for tightening the energy and nutrient cycles, a fundamental concept in sustainable agriculture.

Human Condition: Indicators like nutritional status, food security, and adaptability are linked to the person or household. Even though some of these ideas rely on social interactions (such as those in the home or community), they differ from those in the social domain specifically concerned with interpersonal relationships.

Social: This area focusses on the social ties that exist within agricultural communities or society, including relationships that are equitable across gender and social groupings, the degree of group activity, and the capacity to handle disputes about natural resource management and agriculture.

7.5 Mixed Farming System Typology and Characterization

Mixed farming system patterns in West Africa have distinguishing characteristics, making them highly location-specific. They depend on adaptive strategies devised by farmers to cope with adverse situations and often take advantage of the potential opportunities for intensification and diversification at the household level. Studies have shown that farmers develop strategies to avoid adverse situations vis-à-vis volatile prices, crop failure, food, drought, declining soil fertility, land scarcity, and climate change (Innazent et al. 2022). They also use potential opportunities vis-à-vis the use of new technologies and value addition, which allowed for sustainable production and income (Yaro 2010; Tittonel 2014; Shukla et al. 2019).

7.6 Typologies of Mixed Farming System in West Africa

Fernández-Rivera et al. (2004) provide the most thorough classification of crop–livestock production systems for West Africa, identifying 13 systems based on the leading crops and building on a mix of elements from Seré et al. (1996), Dixon et al. (2001), Manyong (2002), and Thornton et al. (2002). These crop–livestock systems are as follows.

Pearl Millet–Cowpea–Livestock System

The most extensive crop–livestock production system in West Africa is the pearl millet–cowpea–livestock system. According to estimates by Fernández-Rivera et al.

(2004), the mixed farming production system covers 17.7% or 0.64 million km^2 of the land area in the region. Its practice spans the semi-arid zone, from northern Nigeria to south-western Niger, Burkina Faso and southern Mali. Compared to the pastoral system of range grazing, the system has a more substantial potential to support its animal population from usable agricultural wastes. The soils of the region are largely sandy. They are easily stressed by an average cultivated area of 25.2%, a high livestock density of 13.5 cattle/km^2, 37.7 sheep and goats/km^2, and a human population density of 61.3 people/km^2 (Fernández-Rivera et al. 2004). As a result, crop–livestock integration is a common strategy to maintain soil fertility. Moreover, animal traction for cultivation is often used because the agroecological and economic circumstances are suitable to encourage its profitable and intensive usage.

Pearl Millet–Groundnut–Livestock System
This MFS is common in the Gambia, and the northern regions of Senegal, which have comparable agroecological and farming system traits to the pearl millet–cowpea–livestock system. The primary distinctions between this and the Pearl millet–cowpea–livestock system are in the type of cattle bred and the extent to which animals are used as traction for both transportation and crop production (Agyemang et al. 1997). Due to the high prevalence of trypanosomiasis and perhaps due to cultural preferences, N'Dama cattle, known for their trypan tolerance, are often raised for their milk and meat. Animals are herded together during the planting season to protect the crops; the groundnut hay is gathered and kept after harvest to feed labour oxen, while the pearl millet husks are left on the field so that the animals may roam freely and graze anywhere.

Sorghum–Maize–Cowpea–Livestock System
According to Fernandez et al. (2004), this system type is situated in the drier regions of the sub-humid zone, which is also known as the Northern Guinea Savannah; it has a surface area of 0.23 million km^2 and supports 9.4 million sheep and goats in addition to 3.3 million cattle, making it the production system in the area with the highest density of livestock. Crop–livestock integration is probably strongest in this system because it lies in the middle of systems in the drier zones with comparative advantage in livestock production and those in the wetter zones with comparative advantage in crop production. Farmers have been demonstrated to work at a greater production efficiency in the system (86% on average) than farmers in the drier systems, who work at an average of 68% (Okike 2002). The main crops are sorghum, maize, and cowpeas, although other crops, including pearl millet, groundnuts, and soybeans, are also grown. Population expansion, urbanization, and profitable cowpea and meat markets inside and outside the system are the main drivers of this system. There is a sizable market for this product in southern rural and urban parts of Nigeria, where most cowpea is grown and consumed.

Maize–Sorghum–Livestock System
This system is situated in the drier regions of the subhumid zone and covers a comparable area (0.24 million km^2), of which 21.5% is farmed. It is similar to the sorghum–maize–cowpea–livestock system in these regards. According to

Fernández-Rivera et al. (2004), it has a population of 3.9 million Total Livestock Units (TLU), consisting of 2.7 million cattle and 8.5 million sheep and goats. Its livestock population density (16.7 TLU/km^2) is lower than the 19.9 TLU/km^2 in the sorghum–maize–cowpea–livestock system in the Northern Guinea Savanna, a system with a comparable geographic location. Because the industrial need for cereals drives the system the most, it differs from many other production methods. Farmers look for advances and technology in this area to increase their profit margins. For instance, improved seeds, insecticides, herbicides, chemical fertilizers, and mechanical traction are frequently used. The need for meat, policies encouraging the use of cereal grains for industrial purposes, the availability of technology, and the maize–sorghum–livestock system all seem to influence how the system works.

Coconut Oil Palm–Fruits–Livestock System

The coconut oil palm–fruits–livestock system covers an area of 46,581 km^2 (22% cultivated) with 5.1 million people and 0.2 million TLU composed of 79,000 cattle and about one million sheep and goats (Fernández-Rivera et al. 2004). Oil palm, citrus trees, coconut palm, and pineapples are common. In 2000 alone, West Africa produced 12.7 million tons of oil palm fruits, 4.0 million tons of citrus fruits, 0.8 million tons of coconuts, and 1.3 million tons of pineapples (FAO 2003). Fianu et al. (1994) reported in a study of sheep in tree crop plantations in Ghana that plantation sizes ranged from 3 to 16 ha, and plantation owners had about 7–58 sheep. Fianu et al. (1994) cited Wilson and Lansbury (1958) as estimating the average annual green herbage yield of cover crops in the plantations at 20.8 t/ha, thus pointing to the enormous potential for integrating livestock into tree crop plantations. This system can be used in landless urban and peri-urban livestock production systems to integrate cattle, decrease labour expenses for weed management, acquire milk, meat, and manure in return, and offer components for stall feeding.

Rice–Livestock System

The rice–livestock system exists where rice is the predominant crop under the irrigated systems; many of these systems occur along the Niger and Gambia rivers and the flood plains around Lake Chad. The rice–livestock system has an area of 0.34 million km^{2,} of which 12.8% is cultivated. Within the rice–livestock system, there are 15 million people and 4.5 million TLU of 3.2 million cattle and 10 million sheep and goats (Fernández-Rivera et al. 2004). Many of these systems in West Africa occur along the Niger and Gambia rivers and the flood plains around Lake Chad. Dixon et al. (2001) distinguished between large-scale, centrally operated irrigation schemes and small-scale farmer-managed schemes within the irrigated systems. According to their findings, irrigated fields are almost invariably a component of a broader agricultural and subsistence system, including livestock and rainfed agriculture. De Leeuw (1997) claims that roughly 2 months after harvesting the neighbouring fields' sorghum and pearl millet, rice straw from this method starts to appear in harvested fields and community threshing locations. De Leeuw cited Wilson et al. (1989) in the same report, who found that herds grazed upland savannahs during the rainy season but relied on rice straw (57%), pearl millet residues (12%), and harvested rice fields (20%) from November to May in the areas surrounding the extensive irrigation projects along the Niger River in central semi-arid Mali.

Cocoa–Plantain–Cassava–Livestock System

The cocoa–plantain–cassava–livestock system covers 0.19 million km^2 and has 17.6 million people (94.5 persons/km^2) commonly practised in Ghana and Nigeria (Fernández-Rivera et al. 2004). The practice has the lowest livestock population density; the plantain is a regional staple, while cocoa is a major cash crop that contributes significantly to the economies of countries in the region. The integration of livestock into cocoa plantations is limited because the shade of the cocoa does not allow enough light for herbaceous vegetation to grow and offers an opportunity for grazing. However, the dried cocoa pods are combined with cassava, plantain, yam, cocoyam peels and other household wastes as supplementary feed for small ruminants in homesteads.

Rice–Cassava–Maize–Livestock System

In the forests of Côte d'Ivoire, Liberia, Guinea, and Sierra Leone, this system can be found in sporadic locations. The system is reported to cover 0.26 million km^2. It is distinguished by having the lowest agricultural intensity (3.3%) in the whole region as well as the lowest human population density (28.6 persons/km^2) and cattle population density (5.1 TLU/km^2) (Fernández-Rivera et al. 2004). Useable crop residues from the cleared and cultivated areas can sustain all the ruminant livestock within the system for up to 5.8 months/year. There seems to be potential for expanding the livestock population in this system to utilize available crop residues more fully.

Cassava–Yam–Soybean–Livestock System

In this system, cassava and yams predominate, while soybean is of emerging importance, especially in Nigeria. The system covers 0.29 million km^2 and stretches across the middle belts of Nigeria and Côte d'Ivoire, with fair representation in Togo, Ghana, and Benin. Food demand in the region's densely populated coastal districts drives production. Similar to the cassava maize yam livestock system, there are opportunities to improve the system.

Cassava–Maize–Yam–Livestock System

Cassava is arguably the most dominant food staple in the humid zone of West Africa, extending into the wetter parts of the sub-humid zone, from Guinea through Liberia, Côte d'Ivoire, Togo, Benin and Nigeria. This system accounts for 136,518 km^2, with 17,540 km^2 or 12.8% cultivated to crop: about 408,000 cattle and 4.4 million sheep and goats (Fernández-Rivera et al. 2004). Since Tsetse is a major challenge for cattle production in the humid and wetter sub-humid zones where the system is common, the common livestock is N'Dama and Keteku breeds of cattle and the trypan-tolerant West African Dwarf sheep and goats, which are resistant to trypanosomosis. In this system, approximately 74% of households have goats, with an average flock size of 5.1, compared to approximately 13% who have sheep, with an average flock size of 1.9. Goats typically roam freely in the post-harvest season, and their diet is supplemented with household wastes like cassava, yam, and cocoyam peels, as well as cut-and-carry fodder (Onwuka et al. 1992). Maintaining the rising productivity of tubers, especially cassava, and combining crop and animal production are opportunities for realizing the full potential of this system.

Groundnut–Rice–Livestock System

The groundnut–rice–livestock system is found mainly in an area of Senegal known as the Bassin Arachidier (groundnut–producing basin). It covers an area of about 27,407 km^2 (1478 km^2 or 5.4% cultivated) and has 228,000 cattle and 541,000 sheep and goats (Fernández-Rivera et al. 2004). Groundnut production is driven by the commercial value of groundnut oil and the popular food value of groundnut sauce eaten as a regular accompaniment for rice in the zone. With market opportunities and population already playing important roles in the system, its prospects are good, especially with market liberalization and other reforms of the groundnut marketing sector in Senegal, as Badiane et al. (1997) reported.

Cotton–Maize–Sorghum–Livestock System

The system extends across 0.11 million km^2 of what is today called the cotton belt of West Africa, with 2.3 million inhabitants, 1 million cattle, and 1.7 million sheep and goats (Fernández-Rivera et al. 2004). Williams et al. (2000) reported that this MFS became successful in the region due to the introduction of new maize and cotton cultivars, the quick adoption of animal traction, and enhanced crop management techniques such as fertilizer use, higher plant density, and insect control.

High-Value Vegetables–Rice–Livestock System

This system is common along the Niger River, especially between its entry into Mali and exit from Niger and the extensive inland valleys in north-west Nigeria. It differs from the irrigated systems in that high-value vegetables such as onions, peppers, tomatoes, and carrots are planted and irrigated through the dry season after the rice crop has been harvested. In Kano, Nigeria, for instance, satisfies a major part of the demand for vegetables in the urban centres of southern Nigeria. Since these vegetables leave no significant residues for livestock production and occupy the land after the rice is harvested, they limit the grazing period; therefore, the rice straws in the feed are used as feed for the livestock. According to Fernández-Rivera et al. (2004), this system covers an area of 91,270 km^2 with 5 million persons and 1.3 million TLU. Its cultivation intensity (19.4%) is among the highest in the region. Dixon et al. (2001) also reported that the future potential for intensification of this system is good if water shortages and irrigation scheme breakdowns can be avoided and input/output price ratios do not deteriorate. Improving infrastructure will reduce the time spent between harvest and reaching the final product and the associated deterioration in product quality. Better access to market information will also benefit the system.

The characterization of a crop–livestock system in West Africa by Fernández-Rivera et al. (2004) was found to be comparable to Sekaran et al. (2021) characterization for various regions of the world as part of the integrated crop–livestock system (ICLS). It may be noticed from Table 7.1 that cassava, maize, banana, rice, groundnuts, oil palms, millet, beans, and sorghum are common crops, along with

Table 7.1 Different types of crops and livestock used under the integrated crop–livestock system (ICLS) in different parts of the world

Ecological/ climatic zone/ place research carried out	Crops	Livestock	Major livestock output	Location	References
Humid	Cassava, maize, banana, rice groundnut, and oil palms	Cattle, sheep, goat	Meat, milk, and power	Sub-Saharan Africa	Saleem (1998)
Sub-humid	Maize, millet, groundnut, cassava, beans, rice, and sorghum			West Africa	
Arid/Semi-arid/ Sub humid	Millet, banana, beans, and maize				
Arid	Sorghum, millet, and beans				
Cool tropics	Teff, wheat, oats, maize, and pulses				
Cauvery Delta zone, Northwestern zone of Tamil Nadu and Northwestern arid region of Haryana	Paddy, sugarcane, vegetables, and flowers	Cow, buffalo, goat, piggery, sheep, poultry, and fishery	Milk, calf, chicken, fish, and meat	India	Ponnusamy and Devi (2017)
North-West India	Rice and wheat	Buffalo, cattle, and goats	Milk and meat	India	Kumar et al. (2014)
University of Illinois, Pana, Illinois	Corn, oat, cereal rye and turnip	Beef cattle	–	USA	Maughan et al. (2009)
Montana State University-Bozeman	Kohlrabi, spinach, lettuce and cover crops (Buckwheat, beets, sweet clover and pea)	Sheep (grazing)	–	Bozeman, USA	McKenzie et al. (2017)
Southeastern USA	Corn, soybean, and wheat (winter cover Crop)	Cattle		Georgia, USA	Sulc and Franzluebbers (2014)
The Texas Southern High Plains	Old world bluestem (grass), cotton, rye, and wheat	Steer	–	USA	Allen et al. (2007)

(continued)

Table 7.1 (continued)

Ecological/ climatic zone/ place research carried out	Crops	Livestock	Major livestock output	Location	References
Southern Coastal Plain	Cotton, peanut, and rye or wheat or Crimson clover (cover crops)	Steer and heifers	–	Georgia, USA	Franzluebbers (2007)
Southern Piedmont	Sorghum, wheat, and rye and pearl millet or crimson clover (cover crops)	Cattle and broilers	Meat	USA	
U.S. Corn belt, Pana, Illinois	Corn, oats, cereal rye and turnip	Beef cattle		USA	Sulc and Tracy (2007)
Mississippi	Trees, corn and perennial forages (Agroforestry)	Dairy cow	Milk	USA	Sulc and Franzluebbers (2014)
North Dakota	Oat/alfalfa/hairy vetch/red clover, brown midrib sorghum–sudangrass/sweet clover/red clover and corn	Cattle	–	USA	Liebig et al. (2012)
Southern United States	Peanut, cotton, and bahiagrass/ bermudagrass	Cattle			Katsvairo et al. (2006)
Southeastern and North-eastern	Wheat and canola (dual purpose crops)	Beef cattle and	–	Australia	Bell et al. (2014, 2015)
Tablelands and slopes, and the Southern and Western high rainfall zones		Sheep			Dove et al. (2015), Rodriguez et al. (2014) and Villano et al. (2010)

(continued)

Table 7.1 (continued)

Ecological/ climatic zone/ place research carried out	Crops	Livestock	Major livestock output	Location	References
Brazilian subtropical region (Southern Brazil), Tropical region Southeast, Midwest, Northeast and (North of Brazil), Mato Grosso do Sul	Soybean, corn, rice, beans, eucalyptus (Agroforestry), cotton, wheat (winter cover crop) and signal grass (pastures)	Beef cattle, dairy cattle, sheep	Meat, milk, wool	Brazil	de Faccio et al. (2010), Gil et al. (2015), Siebold (2015), Salton et al. (2014)
Coteaux de Gascogne (Hilly region)	Cereals and other cash crops	Beef and dairy cattle	Meat and dairy products	France	Ryschawy et al. (2012)
Charolais suckler cattle farms in central France	Cereals, sunflower, and rapeseed	Beef cattle	Meat	France	Veysset et al. (2014)
Canterbury	Wheat, brassicas, kale, fodder beet, oats, barley, peas, beans, turnips, and rapeseed	Beef cattle, dairy cattle, sheep, and deer	Milk, tallow, potted and salted meat, wool, skins, hides, and lamb	New Zealand	Dynes et al. (2010)
Marlborough	Wine grapes (viticulture)	Sheep	–	New Zealand	Niles et al. (2018)

Source: Sekaran et al. (2021)

various combinations of cattle, sheep, and goats as livestock. This implies that MFS with these crop combinations resonates around the world.

7.7 Productivity of Mixed Farming System in West Africa

Crop–livestock farming occurs in West Africa in socio-economic, climatic, and soil conditions. The scarcity of essential resources like land, plant nutrients, water, finance, and labour significantly impacts the evolution of these systems (van Wijk et al. 2009). The level of development of mixed farming systems across the region has varied greatly due to farmers' preferences about the distribution of resources. Herds are forced to graze on smaller areas as land pressure mounts, which affects the livestock's diet and raises the possibility of overgrazing (Powell and Williams

1995). Rain-fed agriculture will undoubtedly continue to play an important part in ensuring the food and livelihoods of a growing global population, as evidenced by trends in the availability of water in the present and the future (Rockstrom et al. 2010). However, sustaining rainfed agriculture with supplemental irrigation systems by imposing water harvesting and storage practices has become inevitable and essential to reduce terminal water stress, which happens more frequently due to climate change. Including animals in crop production will help to make the most use of available agricultural resources. Although their advantages have not been thoroughly investigated, forage legumes can potentially increase crop and animal output in smallholder agricultural systems. Significant threats to the crop–livestock systems in sub-Saharan Africa are variable rainfall and inadequate soil fertility (Harrington et al. 2009).

7.8 Benefits and Challenges of Mixed Farming Systems in West Africa

To maximize output from MFS, it is essential to maintain diverse agroecosystems with components having complementary functions, increasing soil fertility and preventing nutrient losses. It is vital to provide continuous vegetative cover and limit the use of external pesticides and fertilizers over a long period. MFS is one of the most significant ways to achieve food security in developing countries without hurting the environment (Iiyama et al. 2008; Mekuria and Mekonnen 2018). According to the system, the sequential process is through which farmers raise farm animals and cultivate crops to maintain sustainable agriculture on the same plot of land (Mekuria and Mekonnen 2018). According to several empirical types of research, mixed farming is the most crucial agricultural strategy for emerging economies, especially in Sub-Saharan Africa (SSA), where over 166 million agro-pastoralists live (Iiyama et al. 2008). Farmers can diversify their resources through the system to balance agricultural and animal output. According to Maitland (2019), who cites Stuart's Nuffield research, raising cattle with continuous cropping can increase overall farm profitability by extending pasture productivity, lowering overgrazing, lowering financial risk, and extending pasture productivity. His local experience indicates that planting annual crops expressly for grazing can improve soil quality and subsequent crops' ability to absorb nutrients.

Both Liniger et al. (2011) and Berndes et al. (2010) argued that mixed production systems increase water usage effectiveness and land productivity. According to several studies (Alexandratos and Bruinsma 2012; Darnhofer et al. 2012; Herrero et al. 2012; Liniger et al. 2011), farm diversification reduces risks and provides insurance against crop failures because different farm activities offer a variety of responses to varying conditions and boost household income and resilience (Liniger et al. 2011; Kuria et al. 2014). According to Liniger et al. (2011), families using crop–livestock systems in Ethiopia's highlands had 50% higher productivity and farm income than smallholders that grow and cultivate crops.

Socio-economic factors, including skills and competencies, the role of the agricultural knowledge and innovation system, the economy, and the policy environment, serve as some of the major challenges and obstacles to the growth of MFS (EIP-AGRI Focus Group 2017). Because animals require constant care from those participating in the operation, a mixed farming system necessitates higher expertise (both crop and livestock) and dedication (Peterson and Deiss 2020).

Integrated crop–domestic livestock systems could be more adaptive and sustainable in the drier regions of West Africa. Moreover, unfavourable agricultural conditions and the prevalence of trypanosomiasis in the wetter, more humid zone make it difficult to expand mixed farming, at least for ruminant animals. It is not environmentally preferable to increase livestock numbers by removing trees. Still, options that increase the chances of incorporating trypan-tolerant cattle into long-term tree systems should be explored. Some drawbacks make MFS more challenging to sustain than monoculture farming. Some of these include multiple tasks involved in mixed farming, the challenge of monitoring and maintaining increased operations, and the challenge of investment in mechanization. There is also a chance that a mistake in one crop will affect the other. Given the variety of operations involved in mixed farming, the farmer must be ready to increase his knowledge in mixed farming. Ramnath (2020) noted that the fundamental disadvantage of mixed farming is its constrained ability to produce a crop in commercial quantity because of space constraints. According to Clement et al. (2010), most technical advancements to boost livestock productivity at the household level put more labour on women, particularly when it comes to feeding animals. For instance, stall feeding with no grazing has increased women's workload by 1–2 hours daily in weed-cutting and feed-mixing in the trans.

Crop residues are utilized not just as feed but also as mulch, fuel, and construction material. Thus, the integrated crop–livestock system's fundamental nature sometimes suggests conflicting uses for crop wastes (Erenstein 2002). These trade-offs frequently rely on the socio-economic situation and setting, including the supply and demand for agricultural waste and farmers' preferences (Giller et al. 2011; Erenstein 2011; Andrieu et al. 2015). Intensive pasture grazing may cause species with short life cycles and low productivity to predominate on rangelands. This may impact agricultural nutrient transfers and animal access to crop residue, decreasing livestock output (Rufino et al. 2006). More strain is placed on the feed resources due to the rising demand for fodder, a lack of arable land and water, and resources (Notenbaert et al. 2013). This also suggests that an integrated crop–livestock system might result in an inequitable distribution of nutrients from manure and urine, resulting in uneven crop development. The main drawbacks of integrated crop–livestock systems are continuous labour and infrastructure needs, expensive capital expenditures, and higher nutrient losses due to intense recycling.

7.9 Conclusion

Sustainable intensification (SI) is a production system that increases output without causing significant environmental damage. It emphasizes the end over the means and does not pre-determine technologies, species mix, or design components. Historically, SI was believed to reduce demand for converting natural areas to agriculture. Still, it has also led to adverse effects like biodiversity loss, decreased climate adaptation, and water contamination. The current focus on SI involves enhancing agricultural land production while managing its environmental impact. Evaluation frameworks have expanded to include non-environmental aspects, such as social concerns, economics, and the human condition. The agricultural sustainability assessment now relies on indicator frameworks structured into five domains: productivity, economic, environment, human condition, and social.

Meanwhile, mixed farming systems (MFS) are agricultural production systems where farmers produce crops and animals in the exact location under the same ownership. Sustainable MFS provides enough food for consumers and income for farmers while ensuring soil fertility, biodiversity, and pest control. Several characterizations and typologies of MFS have been identified in West Africa based on leading crop and livestock production systems. However, the level of development varies due to farmers' preferences for resource distribution. Socio-economic factors, such as skills, competencies, agricultural knowledge and innovation, economy, and policy environment, pose significant challenges to MFS growth in West Africa.

References

Agyemang K et al (1997) Village N'Dama cattle production in West Africa: six years of research in The Gambia. ILRI (International Livestock Research Institute)/ITC (International Trypanotolerance Centre), Nairobi/Banjul. 131 pp

Alexandratos N, Bruinsma J (2012) World agriculture towards 2030/2050: the 2012 revision. ESA Working Paper No. 12–03. FAO, Rome

Allen VM et al (2007) Integrated irrigated crop-livestock systems in dry climates. Agron J 99:346–360

Andrieu N et al (2015) From farm scale synergies to village scale trade-offs: cereal crop residues use in an agropastoral system of the Sudanian zone of Burkina Faso. Agric Syst 134:84–96

Badiane O, Gaye M, Mendoza M, Govindan K (1997) Market liberalization and adjustment of the groundnut sector in Senegal. Survey report. IFPRI (International Food Policy Research Institute)/ISRA (Instut Senegalaise des recherches agronomiques), Washington, DC/Dakar, 94 pp

Bell LW, Moore AD, Kirkegaard JA (2014) Evolution in crop-livestock integration systems that improve farm productivity and environmental performance in Australia. Eur J Agron 57:10–20

Bell LW, Harrison MT, Kirkegaard JA (2015) Dual-purpose cropping–capitalizing on potential grain crop grazing to enhance mixed-farming profitability. Crop Pasture Sci 66:i–iv

Berndes G, Bird N, Cowie A (2010) Bio-energy: land use change and climate change mitigation. IEA bioenergy. ExCo, Dallas, p 03

Bockstaller C, Girardin P (2003) How to validate environmental indicators. Agric Syst 76(2):639–653
Clement F et al (2010) Livestock Water Productivity in the Ganga Basin, Completion report of the "Improving water productivity, reducing poverty and enhancing equity in mixed crop-livestock systems in the Indo-Gangetic Basin" CPWF Project (PN 68). CGIAR Challenge Program on Water and Food (CPWF)
Conway GR (1994) Sustainability in agricultural development: trade-offs with productivity, stability and equitability. J Farm Syst Res Ext 4(2):1–14
Conway G (1997) The doubly green revolution: food for all in the twenty-first century. Cornell University Press
Cook S (2015) Sustainable intensification revisited, IIED Issue paper. IIED
Darnhofer I, Gibbon D, Dedieu B (2012) Farming systems research: an approach to inquiry. In: Gibbon D (ed) Farming systems into the 21st century: the new dynamic. Springer, Dordrecht, pp 3–31
de Faccio PC et al (2010) Managing grazing animals to achieve nutrient cycling and soil improvement in no-till integrated systems, nutrient cycle. Agroecosyst 88:259–273
de Leeuw PN (1997) Crop residues in tropical Africa: trends in supply, demand, and use. In: Renard C (ed) Crop residues in sustainable mixed crop/livestock farming systems. CAB International, Wallingford, pp 41–77
Dixon J, Gulliver A, Gibbon D (2001) Global farming systems study: challenges and priorities to 2030. Synthesis and global overview. FAO (Food and Agriculture Organization of the United Nations), Rome, Italy. 98 pp. ESRI (Environmental Systems Research Institute). 1999. ArcView GIS. Version 3.2®
Dove H et al (2015) Integrating dual-purpose wheat and canola into high-rainfall livestock systems in southeastern Australia. 2. Pasture and livestock production. Crop Pasture Sci 66:377–389
Dynes R et al (2010) Canterbury farming: production, processing and farming systems. In: Proceedings of the New Zealand Grassland Association
EIP-AGRI Focus Group (2017) Mixed farming systems: livestock/cash crops Final Report May
Erenstein O (2002) Crop residue mulching in tropical and semi-tropical countries: an evaluation of residue availability and other technological implications. Soil Tillage Res 67:115–133
Erenstein O (2011) The paradox of limited maize stover use in India's smallholder crop-livestmock systems. Exp Agric 47:677–704
FAO (2001) Mixed crop-livestock farming: a review of traditional technologies based on literature and field experience. FAO Animal Production and Health Papers 1521
FAO, Rome, Italy. http://apps.fao.org. Accessed Feb 2003
FAO (2011) Save and Grow A policymaker's guide to the sustainable intensification of smallholder crop production. FAO, Rome
Fernández-Rivera S et al (2004) Classification and description of the major farming systems incorporating ruminant livestock in West Africa. Sustainable crop–livestock production in West Africa, Conference Paper. https://hdl.handle.net/10568/50278
Fianu FK, Addae PC, Adjorlolo L (1994) Sheep rearing under tree crop plantation in Ghana's forest zone: problems and prospects. In: Lebbie SHB, Kagwini E (eds) Small ruminant research and development in Africa (July 1996), Proceedings of the third biennial conference of the African Small Ruminant Research Network, Kampala, Uganda, 5–9 December 1994. ILRI (International Livestock Research Institute), Nairobi, pp 87–91
Franzluebbers A (2007) Integrated crop–livestock systems in the southeastern USA. Agron J 99:361–372
Gil J, Siebold M, Berger T (2015) Adoption and development of integrated crop-livestock-forestry systems in Mato Grosso, Brazil. Agric Ecosyst Environ 199:394–406. https://doi.org/10.1016/j.agee.2014.10.008
Giller KE, Tittonell P, Rufino MC, Van Wijk MT, Zingore S, Mapfumo P, Adjei-Nsiah S, Herrero M, Chikowo R, Corbeels M (2011) Communicating complexity: integrated assessment of

trade-offs concerning soil fertility management within African farming systems to support innovation and development. Agric Syst 104:191–203
Gunton RM (2016) How scalable is Sustainable Intensification? Nat Plants 2:Art no. 16065
Gustafson D (2016) Seven food system metrics of sustainable nutrition security. Sustain For 8(3):196
Harrington L et al (2009) Cross-basin comparisons of water use, water scarcity and their impact on livelihoods: present and future. Water Int 34:144–154. https://doi.org/10.1080/02508060802661584
Herrero M et al (2012) Drivers of change in crop–livestock systems and their potential impacts on agroecosystems services and human wellbeing to 2030. A study commissioned by the CGIAR system wide livestock Programme, Addis Ababa; 2012
Iiyama M et al (2008) Livelihood diversification strategies, incomes and soil management strategies: a case study from Kerio Valley, Kenya. J Int Dev 20(3):380–397
Innazent A et al (2022) Farm typology of smallholders integrated farming systems in Southern Coastal Plains of Kerala, India. Sci Rep 12(333). https://doi.org/10.1038/s41598-021-04148-0
Kumar S et al (2014) Integrating crop and livestock management for enhanced productivity, profitability and sustainability of the rice-wheat system in North West India. Int J Life Sci Biotechnol Pharma Res:374
Kuria A et al (2014) Local knowledge of farmers on opportunities and constraints to sustainable intensification of crop–livestock-trees: mixed systems in Lemo Woreda, Southern Nations Nationalities and People Region (SNNPR), Ethiopian highlands. A field study report
Liebig MD et al (2012) Integrated crops and livestock in central North Dakota, USA: agroecosystem management to buffer soil change. Renew Agric Food Syst 27:115–124
Liniger H et al (2011) Sustainable land management in practice: guidelines and best practices for Sub-Saharan Africa: field application. FAO, Rome
Loos J et al (2014) Putting meaning back into "sustainable intensification". Front Ecol Environ 12:356–361. https://doi.org/10.1890/130157
López-Ridaura S, Masera O, Astier M (2002) Evaluating the sustainability of complex socio-environmental systems. The MESMIS framework. Ecol Indic 2(1):135–148
López-Ridaura S, van Keulen H, van Ittersum MK, Leffelaar PA (2005) Multiscale methodological framework to derive criteria and indicators for sustainability evaluation of peasant natural resource management systems. Environ Dev Sustain 7(1):51–69
Mahon N et al (2017) Sustainable intensification–"oxymoron" or "third-way"? A systematic review. Ecol Indic 74:73–97
Maitland K (2019) Intensive mixed farming: getting the best from cropping and stock. National Land care programme, SANTFA The Cutting Edge WINTERMaitland
Manyong VM (2002) Economic research at IITA for the improvement of agriculture in the sub humid and humid zones of West Africa. In: Sakurai T, Furuya J, Takagi H (eds) Economic analyses of agricultural technologies and rural institutions in West Africa: achievements, challenges, and application to rice farming research. Proceedings of JIRCAS international workshop, 12–13 July 2001. JIRCAS (Japan International Research Center for Agricultural Sciences) Working Paper Report 25, Tsukuba, pp 37–58
Maughan MW et al (2009) Soil quality and corn yield under crop–livestock integration in Illinois. Agron J:101–1503–1510
McKenzie SC et al (2017) Integration of sheep grazing for cover crop termination into market gardens: agronomic consequences of an ecologically based management strategy. Renew Agric Food Syst 32:389–402
Mekuria W, Mekonnen K (2018) Determinants of crop–livestock diversification in the mixed farming systems: evidence from central highlands of Ethiopia. Agric Food Secur 7(1):1–15
Musumba M et al. (2017) Guide for the sustainable intensification assessment framework. Feed the Future/USAID
Niles MT, Garrett RD, Walsh D (2018) Ecological and economic benefits of integrating sheep into viticulture production. Agron Sustain Dev 38:1

Notenbaert A et al (2013) Identifying recommendation domains for targeting dual-purpose maize-based interventions in crop-livestock systems in East Africa. Land Use Pol 30:834–846

Okike I (2002) Transregional analysis of crop-livestock systems: understanding intensification and evolution across three continents – the case of West Africa. A consultant's report submitted to IITA (International Institute of Tropical Agriculture), Ibadan, Nigeria, and ILRI (International Livestock Research Institute), Nairobi, Kenya. 70 pp

Olsson JA et al (2009) A goal-oriented indicator framework to support integrated assessment of new policies for agri-environmental systems. Environ Sci Pol 12(5):562–572

Onwuka CFI, Taiwo BBA, Adu IF (1992) Browse species and supplements used for feeding small ruminants in Ogun State, Nigeria. In: Stares JES, Said AN, Kategile JA (eds) The complementarity of feed resources for animal production in Africa. Proceedings of the Joint Feed Resources Networks Workshop held in Gaborone, Botswana, 4–8 March 1991. African Feeds Research Network. ILCA (International Livestock Centre for Africa), Addis Ababa, pp 173–180

Peterson CA, Deiss ACL (2020) Gaudin Commercial integrated crop-livestock systems achieve comparable crop yields to specialized production systems: a meta-analysis. PLoS One 15:e0231840

Ponnusamy K, Devi MK (2017) Impact of integrated farming system Approach on doubling farmers' income. Agric Econ Res Rev:30

Powell JM, Williams TO (1995) An overview of mixed farming systems in sub-Saharan Africa. In: Powell JM, Fernandez-Rivera S, Williams TO, Renard C (eds) Livestock and sustainable nutrient cycling in mixed farming systems of sub-Saharan Africa. Volume II: Technical papers. ILCA, Addis Ababa, pp 21–36

Pretty J (1997) The sustainable intensification of agriculture. Nat Resour Forum 21(4):247–225

Pretty J, Bharucha ZP (2014) Sustainable intensification in agricultural systems. Ann Bot 114(8):1571–1596). Oxford University Press. https://doi.org/10.1093/aob/mcu205

Ramnath KR, Mukherjee A, Singh DK, Shubha K, Kumar U (2020) Mixed farming: a viable option for sustainable agriculture. Food Sci Rep 1(6):75–78

Reardon T et al (1995) Promoting farm investment for sustainable intensification of african agriculture. Food Security International Development Papers

Rockstrom J, Karlberg L, Wani SP, Barron J, Hatibu N, Oweis T et al (2010) Managing water in rain fed agriculture-the need for a paradigm shift. Agric Water Manag 97:543–550. https://doi.org/10.1016/j.agwat.2009.09.009

Rodriguez DH, deVoil CP, Power B (2014) A participatory whole farm modelling approach to understand impacts and increase preparedness to climate change in Australia. Agric Syst 126:50–61

Rufino MC, Rowe EC, Delve RJ, Giller KE (2006) Nitrogen cycling efficiencies through resource-poor African crop–livestock systems. Agric Ecosyst Environ 112:261–282

Ryschawy J et al (2012) Mixed crop-livestock systems: an economic and environmental-friendly way of farming? Animal 6(10):1722–1730. The Animal Consortium. https://doi.org/10.1017/S1751731112000675

Saleem M (1998) Nutrient balance patterns in African livestock systems. Agric Ecosyst Environ 71:241–254

Salton JC et al (2014) Integrated crop-livestock system in tropical Brazil: toward a sustainable production system. Agric Ecosyst Environ 190:70–79

Sekaran U et al (2021) Role of integrated crop-livestock systems in improving agriculture production and addressing food security – a review. J Agric Food Res 5(2021):100190

Seré C, Steinfeld H, Gronewold J (1996) World livestock production systems: current status, issues and trends. In: FAO animal production and health paper 127. FAO (Food and Agriculture Organization of the United Nations), Rome. 84 pp

Shukla R et al (2019) Farmer typology to understand differentiated climate change adaptation in Himalaya. Sci Rep 9:1–12

Sulc RM, Franzluebbers AJ (2014) Exploring integrated crop–livestock systems in different ecoregions of the United States. Eur J Agron 57:21–30

Sulc RM, Tracy BF (2007) Integrated crop–livestock systems in the US Corn Belt. Agron J 99:335–345

The Montpellier Panel (2013) Sustainable intensification: a new paradigm for African Agriculture, London

The Rise Foundation (2014) The sustainable intensification of European agriculture. A review Sponsored by the RISE Foundation. Brussels

Thornton PK et al (2002) Mapping poverty and livestock in the developing world. ILRI (International Livestock Research Institute), Nairobi. 124 pp. Tittonel 2014

Van Cauwenbergh N et al (2007) SAFE – A hierarchical framework for assessing the sustainability of agricultural systems. Agric Ecosyst Environ 120(2):229–242

Van Wijk M et al (2009) Identifying key entry-points for strategic management of smallholder farming systems in sub-Saharan Africa using the dynamic farm-scale simulation model NUANCES-FARMSIM. Agric Syst 102:89–101. https://doi.org/10.1016/j.agsy.2009.07.004

Veysset P et al (2014) Mixed crop–livestock farming systems: a sustainable way to produce beef? Commercial farms results, questions and perspectives. Animal 8:1218–1228

Villano R, Fleming E, Fleming P (2010) Evidence of farm-level synergies in mixed-farming systems in the Australian Wheat-Sheep Zone. Agric Syst 103:146–152

Williams TO, Hiernaux P, Fernández-Rivera S (2000) Crop- livestock systems in sub-Saharan Africa: determinants and intensification pathways. In: McCarthy N, Swallow B, Kirk M, Hazell P (eds) Property rights, risks, and livestock development in Africa. IFPRI (International Food Policy Research Institute)/ILRI (International Livestock Research Institute), Washington, DC/Nairobi, pp 132–149

Wilson ASB, Lansbury TJ (1958) Centrosema pubescens: ground cover and forage crop in cleared rainforest of Ghana. Empire J Exp Agric 26:351–364

Wilson P, Lhoste J-M, Traore B (1989) Feeding strategies of livestock in the irrigated perimeters of The Niger Inland Delta, Mali. Anim Prod 48(1):57–67

Yaro JA (2010) The Social dimensions of adaptation to climate change in Ghana. Economics of adaptation to climate change. World Bank Discuss Pap Number 15

Chapter 8
The Crop–Livestock–Soil Nutrient–Water Nexus in Mixed Farming System: A Research Gap in West Africa Region

H. E. Igbadun, O. A. Ojeleye, Tafadzwanashe Mabhaudhi, and O. Cofie

Abstract Mixed farming systems are a sustainable closed-loop model for crop–livestock systems, focusing on structure, practices, logic, social, cultural, economic, climatic, and institutional capital interactions. However, literature on crop–livestock–soil–water interactions is scarce, with few separating crop and livestock water productivity in mixed systems. Crop–livestock water productivity is influenced by land management, biophysical, and socioeconomic factors. Integrating thinking in space and time can help understand these interrelationships and analyse their influence on crop–livestock production. Improving livestock water productivity (LWP) is crucial due to rising consumer demand, competition for global freshwater, and water rivalry. Strategies include promoting livestock production, managing grazing, water, livestock marketing, animal health, and minimizing environmental effects.

Keywords Nexus planning · Sustainable development · Inter-cropping · Economic returns · Crop productivity

H. E. Igbadun (✉)
Department of Agricultural and Bioresources Engineering, Ahmadu Bello University, Zaria, Nigeria

O. A. Ojeleye
Department of Agricultural Economics, Ahmadu Bello University, Zaria, Nigeria

T. Mabhaudhi
Centre on Climate Change and Planetary Health, London School of Hygiene and Tropical Medicine, London, UK

Centre for Transformative Agricultural and Food Systems, University of KwaZulu-Natal, Carbis Road, Scottsville, Pietermaritzburg, KwaZulu-Natal, South Africa

O. Cofie
International Water Management Institute (IWMI), West Africa Regional Office, Accra, Ghana

T. Mabhaudhi et al. (eds.), *Enhancing Water and Food Security Through Improved Agricultural Water Productivity*,
https://doi.org/10.1007/978-981-96-1848-4_8

8.1 Introduction

In some regions of the world, mixed farming systems—which combine crop production and animal husbandry—are widespread. In comparison to specialized crop or livestock production systems, the mixed farming system can provide several advantages, such as increased crop yields due to the addition of manure and other nutrients from livestock, an improvement in soil health due to an increase in organic matter content, reduction in soil erosion, provision of a more stable source of income due to diversified production, and a reduction in crop failure risk due to the spreading of risks across multiple enterprises. With food insecurity challenges mounting in the world, exacerbated by the dictates of changing climate, exploiting the potential of an agricultural system for equity and resilience accessible through self-sustaining mixed farming systems is critical for increasing water management, productivity, farmers' economy, and food security. Affirmations from researchers see mixed farming systems as an agricultural system that uses resources more efficiently and a conscientious mechanism by which households can combat poverty and achieve food security. This agricultural system has the potential to provide and supplement up to half of West Africa's food supply (Murendo et al. 2019; Asante et al. 2018; Ryschawy et al. 2012; Hendrickson et al. 2008). Mixed farming systems combine cash crops and livestock on a farm, and it is regarded as a viable strategy for achieving sustainable agricultural system intensification. In the past, these systems were widely used (Mazoyer and Roudart 2006), but they are now attracting more attention on a global scale, especially for economic and environmental reasons. The highest level of interconnectedness across agricultural systems is seen in mixed crop–livestock systems (Ryschawy et al. 2012). This chapter looks at the nexus in mixed farming systems with a particular interest in the crop–livestock–soil–water nexus, which is still primarily a research gap in West Africa.

8.2 Overview of Nexus in Mixed Farming Systems

Nexus in mixed farming systems (MFS) is the sustainable closed-loop in the agricultural production system that uses concepts of structure, practices, logic, social, cultural, economic, climatic and institutional capital to explain the interactions between the principal elements. Four kinds of nexus in MFS which are prominent include the following:

(i) Crop–livestock–market nexus
(ii) Crop–livestock–soil nexus
(iii) Production–population–market–climate change nexus
(iv) Crop–livestock–soil–water nexus

8.3 Crop–Livestock–Market Nexus

This nexus commonly exists in agro-livestock systems emphasizing crop production and raising livestock or animals for economic returns. The production systems for crops and livestock are intertwined, as crops offer feed for livestock, and livestock manure can be used as fertilizer for crops. Additionally, the two systems share resources, including energy, water, and land, and by utilizing these synergies, integrated crop–livestock systems essentially aim to reduce the environmental harm caused by crop and livestock production. Early studies of MFS largely focused on the production mechanism, supply exchange between the crop and livestock, and the economic returns, with little emphasis on the interactive effect on the production environment. Hence, the focus was on using the livestock as a labour force, especially for the crop production system's tillage, haulage, and threshing operations. The crop residues are converted to feed (hays and silages) for the livestock, and the major products from the crops and animals are taken to the market. This is still largely the focal interest of the larger farming population who practice MFS in the West Africa Region.

However, integrating crop and livestock systems could benefit both production and environmental goals. The integration shares the common goal of optimizing the use of resources and minimizing environmental impacts. It could offer the benefits of increased productivity, improvement of soil health, reduction of greenhouse gas emissions, increased water use efficiency, improved animal welfare, and increased economic benefits, hence the need to consider the broader interaction in the production system.

8.4 Crop–Livestock–Soil Nexus

Advancement in MFS brought to the fore the production environment, especially the interaction of crop–livestock production and the consequences on the soil. Crops depend on soil for nutrients, water, and anchorage. Soil, in turn, is affected by crop production practices, such as tillage, fertilization, and crop rotation. Healthy soils are essential for sustainable crop production. The crop-soil nexus involves several key factors that influence crop productivity and soil health, which include nutrient cycling, soil erosion, soil organic matter, and soil water management. Animal grazing on vegetation significantly affects the soil, exposing the land to erosion when not controlled. Animal droppings are manure to the soil, increasing the soil carbon content, aiding water infiltration, enhancing water retention capacity, reducing runoff volume and erosion effect, and increasing nutrient availability and crop utilization. Animal droppings are also potential media for dispersing and spreading weeds, pests and diseases (References). Since soil fertility is frequently historically maintained by fallowing, nutrient availability is a significant constraint for developing countries' agricultural productivity. In mixed farms, nutrient cycling enables various agricultural production processes to reintegrate. Improvements in soil nutrients on mixed farms and nutrient transfers by animals to croplands are possible by controlling soil nutrient materials.

As shown in Fig. 8.1, Wolfert (2002) illustrated a material cycle for a mixed farming system. The soil is grey-coloured and regarded as the pivot on which everything changes or recycles (Neher 1999). When leguminous crops are involved, they absorb nutrients from the soil and extra nitrogen from the air. Animals use crops, and humans use them as food or animal feed. The additional outside feed may be consumed by animals as well. Animal products like milk or meat are taken off the farm. Animals also create manure, which may be added to the soil with external manure. Volatilization is one potential cause of nutrient loss in the agricultural system. The same is true of the processes in the soil known as leaching and denitrification.

It can be seen that mixed farming is favourable for ecological considerations in the farming system. With the farmer's complete material cycle under his control, the management of nutrients is positively stimulated, and the whole process chain or cycle can be optimized. A larger carrying capacity and higher yields will result from this. Production that is both ecologically and economically sustainable can then be combined.

In addition, Wolfert (2002) argued that all farm management activities entail using soil, crops, animals, and waste. These operations can influence the material cycle and nutritional balance in the soil. Potatoes, wheat, and milk are essential agricultural goods, but residual products like manure and crop wastes can become

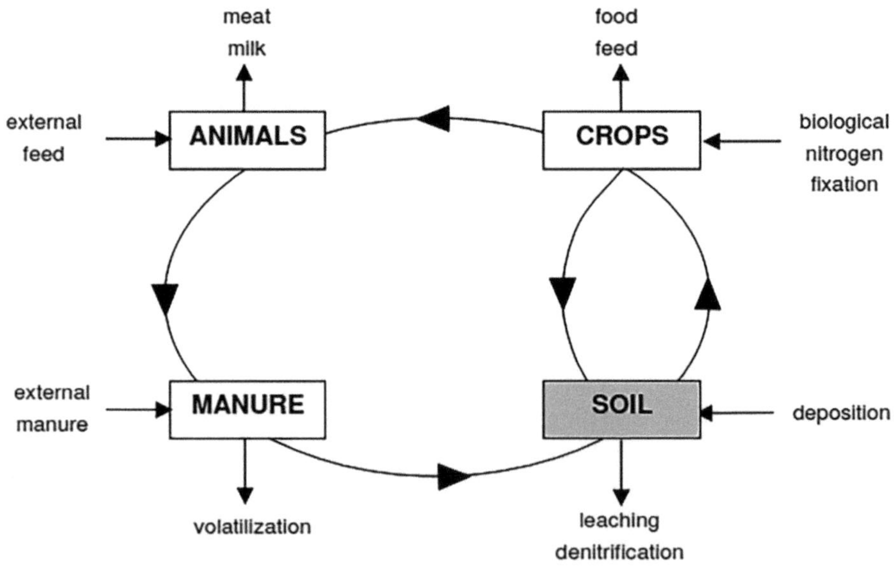

Fig. 8.1 The farm-level system's material cycle is long, including animals and manure, and a short cycle involving crops and soil. (Source: Wolfert 2002)

significant. In the manufacturing cycles that correspond to the ecological material cycles, they are intermediate goods. Product quality, which comprises several characteristics (such as nitrogen content, colour, and structure), impacts how well the cycle works.

The main actors in the mixed agricultural system were expanded by Clement et al. (2010) to include a new variable that had not been mentioned before. This water productivity variable affects the production systems for crops and animals individually and together. They believed enhancing MFS impacted the broader goal of ensuring food and water security. Over 80% of farmers in all field locations reported experiencing water shortage on a seasonal or regular basis, according to their experience in the mixed crop–livestock systems of the Indo-Gangetic Basin.

In the study location with the highest rainfall, over 50% of the population faced water shortage all year round, and the remaining population experienced considerable shortage on a seasonal basis. Similar scenarios exist in West Africa, where preserving food security necessitates access to water in various forms. Cofie (2022) argued that West Africa is exposed to particularly high-water risk, facing hunger, sickness, energy shortages, and poverty owing to many variables caused by water scarcity, pollution, and floods. Clement et al. (2010) thus stated that water shortage is a fundamental restriction for expanding mixed farming systems. Improving the water productivity of such systems through efficient, equitable, and sustainable use of resources thus holds a great potential to contribute to enhanced livelihoods and reduce poverty.

8.5 Production–Population–Market–Climate Change Nexus

This nexus addresses the bio-social, economic, and bio-physical dynamics interconnected in mixed farming systems. These four elements are all connected and have a variety of interactions. This nexus began to gain research attention as more thought was given to the interaction among the elements of mixed farming. For instance, a growing population induces a greater need for food, necessitating increased crop and livestock production. As a result of this rise in production, greenhouse gas emissions may also rise, furthering the effects of climate change. Several other detrimental effects of climate change on production include decreased crop yields, increased extreme weather events, and more challenging access to water sources. Food insecurity, economic instability, and social unrest may follow from these effects (References).

Achieving food and nutritional security seems to conflict with the need to reduce agriculture's detrimental environmental impact due to the numerous demands of the expanding and increasingly prosperous human population. The reduction in the diversity of agricultural systems at the soil, landscape, and farm scales that results from the simplification of agroecosystems may significantly contribute to this paradox (Lemaire et al. 2014). How the animals are raised is to blame for the negative impacts of livestock on the environment, and such problems may be resolved

(Dalibard 1995). Separating crops and animals in various climates and regions is a significant factor in the loss of variety at the farm scale, with the ensuing negative effects on the ecosystem. This is one of the challenges mixed farming is advocated to control. However, such a biodiversity loss also causes carbon cycling to become independent of the cycles of water, nitrogen, phosphorus, and sulphur (Lal 2010). According to Steinfeld et al. (2006), livestock plays a significant role in land use and land use change, and this relationship should be controlled in a sustainable farming system. Concentrating ruminants tend to uncouple the carbon and nitrogen cycles, releasing digestible carbon as CO_2 and CH_4 and digestible nitrogen in waste as N_2O, which increases greenhouse gas emissions (Soussana and Lemaire 2014). Because CH_4 and N_2O have a high potential for global warming, greenhouse gases significantly impact climate change, but they can be reduced by combining cattle with crops and trees (Lal 2020). Goldstein et al. (2012) further suggested establishing vegetation buffers on agricultural fields to boost biodiversity and preserve soil and water. That is, agroforestry or alley cropping can also diminish the environmental footprint of livestock raised on the same land unit.

The paths for reducing the environmental impact of livestock products are shown schematically in Fig. 8.2. In theory, consuming animal products with smaller carbon footprints would positively influence the environment and climate (Lal 2020). In West Africa, livestock is an essential part of agroecosystems. By implementing

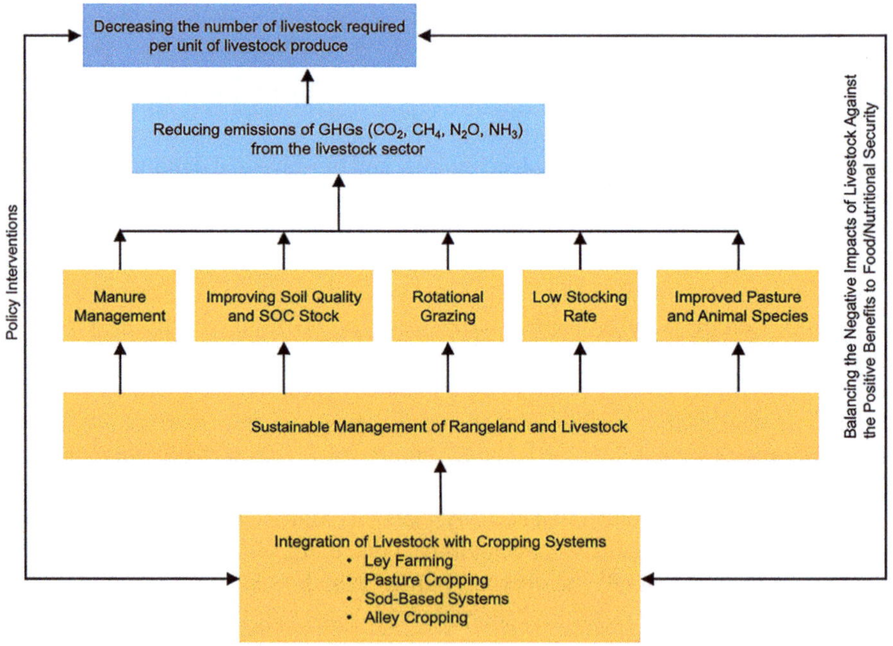

Fig. 8.2 A flow chart depicting livestock integration with arable land use for decreasing the number of livestock required. (Source: Lal 2020)

creative mixed farming practices, productivity may be increased while leaving a smaller environmental imprint. Numerous co-benefits result from skilfully integrating crops and livestock within the same terrain (Gil et al. 2015). The primary contributions of crops are subsistence consumption, household income, animal feed, fuel energy, export revenues, and national returns. Crop production depends on climate, seeds, water, soil nutrients, biodiversity, and farmers' technical skills (Godfray et al. 2010). However, livestock offers a variety of benefits, including food, fuel, manure, draught power, reproduction, societal advantages, and transportation (Belay et al. 2012).

The environmental feature of mixed crop–livestock systems had the most diversified agricultural land usage, Ryschawy et al. (2012) stated, since this is known to improve bird and insect biodiversity through spatial heterogeneity (Fahrig et al. 2011). According to Schiere and Kater (2001) and Russelle et al. (2007), mixed crop–livestock systems have a lower risk of nitrogen contamination than other systems. One technique for reducing external inputs in autonomous systems is nutrient cycling (Schiere et al. 2002).

8.6 Crop–Livestock–Soil–Water Nexus

Crops, livestock, soil, and water are all interdependent, and changes in one can have a knock-on effect on the others. For example, water use for irrigation can lead to soil salinization, reducing crop yields. Similarly, overgrazing of livestock can lead to soil erosion, reducing water infiltration and increasing the risk of flooding.

Literature on crop–livestock–soil–water interactions analysis is scarce, except for those few that separate crop and livestock water productivity in mixed systems. Knowledge gaps in CLWS nexus include mixed farming system's profitability and productivity level under rain-fed, irrigation, or supplementary irrigation; the best combinations of livestock-based nutrients, water application regimes, and crop types/varieties for optimum yield and water use efficiency; the water productivity in livestock production, and how that can be optimized; and the combinations of practices and degree of intensity of mixed farming that can impact water availability vis-à-vis climate variabilities.

Crop or animal production each requires water for efficiency and effectiveness. Combining and integrating both practices will put a demand on water requirement and availability amid climate change vagaries. It is important to understand how MFS strategies respond to combinations of field water management strategies and how such measures can improve production and the overall effect on water use efficiency (WUE).

8.7 Water Productivity in Mixed Farming System Nexus

Soil nutrients, crop yield, pasture production, crop harvest, feed storage, grazing, feeding, and manure handling are a few of the complex operations involved in mixed farming, which depend significantly on water and water productivity. These aspects are all intrinsically tied to each other. Increased water productivity necessitates a thorough understanding of the biophysical, socioeconomic, and environmental aspects of the field, farm, and basin scale, according to Mengistu et al. (2012) in their reviewed study on increasing livestock water productivity under rain-fed mixed crop/livestock farming scenarios of sub-Saharan Africa. Reports from studies on mixed agricultural systems and water production are few. For instance, Khalil and Ahmed (2012) compared crops' water efficiency with dairy buffalo in Egypt. They claimed that milk production uses less water than cash crops do. Kebebe et al. (2014) revealed differences in livestock water productivity (LWP) among agricultural systems and wealth categories in their study on techniques for enhancing the water usage efficiency of livestock production in rain-fed systems in Ethiopia. Therefore, water use efficiency might be increased by removing resource restrictions such as lack of access to agricultural labour and livestock resources, especially oxen.

Furthermore, Bekele and Mengistu (2017) used a multistage selection technique to choose farmers from various wealth statuses to perform a study in the northeast of Addis Ababa to ascertain feed and livestock water production in a mixed farming system managed by smallholders. The study found that most Ethiopian farmers predominantly use the rain-fed mixed farming system to satisfy their subsistence needs; small-scale irrigated farming is only practised by a select few farmers along the Beressa River in sparsely populated areas. They said that ownership of significant agricultural resources, such as cropland and cattle, determines wealth status. These resources control livestock outputs, feed water productivity, and livestock water productivity.

In the dry lands of Eastern Kenya, Esilaba et al. (2020) conducted research to improve soil and water use efficiency to increase crop and livestock productivity. They developed options for managing soil, water, and nutrients in integrated cropping-livestock systems that Kenyan farmers may choose to implement. For on-farm demonstration, field tests in two watersheds were conducted. Various tillage techniques, fertilizer treatment, fertilizer application, and water treatment techniques were used. The daily water needs of livestock and crops differ greatly depending on the kind of animals, crop type, and crop variety. The amount of water utilized each day will be strongly influenced by an animal's size and the stage of crop development. Consumption rates may be affected by environmental and management factors. The amount of water in an animal's diet will affect how much it drinks. The drinking water needed for feed with relatively high moisture content is reduced.

Based on the literature reviewed reviewed, knowledge gaps remain as to what are the best combinations of livestock-based nutrients, water application regimes, and crop varieties for optimum yield and water use efficiency; there is yet to be an established agronomic practice for farmers because of the complex nature of the interactions within factors, coupled with uncontrollable and unpredictable environmental influences which can obscure the anticipated learning outcomes.

8.8 Crop Water Productivity (CWP) in Mixed Crop–Livestock Systems

Numerous factors, which may be divided into plant-specific, climate-related, and management-related factors, affect crop water production (Kilemo 2022). These variables influence crop output and biomass build-up per unit of seasonal evapotranspiration. CWP, as a ratio of yield to water consumed, can be assessed by examining what happens at the numerator (yield) and the size of the denominator (water consumed). If the numerator is greater than the denominator, a low CWP will be attained. Increasing CWP entails decreasing the denominator while increasing the numerator. Remembering that plants must utilize some water to produce a harvest is crucial. The CWP index aims to maximize yield with the least amount of water feasible. Water conservation can boost total water productivity by reducing wasteful water losses (runoff, evaporation, conveyance losses, and deep percolation) and improving the effectiveness of the corresponding system components (Descheemaeker et al. 2010).

Crop water productivity analyses are standard in crop farming systems but not common in mixed farming systems. According to Burke et al. (1999), crop water productivity can be used to examine crop productivity increases that might occur due to greater water availability. It offers a quick way to determine whether water availability or other variables limit yield (Augus and van Herwaarden 2001). The impact and value of additional water supply can be evaluated using the unit increment in yield per unit of water use shown by crop water productivity. Various stakeholders in crop-water concerns employ a variety of definitions and formulations to represent crop-water productivity quantitatively. For instance, Ronald and Marlow (2002) expressed the idea of crop water productivity using three efficiency terms. The definitions of these terms include water use (technical) efficiency, which is the mass of agricultural produce per unit of water consumed; water use (economic) efficiency, which is the value of the product(s) produced per unit of water volume consumed; and water use (hydraulic) efficiency, which is the proportion of water used by irrigated agriculture for instance, to the volume of water supplied.

Other terms used in the concept of crop water productivity also include agronomic water use index (AWUI), defined as crop yield per volume of water input; crop water use index (CWUI), defined as crop yield per volume of water used by the crop (evapotranspiration) in production; and economic water use index (EWUI), defined as gross revenue per water input (Igbadun et al. 2006).

8.9 Livestock Water Productivity (LWP)

Globally, consumer demand for fresh water is increasingly becoming a concern. In developing countries, for instance, rain-fed mixed crop–livestock systems provide the mainstream of economic development. Improving livestock water productivity (LWP) has become imperative due to rising consumer demand for animal products, increased competition for the dwindling global freshwater, and within and trans-boundary water rivalry. In agricultural production, the idea of water productivity is well established. Although valuable in enhancing livestock output with more efficient water usage, livestock water productivity is still a relatively new and growing idea in the production of animals (Daniel and Kevin 2019; Herrero et al. 2013; Thornton and Herrero 2010).

According to Peden et al. (2007), LWP is a reliable indicator of a livestock production system's capacity to transform available water into advantages for livestock. It is the ratio of livestock products and services to water depleted in production (Descheemaeker et al. 2011). The type, quality, and quantity of forage/feed crops produced, the amount of water used to grow these feeds, the productivity level of the animal using these feeds, which may be influenced by breed, the health and management conditions of the animals, the standard of veterinary services, and other socioeconomic incentives, have been identified as the main factors that directly affect LWP (Peden et al. 2007).

A conceptual framework for comprehending livestock water productivity in mixed crop–livestock systems is shown in Fig. 8.1. The framework considers the several advantages of livestock, the various water flows involved, and the numerous factors affecting LWP, which are biophysical but also institutional and socioeconomic. Based on the model developed by Peden et al. (2007), the modified framework for mixed crop–livestock systems in Fig. 8.3 combined the production system with hydrology. The framework is provided to better understand water consumption in mixed agricultural systems and systematically examine livestock–water interactions.

According to Descheemaeker et al. (2010), the system receives water from precipitation, surface water, and groundwater. Water is utilized for servicing, drinking, processing, and producing biomass. It enables the system to generate animal outputs while utilizing various feeds and other natural resources and inputs. Then, animal by-products help provide livelihoods and environmental services. If handled well, e.g. if sufficient amounts of manure are put into fields, its contribution is beneficial; nevertheless, if poorly managed, it might be detrimental (e.g. if manure and urine contaminate watering points). In some way, the water that enters the system also exits it. Depleted water flows are those caused by transpiration, evaporation, polluted water, and degraded runoff water and cannot be replenished by the system. However, deep percolation under the root zone and non-degraded discharge may be utilized by other systems later or after recycling.

The green area represents the livestock portion of crop–livestock farming systems, emphasizing animal outputs and feeds. The many water inputs and outflows

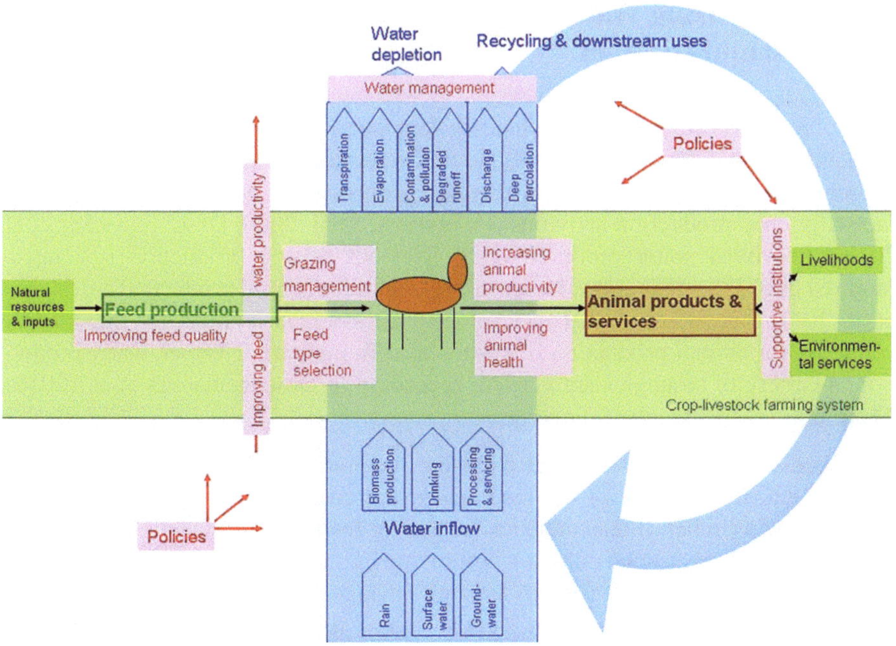

Fig. 8.3 Foundation for livestock water productivity in mixed crop and livestock production systems. (Built upon Peden et al. 2007)

that make up the system's water balance are denoted by blue arrows. The pink is the strategy that can assist in boosting LWP in the nexus.

Descheemaeker et al. (2010) identified nine strategies for improving LWP using the LWP framework (Fig. 8.3). These include (i) water management, (ii) choosing the right type of feed, (iii) feed quality, (iv) feed water productivity, (v) management of grazing, (vi) raising animal productivity, (vii) maintaining animal health, (viii) supportive institutions, and (ix) enabling policies. The strategies aimed at the biophysical elements of farming systems may also be divided into three categories: (a) animal management (increasing animal productivity and management of animal health), (b) water management, and (c) editing management (which includes enhancing feed quality, feed water productivity, feed type selection, and grazing management). Increased LWP will reverse land degradation and protect against environmental vulnerability while enhancing household nutrition, food security, and livelihoods (Bossio 2009; Peden et al. 2009).

Biophysical, socio-political, and economic factors influence the innovations required for LWP gains. To enhance biophysical circumstances, technical interventions must be linked with institutional and regulatory measures to assure acceptance. An evaluation of these initiatives is required; knowing the baseline situation will assist policymakers, development actors, and investors in making appropriate decisions to increase livestock water productivity.

8.10 Optimizing Water Productivity in a Mixed Farming System

Lack of feed, whose production is heavily dependent on scarce freshwater resources (Falkenmark and Rockstrom 2004), is the main barrier to livestock production, especially in developing countries. It is also progressively becoming a determining factor for crop production activities (Haileslassie et al. 2009; Descheemaeker et al. 2013). Improved water productivity of livestock-related production systems is required due to the anticipated increase in demand for animal products (Haileslassie et al. 2011; Kebebe et al. 2014). Exploring all available options for maximizing water use for animal production is worthwhile. To increase livestock water production, a variety of strategies are needed, according to Descheemaeker et al. (2010). These options include the following:

(i) Feed-related, which includes choosing the suitable types of feed, enhancing the nutritional value of the feed, and using crops that can be used for food, feed, and timber. It also includes increasing feed water productivity by choosing suitable crops and cultivars, improving agronomic management, and using more environmentally friendly grazing management techniques.
(ii) Water management practices include saving, harvesting, strategically placing and maintaining watering sites, and incorporating animal farming into irrigation plans.
(iii) Animal management comprises breeding for better health, disease prevention and control, and proper animal husbandry; also supported by educating livestock owners on how smaller, more productive herds may provide the same value.

8.11 Conclusion

Crop–livestock water productivity is influenced by several inextricably linked components, including land management, biophysical, and socioeconomic factors. So, addressing these issues with integrated thinking in space and time is necessary to comprehend the interrelationships and analyse the influence on crop–livestock production. Such an integration paradigm offers significant promise for analysing integrated crop–livestock farming systems. Attention to improving crop–livestock water productivity (CLWP) has become imperative due to rising consumer demand for crop and animal products, increased competition for the dwindling global freshwater, and within and transboundary water rivalry. Promoting MFS production, managing grazing land better, managing water better, increasing livestock marketing, enhancing soil health, and minimizing adverse environmental effects are all strategies that can increase CLWP. The MFS interalationship, which focuses on feed-related practices involving selecting appropriate feed types, enhancing nutritional value, and using suitable crops for food, feed, and water management involving water conservation, harvesting, and strategically placing watering sites, are recommended strategies.

References

Asante BO et al (2018) Determinants of farm diversification in integrated crop-livestock farming systems in Ghana. Renew Agric Food Syst 33(2):131–149

Augus JF, van Herwaarden AF (2001) Increasing water use and water use efficiency in dryland wheat. Agron J 93(2001):290–298

Bekele AM, Mengistu TB (2017) Livestock and feed water productivity in the mixed crop-livestock system. Animal, page 1 of 9 © The Animal Consortium 2017. https://doi.org/10.1017/S1751731117000416

Belay D, Azage T, Hegde BP (2012) Smallholder livestock production system in Dandi District, Oromia regional state. Central Ethiopia Glob Vet 8(5):472–479

Bossio D (2009) Livestock and water: understanding the context based on the 'comprehensive assessment of water Management in Agriculture'. Rangel J 31:179–186

Burke S, Mulligan M, Thornes JB (1999) Optimal irrigation efficiency for maximum plant productivity and minimum water loss. Agric Water Manag 40(2/3):377–391

Clement F et al (2010) Livestock water productivity in the Ganga Basin, completion report of the "improving water productivity, reducing poverty and enhancing equity in mixed crop-livestock systems in the Indo-Gangetic Basin" CPWF project (PN 68), CGIAR challenge program on water and food (CPWF)

Cofie O (2022) Exploring agricultural water challenges and opportunities in West Africa as climate change signals a new normal. Available at https://www.iwmi.cgiar.org/2022/05/exploring-agricultural-water-challenges-and-opportunities-in-west-africa-as-climate-change-signals-a-new-normal/

Dalibard C (1995) Livestock's contribution to the protection of the environment. In: Diouf J (ed) World animal review. Food and Agriculture Organization of the United Nations, pp 104–112. Available online at: http://www.fao.org/3/v8180t/v8180T14.htm

Daniel W, Kevin M (2019) Water requirements of livestock. Factsheet water requirements of livestock (gov.on.ca). Accessed 13 Dec 2021

Descheemaeker K, Amede T, Haileslassie A (2010) Improving water productivity in sub-Saharan Africa's mixed crop–livestock farming systems. Agric Water Manag 97:579–586. https://doi.org/10.1016/j.agwat.2009.11.012

Descheemaeker K et al (2011) Analysis of water productivity gaps and effects of interventions on livestock water productivity in mixed crop-livestock systems. Exp Agric 47:21–38

Descheemaeker K et al (2013) Increasing water productivity in agriculture. In: Boelee E (ed) Managing water and agroecosystems for food security. CAB International, Wallingford, pp 104–123. https://doi.org/10.1079/9781780640884.0104

Esilaba AO et al (2020) Enhancing soil productivity and water and nutrient use efficiencies in integrated cropping-livestock production systems in Kenya. Kenya Agricultural and Livestock Research Organization, Nairobi

Fahrig L et al (2011) Functional landscape heterogeneity and animal biodiversity in agricultural landscapes. Ecol Lett 14:101–112

Falkenmark M, Rockstrom J (2004) Balancing water for human and nature: the new approach to ecohydrology. Earthscan, London, pp 74–89

Gil J, Siebold M, Berger T (2015) Adoption and development of integrated crop-livestock-forestry systems in Mato Grosso. Brazil Agric Ecosyst Environ 199:394–406. https://doi.org/10.1016/j.agee.2014.10.008

Godfray HCJ et al (2010) Food security: the challenge of feeding 9 billion people. Science 327(812):812–818. https://doi.org/10.1126/science.1185383
Goldstein JH et al (2012) Integrating ecosystem-service tradeoffs into land-use decisions. Proc Natl Acad Sci USA 109:7565–7570. https://doi.org/10.1073/pnas.1201040109
Haileslassie A et al (2009) Livestock water productivity in mixed crop–livestock farming systems of the Blue Nile basin: assessing variability and prospects for improvement. Agric Syst 102:33–40
Haileslassie A et al (2011) Assessment of the livestock-feed and water nexus across a mixed crop-livestock system's intensification gradient: an example from the Indo-Ganga basin. Exp Agric 47:113–132
Hendrickson JR et al (2008) Principles of integrated agricultural systems: introduction to processes and definition. Renew Agr Food Syst 23:265–271
Herrero M et al (2013) The roles of livestock in developing countries. Animal 7(Supplement 1):3–18. https://doi.org/10.1017/S1751731112001954
Igbadun HE, Mahoo HO, Tarimo AA, Salim BA (2006) Irrigation scheduling scenarios studies for a maize crop in Tanzania using a computer-based simulation model. Agric Eng Int: CIGR J 8(1):1–14
Kebebe EG et al (2014) Strategies for improving water use efficiency of livestock production in rain-fed systems. Animal 9(5):908–916. © The Animal Consortium 2014. https://doi.org/10.1017/S1751731114003115
Khalil MAI, Ahmed AIM (2012) Water efficiency under Egypt's mixed (crop/livestock) farming system: 1. Water efficiency of milking buffalo compared to cash crops. Egyptian J Anim Prod 49(Suppl):119–129
Kilemo DB (2022) The review of water use efficiency and water productivity metrics and their role in sustainable water resources management. OALib J 9:1–21. https://doi.org/10.4236/oalib.1107075
Lal R (2010) Managing soils and ecosystems to mitigate anthropogenic carbon emissions and advance global food security. Bioscience 60:708–721. https://doi.org/10.1525/bio.2010.60.9.8
Lal R (Ed.) (2020) Soil and fertilizers: Managing the environmental footprint (1st ed.). CRC Press. https://doi.org/10.1201/9780429471049
Lemaire G et al (2014) Integrated crop-livestock systems: strategies to achieve synergy between agricultural production and environmental quality. Agric Ecosyst Environ 190:4–8. https://doi.org/10.1016/j.agee.2013.08.009
Mazoyer M, Roudart L (2006) A history of world agriculture from the neolithic age to the current crisis. Earthscan, New York
Mengistu A et al (2012) Increasing livestock water productivity under rain fed mixed crop/livestock farming scenarios of Sub-Saharan Africa: a review. J Sustain Dev 5(7) ISSN 1913-9063 E-ISSN 1913-9071 Published by Canadian Center of Science and Education. https://doi.org/10.5539/jsd.v5n7p1
Murendo C et al (2019) Linking crop and livestock diversification to household nutrition: evidence from Guruve and Mt Darwin districts Zimbabwe. World Dev Perspect 14:100104
Neher DA (1999) Soil community composition and ecosystem processes. Agrofor Syst 45:159–185
Peden D, Tadesse G, Misra A (2007) Water and livestock for human development. In: Molden D (ed) Water for food, water for life: a comprehensive assessment of water management in agriculture. Earthscan, IWMI, London, Colombo, pp 485–514
Peden D, Taddesse G, Haileslassie A (2009) Livestock water productivity: implications for sub-Saharan Africa. Rangel J 31:187–193. https://doi.org/10.1071/RJ09002
Ronald L, Marlow PE (2002) Agriculture water use efficiency in the United States. Paper presented at the US/China water resources management conference, National Water Management Engineer. Nat Res Conserv Serv, USDA, Washington. http://www.lanl.gov/chinawater/documents/usagwue.pdf site visited on 03/07/2004
Russelle MP, Entz MH, Franzluebbers AJ (2007) Reconsidering integrated crop-livestock systems in North America. Agron J 99:325–334

Ryschawy J et al (2012) Mixed crop-livestock systems: an economic and environmental-friendly way of farming? Animal 6(10):1722–1730. The Animal Consortium. https://doi.org/10.1017/S1751731112000675

Schiere JB, Ibrahim MNM, van Keulen H (2002) The role of livestock for sustainability in mixed farming: criteria and scenario studies under varying resource allocation. Agric Ecosyst Environ 90:139–153

Schiere H & Kater L (2001). Mixed crop-livestock farming: A review of traditional technologies based on literature and field experiences. FAO. https://books.google.com.ng/books/about/Mixed_Crop_Livestock_Farming.html?id=5DvbAAAAMAAJ&redir_esc=y

Soussana JF, Lemaire G (2014) Coupling carbon and nitrogen cycles for environmentally sustainable intensification of grasslands and crop-livestock systems. Agric Ecosyst Environ 190:9–17. https://doi.org/10.1016/j.agee.2013.10.012

Steinfeld H et al (2006) Livestock's long shadow: environmental issues and options. Food and Agriculture Organization of the United Nations, Rome

Thornton PK, Herrero M (2010) The inter-linkages between rapid growth in livestock production, climate change, and the impacts on water resources, land use, and deforestation. The World Bank

Wolfert S (2002) Sustainable agriculture: how to make it work? A modelling approach to support management of a mixed ecological farm, PhD thesis Wageningen Universiteit

Chapter 9
Using AquaCrop, DSSAT and the SIMPLE to Estimate Water Use of Underutilised Cereal in South Africa

T. N. M. Nzimande, Vimbayi G. P. Chimonyo, E. M. Wimalasiri, and Tafadzwanashe Mabhaudhi

Abstract The study compares yield, biomass and water use (WU) for maize, sorghum and millet simulated using three crop models of varying complexity: AquaCrop, DSSAT and the SIMPLE model. A standard set of crop parameters was used to develop crop files for all three models. Similar soil, climate and management descriptions from the Ukulinga Research Farm were used across the models. The performance of the three models was observed to be statistically different. Based on the mean bias error, all models overestimated yield, but the lowest overestimation was with AquaCrop (0.22 t/ha), followed by DSSAT (0.24 t/ha) and the SIMPLE model (0.69 t/ha). Other statistical indicators, namely, RMSE and R^2, illustrated that the simulation of yield and WP in AquaCrop was more satisfactory than DSSAT and the SIMPLE model. The study confirms that DSSAT requires relatively more input data but does not always perform more satisfactorily. Before their application, it is essential to calibrate crop growth parameters for local conditions or

T. N. M. Nzimande
Centre for Transformative Agricultural and Food Systems, School of Agricultural, Earth & Environmental Sciences, Pietermaritzburg, South Africa

V. G. P. Chimonyo (✉)
Centre for Transformative Agricultural and Food Systems, School of Agricultural, Earth & Environmental Sciences, University of KwaZulu-Natal, Pietermaritzburg, South Africa

International Maize and Wheat Improvement Centre (CIMMYT), Harare, Zimbabwe
e-mail: v.chimonyo@cgiar.org

E. M. Wimalasiri
Crops For the Future Research Centre, Semenyih, Malaysia

T. Mabhaudhi
Centre on Climate Change and Planetary Health, London School of Hygiene and Tropical Medicine, London, UK

Centre for Transformative Agricultural and Food Systems, University of KwaZulu-Natal, Carbis Road, Scottsville, Pietermaritzburg, KwaZulu-Natal, South Africa

T. Mabhaudhi et al. (eds.), *Enhancing Water and Food Security Through Improved Agricultural Water Productivity*,
https://doi.org/10.1007/978-981-96-1848-4_9

use parameters from local field studies when applying complex crop models such as DSSAT specifically for marginal environments, such as South Africa. On the other hand, AquaCrop performed reasonably well with minimal input requirements, confirming its application in data-limited and marginal environments. However, it is recommended that there must be calibration for all the models using inputs specific to locations.

Keywords Indigenous crops · Water productivity · Sustainable development · Water-use efficiency · Smart technologies

9.1 Introduction

Neglected and underutilised (NUS) species play a crucial role in food security and nutrition, income generation of rural people and sustainable production in marginal environments (Magbagbeola et al. 2010; Mal 2007). However, due to the lack of data on growth and yield responses to different environments and management strategies, current mainstreaming efforts have not matched those for commercially important crops. Numerous computer-based tools, such as crop simulation models (CSMs), are presently being used to generate much-needed data in crop research to bridge the gap in NUS research (Ewert et al. 2011). These tools are effective since they minimise the need for expensive and time-consuming field experiments. While crop simulation models as decision support tools (DSTs) offer users and policymakers data for best management practices, their uptake and use are low primarily because there are huge uncertainties and complexities related to model structure and parameters (Palosuo et al. 2011). Further, it is uncertain whether the current ensemble of models is suited for modelling NUS. Due to the lack of data on NUS to parameterise and calibrate, many researchers have been calling for developing and using simpler but equally robust models (Mabhaudhi et al. 2014, 2019).

Currently, the most common crop models implemented in NUS studies include DSSAT (Jones et al. 2003) and AquaCrop (Steduto et al. 2009), which differ in complexity. There are simpler models, such as the SIMPLE Model developed by Zhao et al. (2019). Although the SIMPLE model has been described as simple by several researchers (Manschadi et al. 2021; Soltani et al. 2020; Zhao et al. 2019), no published studies on predicting the growth, yield and resource use of any NUS exists. Furthermore, there is an ongoing debate on whether a model's simplicity may be appropriate to depict crop responses under observed climate and management options (Palosuo et al. 2011; Zhao et al. 2019). The question then becomes, should there be simple models to simulate crop growth and yield, or focus should be given to complex models that require a high level of expertise and data? This is important with the current drive for sustainable intensification in the wake of climate change using neglected and underutilised crops (NUS). As such, there is a need to assess the appropriateness of using models with varying degree of complexity. Thus, this study compares the performance of three crop simulation models, namely, AquaCrop, DSSAT and the SIMPLE model, in predicting yield, biomass and water use of selected cereal NUS. It was hypothesised that all models were best suitable for predicting yield, biomass and water use of selected cereal NUS regardless of complexity.

9.1.1 *Materials and Methods*

Simulations for maize, sorghum and millet were performed using AquaCrop Version 6.1, DSSAT version 4.7.5 and SIMPLE version 1.1. These models required inputs of climate data, crop characteristics, soil characteristics and description of management practices to run simulations. These inputs were attained from the Ukulinga Research Farm. Maize was used as a base crop since it is a commercially important crop compared to millet and sorghum, which are identified as generally underutilised.

Study Site

This study used climate and soil data from the Ukulinga Research Farm to create input data files for AquaCrop, DSSAT and SIMPLE model. Ukulinga Research Farm is in Pietermaritzburg in the subtropical hinterland of KwaZulu-Natal Province, South Africa (Fig. 9.1). The farm is mainly used for training and research by the University of KwaZulu-Natal (UKZN) (Everson et al. 2012). The farm lies 29°37′ S, 30°16′ E, with an elevation of approximately 775 m above sea level. Rain falls mostly in summer, between September and April. Rainfall distribution varies during the growing season (Swemmer et al. 2007), with most rain falling in November, December, early January and March (Fig. 9.2). Occasionally, light to moderate frost occurs in winter (May–July). Chibarabada et al. (2020) reported that Ukulinga receives an average annual rainfall of 694 mm, mainly during the summer months (mid-October to mid-February). During the summer, average maximum temperatures are between 26 and 28 °C, while minimum temperatures can be as low as 10 °C. The landform at Ukulinga is a colluvial fan, and soils are derived from marine shales.

9.1.2 *Climate Description*

The daily climate data were obtained from an automatic weather station (AWS) at the Ukulinga Research farm (Fig. 9.2). The AWS is part of the Agricultural Research Council-Institute for Soil, Climate and Water (ARC-ISCW) network of automatic weather stations. The climate data for all models comprised rainfall (mm), minimum and maximum air temperature (°C), solar radiation (MJ day^{-1}) and reference evapotranspiration (ET_0). ET_0 was based on the FAO Penman-Monteith equation from full daily weather datasets described by Allen et al. (1998). ET_0 was calculated

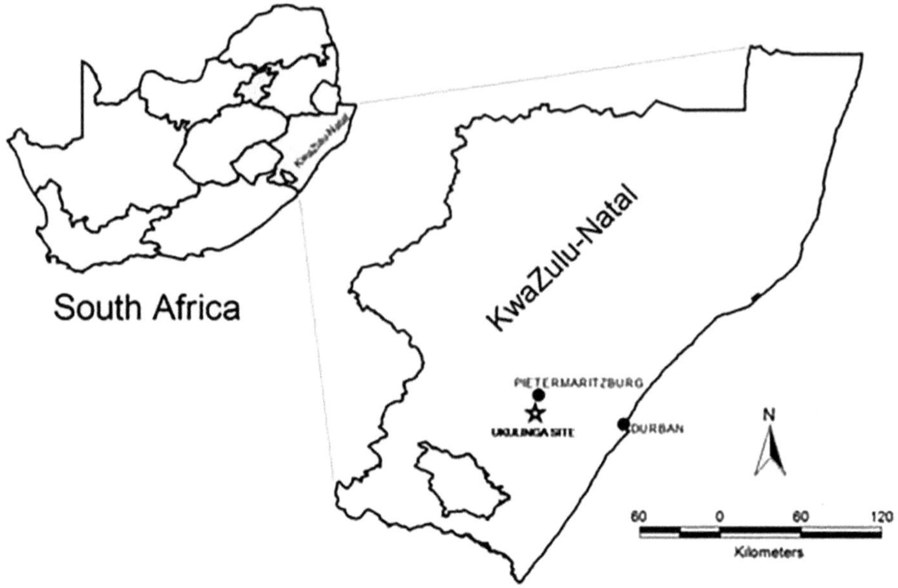

Fig. 9.1 Geographical view of Ukulinga Research Farm (Everson et al. 2012)

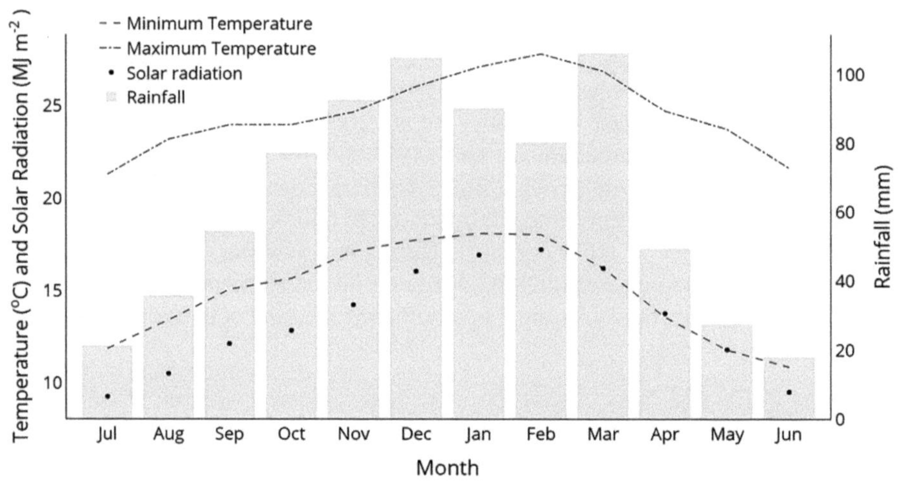

Fig. 9.2 Long-term climate data for Ukulinga Research Farm from 2004 to 2019 obtained from a nearby weather station

using FAO's ET_0 calculator (Raes et al. 2009) using air temperature, solar radiation, wind speed and relative humidity from the meteorological station. For DSSAT and the SIMPLE model, ET_0 was calculated using the Priestley and Taylor (1972) approach. This approach was used because it is simple and requires less data. For AquaCrop, a default file (.CO2) of the mean annual CO_2 concentration of 369.41 ppm as a reference in 2000 and 390 ppm in 2020, measured at the Mauna Loa Observatory in Hawaii, was used. DSSAT and the SIMPLE model also used this atmospheric CO_2 concentration as input data for simulations. To calculate growing degree days (GDD), AquaCrop used daily minimum and maximum temperature. Temperature is equally important in both DSSAT and the SIMPLE model for phenological development and crop growth.

Soil Description

According to Chimonyo et al. (2016), the dominant soils at Ukulinga are chromic luvisols (FAO soil classification), generally characterised as shallow brown acidic soils with low to moderate fertility. The soil textural class of the Ukulinga Research farm soil profile was classified as clay to clay-loam (USDA Taxonomic System) with an effective rooting depth of 0.6 m and three horizons (Table 9.1). No stress (water, soil fertility and soil salinity) was considered in creating the soil file. The Soil file (.SOL) for AquaCrop and SIMPLE model was created using Ukulinga Research farm soil data. In DSSAT, the clay loam soil file embedded in the model was selected to best resemble the Ukulinga soils. Details of the actual soil parameters used in each model are presented Tables 9.1 and 9.2.

Crop File

The creation of crop files entailed matching phenology (GDD) and yield potential on pre-existing cultivars across the models. The study used cultivars described by Akumaga et al. (2017), Hadebe et al. (2017) and Bello and Walker (2016) for maize, sorghum and pearl millet, respectively (Table 9.3).

AquaCrop Model Crop File

The AquaCrop model comprises two types of crop parameters: conservative and non-conservative (cultivar-specific) parameters (Raes et al. 2009). Conservative parameters were used as presented in the model because they do not change substantially with time, management practices, geographic location or climate (Raes et al. 2009). Additionally, conservative parameters are assumed not to change with cultivars unless shown otherwise. In this study, the fine-tuning of non-conservative parameters was not required because maize, sorghum and pearl millet were calibrated in the AquaCrop model. Hence, crop parameters used in this study are similar to those used to calibrate the crops in the selected studies where they were obtained (Table 9.4).

Table 9.1 Soil water properties from the experimental site at Ukulinga Research Farm (Mabhaudhi 2012)

Depth (m)	Bulk density (g/cm^3)	Permanent wilting point (mm/mm)	Field capacity (mm/mm)	Total available water (mm/mm)	Saturation (mm/mm)	Saturated hydraulic conductivity (mm/day)
0.60	1.20	283.00	406.00	123.00	481.00	25

The soil's physical characteristics, such as the soil texture, bulk density and porosity, are considered (Table 9.2)

Table 9.2 The soil file from Ukulinga Research Farm

Parameters	Units	Values	AquaCrop	DSSAT	SIMPLE
Upper horizon depth	Cm	30	X	X	
Lower horizon depth	Cm	10	X	X	
Number of soil horizons	–	3	X		
Sand content	%	33	X	X	
Silt content	%	33	X	X	
Clay content	%	34	X	X	
Bulk density	g/cm^3	1.20	X	X	
Organic carbon	%	2.90		X	
Organic nitrogen	%	0.24		X	
pH	–	4.51		X	
Saturated water content	%	46.73	X	X	
Field capacity	mm	46.32	X	X	
Wilting point	mm	23.03	X	X	
Plant available water-holding capacity	mm[b]	233	X		X
Runoff-curve number	–	75	X	X	X
Deep drainage	–	0.27	X	X	X
Total pore space	mm	36	X	X	
Saturated hydraulic conductivity	cm/h[a]	25	X	X	
Maximum rooting depth	mm	1	X	X	X

[a]For AquaCrop, units are mm/d
[b]For SIMPLE, units are mm/m

Table 9.3 Provincial potential, observed and mean simulated yield for maize, millet and sorghum

Crops	Crop type (maturity)	Provincial potential yield (t/ha)	Observed yield (t/ha)
Maize	Early to medium	8–11 (9.50)	5.51
Sorghum	Medium to late	4	5.31
Pearl millet	Early to medium	4.3–5.6 (4.95)	6.83

DSSAT Model Crop File

In the DSSAT model, the coefficients for one crop species are stored in three different files, namely cultivar (.CUL), ecotype (.ECO) and species (.SPE). Cultivar coefficients are for a single cultivar (traits differ among cultivars), ecotype coefficients are common to a group of cultivars, and species coefficients are common to all cultivars (crop-specific traits) (Jones et al. 2003). Hence, maize, sorghum and millet cultivars were selected in DSSAT. Cultivars chosen for maize, sorghum and millet were 2500–2600 GDD, PIONEER 8333 and BJ104, respectively (Table 9.5), as they best described the cultivars used by Akumaga et al. (2017), Hadebe et al. (2017) and Bello and Walker (2016). The cultivars were selected based on their GDD similarity with the GDD of crops used in the AquaCrop model.

Table 9.4 Crop parameters of maize, sorghum and pearl millet for AquaCrop model

Crop parameters	Pearl millet	Sorghum	Maize
Base temperature (°C) below which crop development does not progress	8	8	8
Upper temperature (°C) above which crop development no longer increases with an increase in temperature	32	30	30
Soil water depletion factor for canopy expansion (p-exp) - Upper threshold	0.30	0.15	0.14
Soil water depletion factor for canopy expansion (p-exp) - Lower threshold	0.65	0.70	0.72
Shape factor for water stress coefficient for canopy expansion (0.0 = straight line)	3	3	2.9
Soil water depletion fraction for stomatal control (p - sto) - Upper threshold	0.70	0.70	0.69
Shape factor for water stress coefficient for stomatal control (0.0 = straight line)	3	6	6
Soil water depletion factor for canopy senescence (p - sen) - Upper threshold	0.75	0.70	0.69
Shape factor for water stress coefficient for canopy senescence (0.0 = straight line)	3	3	2.70
Sum (ET_o) during stress period to be exceeded before senescence is triggered	0	0	0
Soil water depletion factor for pollination (p - pol) - Upper threshold	0.92	0.80	0.80
Vol% for Anaerobiotic point (* (SAT - [vol%]) at which deficient aeration occurs *)	10	5	5
Soil fertility stress at calibration (%)	50	50	50
Crop coefficient when canopy is complete but prior to senescence (Kcb,x)	1.10	1.07	1.03
Decline of crop coefficient (%/day) due to ageing, nitrogen deficiency, etc.	0.15	0.30	0.30
Minimum effective rooting depth (m)	0.30	0.30	0.30
Maximum effective rooting depth (m)	1.75	0.60	1
Shape factor describing root zone expansion	18	13	13
Maximum root water extraction (m^3 water/m^3 soil/day) in top quarter of root zone	0.05	0.03	0.08
Maximum root water extraction (m^3 water/m^3 soil/day) in bottom quarter of root zone	0.01	0.01	0.02
Effect of canopy cover in reducing soil evaporation in late season stage	60	50	50
Soil surface covered by an individual seedling at 90% emergence (cm^2)	5	3	6.50
Number of plants per hectare	55,556	44,444	53,333
Canopy growth coefficient (CGC): Increase in canopy cover (fraction soil cover per day)	0.24	0.13	1.30
Maximum canopy cover (CCx) in fraction soil cover	0.95	0.89	0.96

(continued)

Table 9.4 (continued)

Crop parameters	Pearl millet	Sorghum	Maize
Canopy decline coefficient (CDC): Decrease in canopy cover (in fraction per day)	0.10	1.70	1.06
Calendar Days: from sowing to emergence	4	14	7
Calendar Days: from sowing to maximum rooting depth	45	97	65
Calendar Days: from sowing to start senescence	80	98	91
Calendar Days: from sowing to maturity (length of crop cycle)	120	140	120
Calendar Days: from sowing to flowering	39	70	67
Length of the flowering stage (days)	20	77	30
Crop determinancy linked with flowering	1	1	1
Excess of potential fruits (%)	50	50	50
Building up of Harvest Index starting at flowering (days)	50	70	56
Water productivity normalised for ET_o and CO_2 (WP*) (gram/m^2)	32	33.70	33.70
Water productivity normalised for ET_o and CO_2 during yield formation (as % WP*)	100	100	100
Reference Harvest Index (HIo) (%)	30	45	40
Possible increase (%) of HI due to water stress before flowering	10	4	0
Coefficient describing positive impact on HI of restricted vegetative growth during yield formation	10	1	7
Coefficient describing negative impact on HI of stomatal closure during yield formation	8	3	3
Allowable maximum increase (%) of specified HI	15	25	15

Table 9.5 Cultivar coefficients of maize, sorghum and millet for the DSSAT model

Cultivar coefficients		Maize (2500–2600 GDD)	Sorghum (PIONEER 8333)	Millet (BJ104)
P1	Thermal time from seedling emergence to the end of the juvenile phase (expressed in degree days above a base temperature of 8 °C) during which the plant is not responsive to changes in photoperiod	160.00	325.00	120.00
P2	Extent to which development (expressed as days) is delayed for each hour increase in photoperiod above the longest photoperiod at which development proceeds at a maximum rate (which is considered to be 12.5 h)	0.75		
P2O	Critical photoperiod or the longest day length (in hours) at which development occurs at a maximum rate. At values greater than P2O, the rate of development is reduced		15.50	13.40
P2R	Extent to which phasic development leading to panicle initiation (expressed in degree days) is delayed for each hour increase in photoperiod above P2O		30.00	145.09

(continued)

Table 9.5 (continued)

Cultivar coefficients		Maize (2500–2600 GDD)	Sorghum (PIONEER 8333)	Millet (BJ104)
PANTH	Thermal time from the end of tassel initiation to anthesis (degree days above TBASE)			
P3	Thermal time from to end of flag leaf expansion to anthesis (degree days above TBASE)		152.50	
P4	Thermal time from anthesis to beginning grain filling (degree days above TBASE)		81.50	
P5	Thermal time (degree days above a base temperature of 8 °C) from beginning of grain filling (3–4 days after flowering) to physiological maturity	780.00	540.00	340.00
G1	Scaler for relative leaf size		11.00	0.60
G2	Maximum possible number of kernels per plant	750.00	6.00	
G3	Kernel filling rate during the linear grain filling stage and under optimum conditions (mg/day)	8.50		
G4	Scaler for partitioning of assimilates to the panicle (head)			1.00
PHINT	Phylochron interval; the interval in thermal time (degree days) between successive leaf tip appearances	49.00	49.00	43.00
GT	Tillering coefficient, equivalent to G1, but on tillers			
G5	Potential grain size, mg			
AX	Leaf surface area (cm^2/leaf) of largest leaf			
ALL	Leaf longevity (degree days) of the most longevous leaf			

SIMPLE Model Crop File

The SIMPLE model included nine species parameters to specify crop types and four cultivar parameters characterising cultivar differences. Species parameters were derived from accepted values in the literature (Zhao et al. 2019). Only cultivar parameters were calibrated within a reasonable range, but cultivar parameters are kept constant for the same cultivar when grown in different years or locations. The SIMPLE model calibrated only maize; however, it is not the cultivar used in this study (Table 9.4). Thus, calibrated maize values for species parameters (CO2_RUE and S_Water) were used to simulate maize, millet and sorghum results. I50A and I50B were not readily available for cultivar parameters; hence, these were calculated as I50A (thermal time for flowering*50%) and I50B (thermal time for the start of senescence*50%) (Table 9.6).

Table 9.6 Crop parameters of maize, sorghum and pearl millet for SIMPLE model

	Crop parameters	Maize	Sorghum	Millet
Tsum	Cumulative temperature requirement from sowing to maturity (°C d)	1419	1648	1337
HI	Potential harvest index	0.40	0.45	0.30
I50A	Cumulative temperature requirement for leaf area development to intercept 50% of radiation (°C d)	402	420	234
I50B	Cumulative temperature till maturity to reach 50% radiation interception due to leaf senescence (°C d)	546	588	480
Tbase	Base temperature for phenology development and growth (°C)	8	8	8
Topt	Optimal temperature for biomass growth (°C)	30	30	32
RUE	Radiation use efficiency (above ground only and without respiration) (g MJ^{-1} m^{-2}).	4.20	3.20	4
I50maxH	The maximum daily reduction in I50B due to heat stress (°C d)	100	100	100
I50maxW	The maximum daily reduction in I50B due to drought stress (°C d)	12	12	12
MaxT	Threshold temperature to start accelerating senescence from heat stress (°C)	44	44	48
ExtremeT	The extreme temperature threshold when RUE becomes 0 due to heat stress (°C)	50	50	50
CO_2_RUE	Relative increase in RUE per ppm elevated CO_2 above 350 ppm	0.01	0.01	0.01
S_Water	Sensitivity of RUE (or harvest index) to drought stress (ARID index)	1.50	1.50	1.50

Management File

In AquaCrop, DSSAT and SIMPLE models, the three crops were sown by the direct planting method. The planting period for simulation in all three models ranged from 10/2004 to 04/2019, and the planting dates chosen were 25 October for maize, 15 October for sorghum and 1 October for pearl millet. These dates represented the recommended planting dates for KZN for the different cereal crops. However, AquaCrop started the simulation one day after planting. The optimal planting date was based on Department of Agriculture, Forestry and Fisheries (DAFF) recommendations and historical weather data at Ukulinga. In all the models, the plant population for maize, sorghum and pearl millet were 53,333, 44,444 and 55,556 plants/ha, respectively. Management practices undertaken were similar for AquaCrop, DSSAT and SIMPLE models. Due to the SIMPLE model having only irrigation as a management variable, all three models considered irrigation solely as a management practice. The assumption was that all the models were free from weeds and stress (water, soil fertility and soil salinity).

Statistical Analyses and Model Evaluation Statistics

Descriptive statistics such as means, standard deviations, line graphs, and box and whisker plots were used to analyse outputs. Box and whisker plots can show stability and general distribution of the data sets. A Mann–Kendall trend test was also used to perceive statistically significant decreasing or increasing data trends. A p-value <0.05 indicates a trend, and if τ is +ve, increasing trend, and if τ is −ve, decreasing trend. Also, there was a limited number of data points to run all statistic tests; hence, statistical analysis was not done for the individual crops but considered maize, millet and sorghum as sub-factors.

The methods of assessing and comparing the performance of models have been discussed widely (see, e.g. Bellocchi and Rivington (2009), Kobayashi and Salam (2000), Wallach et al. (2006), Willmott (1981)). We extracted data for observed yield from the articles used for model calibration. Arathoon and Mtumtum (2013), GAIN Report (2019) and GRAINSA (2021) were also used for the provision of provincial potential yield for the three crops. The provincial data was used to benchmark the yield potentials for each crop and each model's performance for the three crops. The calibrated parameters were used to simulate the outputs of the three models, including grain yield, aboveground biomass and evapotranspiration (water use). Since AquaCrop, DSSAT and the SIMPLE model does not calculate WUE directly, simulated outputs of water use (WU in mm) and yield (Y in kg ha^{-1}) were used to determine water use efficiency (WUE in kg mm^{-1} ha^{-1}) as follows:

$$\text{WUE}_{\text{s}} = \frac{Y\ (\text{kg}/\text{ha})}{\text{WU}(\text{mm})} \tag{9.1}$$

The root mean square error (RMSE; Eq. 9.2), normalised root mean square error (NRMSE; Eq. 9.3), coefficient of determination (R^2; Eq. 9.4), mean bias error (MBE; Eq. 9.5) and index of agreement (d; Eq. 9.6)

$$\text{RMSE} = \sqrt{\frac{1}{N}\sum_{i=1}^{N}(Pi - Oi)(1)^2} \tag{9.2}$$

where N is the number of estimate–observation pairs, P_i is the model prediction and O_i is the observed value of the model i.

$$\text{NRMSE} = \frac{\text{RMSE}}{\bar{O}} \tag{9.3}$$

where $\bar{O}$ is the average of observation value

$$r = \frac{n(\sum xy) - (\sum x)(\sum y)}{\sqrt{\left[n\sum x^2 - (\sum x)^2\right]\left[n\sum y^2 - (\sum y)^2\right]}} \tag{9.4}$$

MBE was taken as an indicator of under- or overestimation, i.e. the direction and magnitude of bias.

$$\mathrm{MBE} = N^{-1}\sum_{i=1}^{N}\left(Pi - Oi\right) \tag{9.5}$$

IA was used as a more general indicator of model efficiency.

$$\mathrm{IA} = 1 - \frac{N x \mathrm{MSE}}{\mathrm{PE}} \tag{9.6}$$

Where $\mathrm{PE} = \sum_{i=1}^{N}\left(\left|\dot{P}\right| + \left|\dot{O}\right|\right)\left(1^2\right)$ and where $\dot{P} = P_i - \bar{O}$ and $\dot{O} = O_i - \bar{O}_i$ and $\bar{O}$ is the mean of the observed variable. The main objective of the study was to evaluate model performance. However, in this study, all three models were compared against each other (i.e. AquaCrop vs. DSSAT, SIMPLE vs. DSSAT and AquaCrop vs. SIMPLE) because observed data was unavailable.

9.2 Results

Yield

The highest mean yield was simulated by the AquaCrop model for maize (6.55 t/ha) and millet (7.47 t/ha). Meanwhile, for sorghum, SIMPLE simulated the highest mean yield of 5.74 t/ha. Yield simulations in AquaCrop showed significant variability for maize and sorghum. This was in line with the large standard deviation (±1.48 t/ha) observed for maize under AquaCrop. DSSAT had minimal yield variability for both crops. Contrary to this, for millet, greater yield variability was observed under DSSAT, while the least variability was observed under the SIMPLE model (Fig. 9.3). The DSSAT yield simulation for millet showed the most deviation from the mean by ±1.21 t/ha than other models. The highest standard deviation for sorghum yield simulated by the SIMPLE model was ±1.59 t/ha (Fig. 9.4). Across the simulation years, the Mann–Kendall trend analysis showed a significant ($P < 0.05$) and positive trend (0.30) in yield for sorghum and a negative trend (−0.40 and −0.63) for maize and millet, respectively.

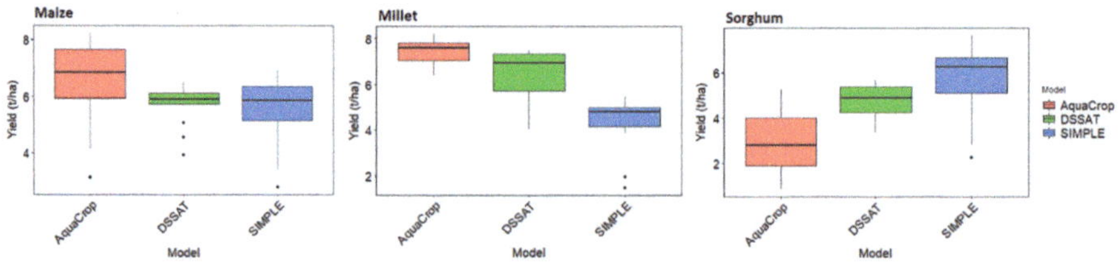

Fig. 9.3 Yield for maize, millet and sorghum simulated by AquaCrop, DSSAT and SIMPLE

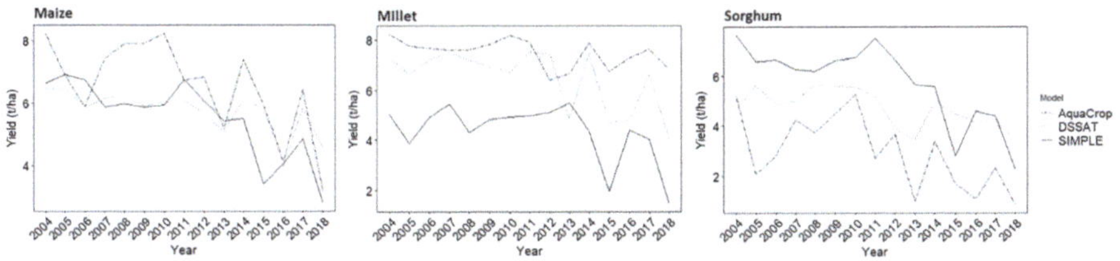

Fig. 9.4 Yield for maize, millet and sorghum simulated by AquaCrop, DSSAT and SIMPLE from 2004 to 2018

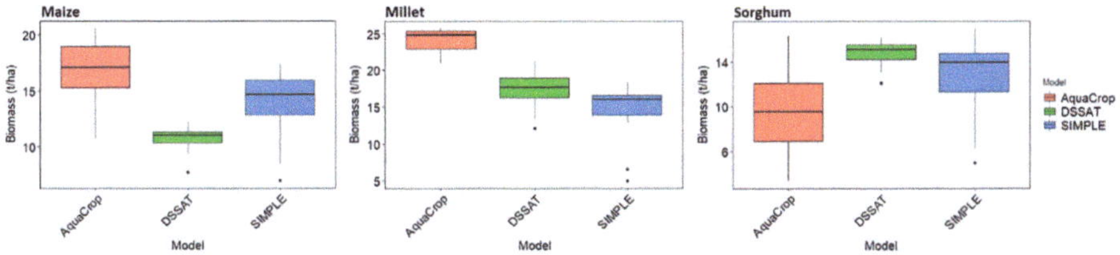

Fig. 9.5 Descriptive statistics for simulated biomass for maize, millet and sorghum simulated by AquaCrop, DSSAT and SIMPLE

Biomass

The calculated mean for biomass showed that AquaCrop simulations attained the highest value for maize (16.78 t/ha) and millet (24.08 t/ha) and DSSAT for sorghum (14.69 t/ha). In (Fig. 9.5), the wider box plots shown by the AquaCrop model across maize and sorghum indicated more biomass variation. For millet, the variation in biomass was similar across the three models. Compared to other models, DSSAT showed less variability in biomass for maize and sorghum. Analysing simulated biomass results showed large standard deviations for maize (3.11 units) and millet in SIMPLE (3.87 units) and for sorghum (4.03 units) in AquaCrop (Fig. 9.6). Overall, the Mann–Kendall trend analysis did not detect a trend in biomass for sorghum. Overall significant ($p < 0.05$) and negative trend was observed in simulated biomass for maize ($\tau = -0.66$) and millet ($\tau = -0.37$). For maize, the AquaCrop model depicted a negative trend for biomass from 2007 to 2011, while no clear trend was observed for biomass simulations using DSSAT and the SIMPLE models (Fig. 9.6).

Water Use Efficiency

The calculated mean WUE showed that SIMPLE simulations attained the highest value for maize (13.77 t/ha) and, sorghum (12.89 t/ha) and DSSAT for millet (14.78 t/ha) (Fig. 9.9). This observation was inconsistent with the yield trends where the AquaCrop model simulated higher yield for maize and millet while, for sorghum, SIMPLE simulated the highest mean yield. The maize and sorghum AquaCrop box plots suggested more variability in WUE. Whereas for millet, more variability in WUE was observed under DSSAT. DSSAT depicted thinner box plots compared to other models, suggesting less variability in WUE for maize and

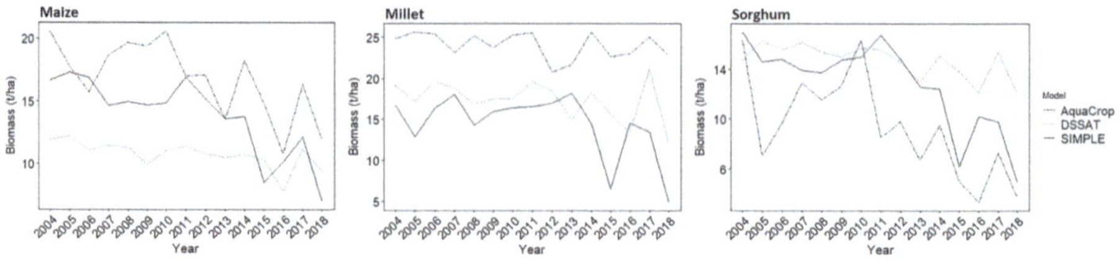

Fig. 9.6 Biomass trends for maize, millet and sorghum simulated by AquaCrop, DSSAT and SIMPLE from 2004 to 2018

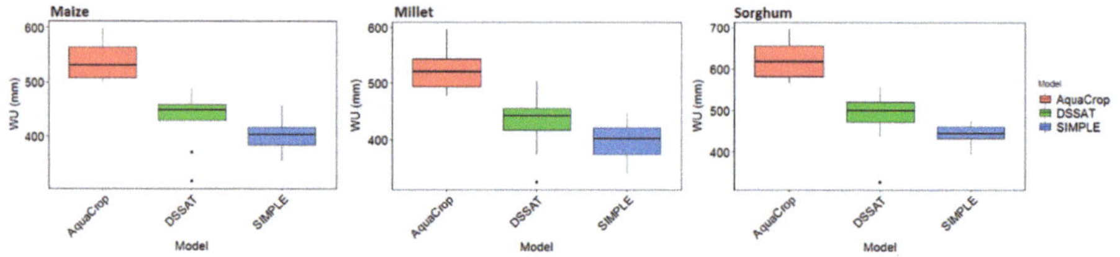

Fig. 9.7 Descriptive statistics for the simulated WUE for maize, millet and sorghum under AquaCrop, DSSAT and SIMPLE

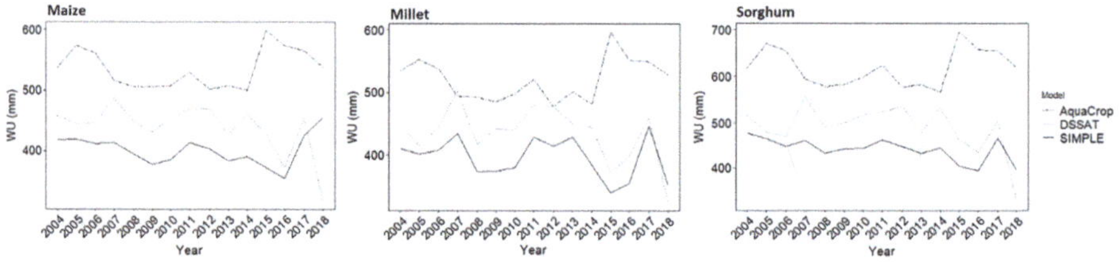

Fig. 9.8 WUE trends for maize, millet and sorghum simulated by AquaCrop, DSSAT and SIMPLE from 2004 to 2018

sorghum. This is consistent with the observed low variability in yield (Figs. 9.7 and 9.8). The standard deviation calculated for WUE infers that there was a high degree of standard deviation for millet (3.21 units) and sorghum (2.65 units) under the SIMPLE model and for maize (3.16 units) under AquaCrop. The WUE trend across all three models was consistent for maize from 2004 to 2018 (Figs. 9.9 and 9.10). The Mann–Kendall trend analysis showed a significant ($p < 0.05$) and negative trend in WUE of $\tau = -0.45$ for millet and a positive trend in WUE of $\tau = 0.42$ for sorghum.

Provincial potential observed and mean simulated yield were compared (Table 9.7) using statistical indicators. Based on MBE, AquaCrop underestimated the yield by 0.22 t/ha. In contrast, in the case of DSSAT, the underestimation is observed to be 0.24 t/ha and 0.69 t/ha for SIMPLE when comparing the observed yield and simulated yield. Moreover, a higher coefficient of determination (R^2) is observed in AquaCrop (0.99) compared to DSSAT (0.92) and the SIMPLE (0.51)

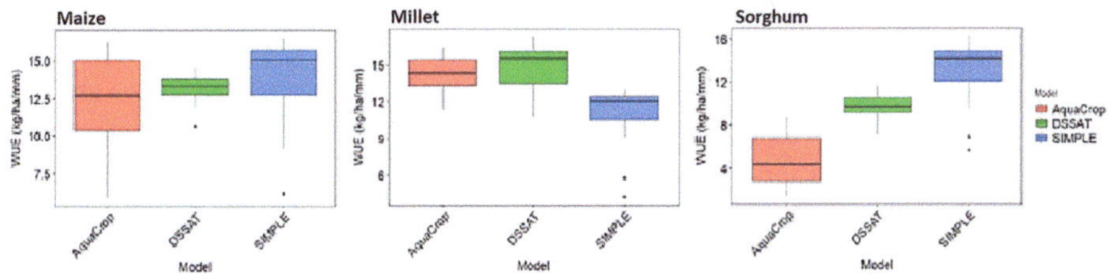

Fig. 9.9 WUE for maize, millet and sorghum simulated by AquaCrop, DSSAT and SIMPLE

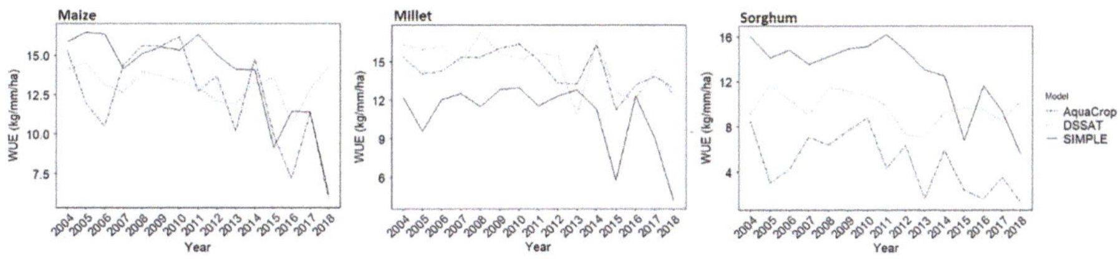

Fig. 9.10 WUE for maize, millet and sorghum simulated by AquaCrop, DSSAT and SIMPLE from 2004 to 2018 Model performance

Table 9.7 Provincial potential, observed and mean simulated yield for maize, millet and sorghum

Crops	Crop type (maturity)	Provincial potential yield (t/ha)	Observed yield (t/ha)	Simulated yield (t ha^{-1})		
				AquaCrop	DSSAT	SIMPLE
Maize	Early to medium	8–11 (9.50)	5.51	6.55 (1.49)	5.75 (0.70)	5.52 (1.24)
Sorghum	Medium to late	4	5.31	7.47 (0.57)	6.45 (1.21)	4.33 (1.16)
Pearl millet	Early to medium	4.3–5.6 (4.95)	6.83	2.97 (1.47)	4.72 (0.77)	5.74 (1.59)

Table 9.8 Statistical comparison of the observed, provincial potential and simulated yield for AquaCrop, DSSAT and the SIMPLE model

Model performance	R^2	D-index	RMSE	NRMSE	MBE
Observed vs. AquaCrop	0.99	−3.15	2.62	0.45	−0.22
Observed vs. DSSAT	0.92	−18.13	1.39	0.24	−0.24
Observed vs. SIMPLE	0.51	0.62	0.85	0.14	−0.69
Potential vs. AquaCrop	0.03	0.27	2.87	0.47	−0.49
Potential vs. DSSAT	0.003	−0.06	2.59	0.42	−0.51
Potential vs. SIMPLE	0.27	0.36	2.35	0.38	−0.95
Observed vs. Potential	0.06	−0.28	2.66	0.45	0.27

model (Table 9.8). Considering the provincial potential yield as a comparative factor, the three models underestimated yield. The statistical comparison of observed and provincial potential yield showed that the provincial potential yield overestimated yield by 0.27 (Table 9.8).

Table 9.9 Statistical comparison of simulated output variables for AquaCrop, DSSAT and the SIMPLE model

Output variables	Model comparison	R^2	D-index	RMSE	NRMSE	MBE
Yield	AquaCrop-DSSAT	0.95	0.77	1.26	0.22	−0.80
	AquaCrop-SIMPLE	0.58	−1.44	2.48	0.44	1.03
	DSSAT-SIMPLE	0.78	−5.06	1.37	0.24	0.80
Biomass	AquaCrop-DSSAT	0.16	0.42	6.04	0.36	−6.06
	AquaCrop-SIMPLE	0.98	0.31	6.14	0.37	−2.98
	DSSAT-SIMPLE	0.07	0.22	2.71	0.19	3.08
Water use	AquaCrop-DSSAT	0.99	0.27	105.79	0.19	−96.61
	AquaCrop-SIMPLE	1.00	0.14	146.59	0.26	−132.87
	DSSAT-SIMPLE	0.99	0.54	40.95	0.09	−36.26

Except for NRMSE and according to the rest of the statistical indicators for yield, results suggested that there was a satisfactory agreement between AquaCrop-DSSAT and DSSAT-SIMPLE. There was a positive correlation for all three models being compared against each other. The correlations (R^2) between AquaCrop-DSSAT, AquaCrop-SIMPLE and DSSAT-SIMPLE were 0.95, 0.58 and 0.78, respectively. There were slight differences in model agreement across the different statistical indicators. For instance, good agreement was observed for yield simulated by AquaCrop-DSSAT ($D^2 = 0.77$). RMSE and NRMSE of 2.48 t/ha and 0.44 t/ha, respectively, for the comparison of AquaCrop and SIMPLE, showed a higher deviation in this combination when compared with other comparisons in this study (Table 9.9).

There was a positive correlation in biomass for all three models compared to each other. The comparison between AquaCrop and SIMPLE SHOWED the highest R^2 of 0.98, and the comparison between DSSAT and SIMPLE had the lowest R^2 of 0.07. All the models simulated biomass well in agreement of 0.42 for AquaCrop-DSSAT, 0.31 for AquaCrop-SIMPLE and 0.22 for DSSAT-SIMPLE. However, due to the high value of AI for AquaCrop and DSSAT, it was evident that this comparison agreed with simulating biomass. Biomass was overestimated in AquaCrop comparison to SIMPLE by 1.03 t/ha, similarly with the comparison between DSSAT and SIMPLE by 0.80 t/ha. Meanwhile, AquaCrop underestimated biomass compared to DSSAT by 0.80 t/ha. RMSE and NRMSE of 6.14 t/ha and 0.37 t/ha, respectively, for the comparison of AquaCrop and SIMPLE, showed a higher deviation in this combination when compared with other comparisons in this study (Table 9.9).

The results showed a perfect coefficient ($R^2 = 1$) in water use simulated by AquaCrop-SIMPLE. The same positive relationship of $R^2 = 0.99$ was observed for comparing AquaCrop-DSSAT and DSSAT-SIMPLE. AI as a statistical indicator showed that there was an agreement for water use simulated by AquaCrop-DSSAT (0.27), AquaCrop-SIMPLE (0.14) and DSSAT-SIMPLE (0.54). There was an underestimation of water use for AquaCrop-DSSAT (−96.61 t/ha), AquaCrop-SIMPLE (−132.87 t/ha) and DSSAT-SIMPLE (−36.26 t/ha). RMSE and NRMSE of 146.59 t/ha and 0.26 t/ha, respectively, for the comparison of AquaCrop and

SIMPLE, showed a higher deviation in this combination when compared with other comparisons in this study. The least deviations (RMSE = 40.95 and NRMSE = 0.09) for water use were observed for the DSSAT-SIMPLE comparison (Table 9.8).

9.3 Discussion

AquaCrop, DSSAT and SIMPLE could simulate yield, biomass and water use for selected NUS. However, the performance of the three models was observed to be statistically different across the simulated years and for the different crop species. The statistical indicators (R^2 and MBE) for observed yield compared with simulated yield suggest that yield, biomass and WU simulated by AquaCrop across the selected NUS were more satisfactory than when DSSAT and the SIMPLE model did simulations. High yield and biomass variability simulated for maize and sorghum under AquaCrop and millet under DSSAT could suggest that both these models were more sensitive to input parameters, and this sensitivity was crop-specific. Moreover, in line with the statement by Manschadi et al. (2021), different simulated results in this study could be attributed to secondary data, different parameters, number of parameters, model types and algorithms. Similar to a study conducted by Timsina et al. (2008), the crop, soil and climate inputs in this study have a degree of uncertainty due to random errors and bias in their measurement and calibration. Also, the three models used in the study were calibrated using secondary data from different sources; hence, the simulated and observed results differ. The disadvantage of using secondary data is that there is a high likelihood that not all parameters were captured that were used in calibrating the models. Hence, when interpreting and extrapolating the model results, due consideration should be given to uncertainties arising from model structure, model parameters and inputs, and the experimental data used for model calibration, validation and application (Timsina et al. 2008).

On the other hand, AquaCrop and the SIMPLE model were calibrated, and this means that the confidence in results for these models was improved, as the need to use default input values was minimised. The SIMPLE model used maize values embedded in the model for species parameters (CO_2_RUE and S_Water) of the three crops; DSSAT used default files for the soil and crop data, suggesting that the confidence in results was low compared to AquaCrop and the SIMPLE model. Nevertheless, scientific processes and coefficients in AquaCrop, DSSAT and the SIMPLE model can never be complete, leading to prediction uncertainties (Timsina et al. 2008). This is the case as perfect modelling can never be attained for natural systems because models abstract reality.

The results indicate that AquaCrop, with a water-driven growth engine, is better than DSSAT and the SIMPLE model, with a solar energy-driven growth engine for biomass and yield simulation. DSSAT and the SIMPLE model differ from AquaCrop by calculating biomass accumulation based on RUE rather than normalised water productivity (WP*). Albrizio and Steduto (2005) showed high variability in RUE values but failed to normalize RUE by vapour pressure deficit to reduce climate variability. The conclusion was that RUE's robustness to simulate biomass in crop models was constrained. The follow-up studies also indicated that calculating biomass through WP* was more robust than RUE (Albrizio and Steduto 2005; Steduto et al. 2007, 2009), which agrees with the findings of this study.

In this study, AquaCrop was the only model previously calibrated for sorghum using soil and climate data for Ukulinga Research Farm. For DSSAT and the SIMPLE model, current simulations are done for climatic and soil conditions that might have been similar in agroecological classification but differed based on the year the studies were performed. Asseng et al. (2013) state that conducting both model calibration and validation before their comparison provides more robust and reliable results. Furthermore, good quality data must be set to give researchers full confidence for calibration and validation. Furthermore, previous model comparisons have shown that minimal calibration of crop models can lead to high uncertainty in yield estimates (Asseng et al. 2013; Palosuo et al. 2011; Rötter et al. 2012). Also, providing more detailed data for model calibration did not necessarily result in high model performance when applied to new situations.

Many input parameters resulted in large and confounding impacts on overall outputs. According to Babel et al. (2019), the performance of a model in any specific site depends on the fine-tuning of the parameters and sound validation under a range of conditions. This study confirms that DSSAT requires relatively more site-specific and crop variety-associated data, in which default files were used to accommodate data unavailability. In contrast, AquaCrop is a simpler model with lesser soil and crop management data required as input. Meanwhile, SIMPLE requires fewer input requirements than AquaCrop and DSSAT. There is limited data availability for NUS to parameterise complex models fully. Since simpler models are found to perform equally and the complex ones, this suggests that going forward, less complex models can be adopted to advance modelling on NUS. Nonetheless, parameters are still needed to be used in modelling NUS. Thus, standards and protocols for data collection must be formulated to attain the required quantity and quality of data.

9.4 Conclusion

Three crop simulation models—AquaCrop (v 6.1), DSSAT (v 4.7.1) and SIMPLE (v 1.1)—were evaluated for their comparative performance for maize, millet and sorghum at Ukulinga Research Farm in Pietermaritzburg, South Africa. AquaCrop, DSSAT and the SIMPLE model simulated yield, biomass and water use for selected NUS. The presented results are based on several assumptions, and the predictions by the three models may be affected by a degree of uncertainty. These assumptions may have affected results and biased conclusions regarding yield, biomass, water use and WUE estimates. Further, the interactions between weather, soil characteristics, plant growth dynamics and management alternatives may have affected simulation results. Despite the potential limitations, the AquaCrop, DSSAT and SIMPLE models can be used as decision support tools to assist farmers in producing

NUS. However, based on the statistical differences, AquaCrop was observed as the better suitable model for simulating yield, biomass and water use for selected NUS. Further, the model can be used, and the results of this study can be extrapolated to other areas with similar climatic and soil environments in South Africa where crop, soil, weather and management data are available.

References

Akumaga U, Tarhule A, Yusuf AA (2017) Validation and testing of the FAO AquaCrop model under different levels of nitrogen fertilizer on rainfed maize in Nigeria, West Africa. Agric For Meteorol 232:225–234. https://doi.org/10.1016/j.agrformet.2016.08.011

Albrizio R, Steduto P (2005) Resource use efficiency of field-grown sunflower, sorghum, wheat and chickpea I. Radiat Use Efficien 130:254–268. https://doi.org/10.1016/j.agrformet.2005.03.009

Allen RG, Pereira LS, Raes D, Smith M (1998) Crop evapotranspiration - guidelines for computing crop water requirements - FAO irrigation and drainage paper 56. FAO - Food and Agriculture Organization of the United Nations

Arathoon J, Mtumtum N (2013) Maize cultivar recommendations

Asseng S, Ewert F, Rosenzweig C, Jones JW, Hatfield JL, Ruane AC, Boote KJ, Thorburn PJ, Rotter RP, Cammarano D, Brisson N, Basso B (2013) Uncertainty in simulating wheat yields under climate change. Nat Clim Chang 3. https://doi.org/10.1038/ncliamte1916

Babel MS, Deb P, Soni P (2019) Performance evaluation of AquaCrop and DSSAT-CERES for maize under different irrigation and manure application rates in the Himalayan region of India. Agric Res 8(2):207–217. https://doi.org/10.1007/s40003-018-0366-y

Bello ZA, Walker S (2016) Calibration and validation of AquaCrop for pearl millet (Pennisetum glaucum). Crop Pasture Sci 67(9):948–960. https://doi.org/10.1071/CP15226

Bellocchi G, Rivington M (2009) Validation of biophysical models: issues and methodologies. A review. Agron Sustain Dev. https://doi.org/10.1051/agro/2009001

Chibarabada TP, Modi AT, Mabhaudhi T (2020) Calibration and evaluation of AquaCrop for groundnut (Arachis hypogaea) under water deficit conditions. Agric For Meteorol 281(2020):107850. https://doi.org/10.1016/j.agrformet.2019.107850

Chimonyo VGP, Modi AT, Mabhaudhi T (2016) Simulating yield and water use of a sorghum–cowpea intercrop using APSIM. Agric Water Manag 177(August):317–328. https://doi.org/10.1016/j.agwat.2016.08.021

Everson CS, Mengistu MG, Gush MB (2012) A field assessment of the agronomic performance and water use of Jatropha curcas in South Africa. Biomass Bioenergy 59:59–69

Ewert F, Van Ittersum MK, Heckelei T, Therond O, Bezlepkina I (2011) Scale changes and model linking methods for integrated assessment of agri-environmental systems. Agric Ecosyst Environ 142:6–17. https://doi.org/10.1016/j.agee.2011.05.016

GAIN, R (2019) The South African sorghum market

GRAINSA (2021) Integrated crop and pasture-based livestock production systems - Part 10: Pearl millet (Pennisetum glaucum). https://www.grainsa.co.za/integrated-crop-and-pasture-based-livestock-production-systems%2D%2D-part-10

Hadebe ST, Modi AT, Mabhaudhi T (2017) Calibration and testing of AquaCrop for selected sorghum genotypes. Water SA 43(2):209. https://doi.org/10.4314/wsa.v43i2.05

Jones JW, Hoogenboom G, Porter CH, Boote KJ, Batchelor WD, Hunt LA, Wilkens PW, Singh U, Gijsman AJ, Ritchie JT (2003) The DSSAT cropping system model. Eur J Agron 18(3–4). https://doi.org/10.1016/S1161-0301(02)00107-7

Kobayashi K, Salam MU (2000) Comparing simulated and measured values using mean squared deviation and its components. Semingroup Forum. https://doi.org/10.1007/s100870050043

Mabhaudhi T (2012) Drought tolerance and water use of selected South African landraces of taro (Colocasia esculenta L. Schott) and bambara groundnut (Vigna Subterranea L. Verdc). Doctoral dissertation, University of KwaZulu-Natal

Mabhaudhi T, Chibarabada TP, Chimonyo GVP, Murugani VG, Pereira LM, Sobratee N, Govender L, Slotow R, Modi AT (2019) Mainstreaming underutilized indigenous and traditional crops into food systems: a South African perspective. Sustainability. https://doi.org/10.3390/su11010172

Mabhaudhi T, Modi AATA, Beletse YYGY (2014) Parameterisation and evaluation of the FAO-AquaCrop model for a South African taro (Colocasia esculenta L. Schott) landrace. Agric For Meteorol 192–193:132–139. https://doi.org/10.1016/j.agrformet.2014.03.013

Magbagbeola JAO, Adetoso JA, Owolabi OA (2010) Neglected and underutilized species (NUS): a panacea for community focused development to poverty alleviation/poverty reduction in Nigeria. J Econ Int Financ 2(10):208–211

Mal B (2007) Neglected and underutilized crop genetic resources for sustainable agriculture. Indian J Plant Genet Resour 20(1):1–14

Manschadi AM, Eitzinger J, Breisch M, Fucchs W, Neubauer T, Soltani A (2021) Full parameterisation matters for the best performance of crop models: inter-comparison of a simple and a detailed maize model. Int J Plant Prod 15(1):61–78. https://doi.org/10.1007/s42106-020-00116-2

Palosuo T, Christian K, Angulo C, Hlavinka P, Moriondo M, Olesen JE, Patil RH, Rumbaur C, Takáˇ J, Bindi M, Ewert F, Ferrise R, Mirschel W, Levent S, Siˇ B (2011) Simulation of winter wheat yield and its variability in different climates of Europe: a comparison of eight crop growth models. Eur J Agron 35:103–114. https://doi.org/10.1016/j.eja.2011.05.001

Priestley CHB, Taylor RJ (1972) On the assessment of surface heat flux and evaporation using large-scale parameters. Mon Weather Rev 100(2):81–92

Raes D, Steduto P, Hsiao TC, Fereres E (2009) AquaCrop — The FAO crop model to simulate yield response to water: II. Main algorithms and software description. Agron J. https://doi.org/10.2134/agronj2008.0140s

Rötter RP, Palosuo T, Christian K, Angulo C, Bindi M, Ewert F, Ferrise R, Hlavinka P, Moriondo M, Nendel C, Olesen JE, Patil RH, Takáˇ J, Trnka M (2012) Simulation of spring barley yield in different climatic zones of Northern and Central Europe: a comparison of nine crop models. Field Crops Res 133:23–36. https://doi.org/10.1016/j.fcr.2012.03.016

Soltani A, Alimagham SM, Nehbandani A, Torabi B, Zeinali E, Dadrasi A, Zand E (2020) SSM-iCrop2: a simple model for diverse crop species over large areas. Agric Syst 182(April):102855. https://doi.org/10.1016/j.agsy.2020.102855

Steduto P, Hsiao TC, Fereres E (2007) On the conservative behavior of biomass water productivity On the conservative behavior of biomass water productivity. Irrig Sci. https://doi.org/10.1007/s00271-007-0064-1

Steduto P, Hsiao TC, Raes D, Fereres E (2009) AquaCrop—The FAO crop model to simulate yield response to water: I. concepts and underlying principles. Agron J. https://doi.org/10.2134/agronj2008.0139s

Swemmer AM, Knapp AK, Snyman HA (2007) Intra-seasonal precipitation patterns and aboveground productivity in three perennial grasslands. J Ecol 95:780–788. https://doi.org/10.1111/j.1365-2745.2007.01237.x

Timsina J, Godwin D, Humphreys E, Kukal SS, Smith D (2008) Evaluation of options for increasing yield and water productivity of wheat in Punjab, India using the DSSAT-CSM-CERES-Wheat model. Agric Water Manag 95:1099–1110. https://doi.org/10.1016/j.agwat.2008.04.009

Wallach D, Makowski D, Jones JW (2006) In: Wallach D, Makowski D, Jones JW (eds) Working with dynamic crop models, 1st edn. Elsevier Science Inc.

Willmott CJ (1981) On the validation of models. Phys Geogr 2(2):37–41. https://doi.org/10.1080/02723646.1981.10642213

Zhao C, Liu B, Xiao L, Hoogenboom G, Boote KJ, Kassie BT, Pavan W, Shelia V, Kim KS, Hernandez-Ochoa IM, Wallach D, Porter CH, Stockle CO, Zhu Y, Asseng S (2019) A SIMPLE crop model. Eur J Agron 104:97–106. https://doi.org/10.1016/j.eja.2019.01.009

Chapter 10
Water Productivity in South Asia: Spatial and Temporal Variations

Upali Amarasinghe, Alok Sikka, Madhusha Perera, and Niranga Alahakoon

Abstract This chapter assesses the variations and options for improving water productivity to address water risks and insecurity in South Asian countries. The water productivity indicators of focus are physical water productivity (PWP)—the production per unit of water use, and economic water productivity (EWP), the value of production per unit of water use. A significant potential exists to increase PWP in many South Asian countries and regions with no water scarcity. These regions require increased access to water. However, increasing EWP should take precedence under water-scarce conditions. The latter may require reducing water-intensive crop areas and diversifying to less water-intensive crops.

Keywords Climate change · Rainfall variability · Mixed farming · Vulnerability assessment · Water scarcity

10.1 Introduction

Water is central to South Asia's (SA) social vulnerabilities of poverty, water insecurity, and climate change (Adger 1999; Soussan and Arriens 2004; Scott et al. 2021; Paegelow et al. 2022). Multidimensional poverty, defined as inadequate education, health, and living standards (UNDP and OPHI 2021), is very high in SA. In 2021, 9% of the SA population was income poor, earning less than $2 daily (World Bank 2022). In contrast, the multidimensional poor include 29% of the population (UNDP and OPHI 2021), and many of them live in rural areas, which in 2021 was 34% of the rural population (World Bank 2022).

U. Amarasinghe (✉) · M. Perera · N. Alahakoon
International Water Management Institute, Colombo, Sri Lanka
e-mail: u.amarasinghe@cgiar.org

A. Sikka
International Water Management Institute, New Delhi, India

T. Mabhaudhi et al. (eds.), *Enhancing Water and Food Security Through Improved Agricultural Water Productivity*,
https://doi.org/10.1007/978-981-96-1848-4_10

Another key development challenge is vulnerability to water insecurity due to highly variable rainfall patterns and water scarcity and quality issues. Water scarcity due to overexploitation (physical water scarcity) or due to lack of development (economic water scarcity) affects most countries in SA (Seckler et al. 1999; Molden 2013). Pakistan and large parts of India are physically water scarce. Economic water scarcity dominates Nepal and Bangladesh. With monsoonal weather patterns, where most precipitation falls in a few months, seasonal water scarcity is recurrent in many SA countries (Mekonnen and Hoekstra 2016). Along with water scarcity, critical water quality issues are also emerging, making SA one of the water insecurity hotspots (Habiba et al. 2014).

Water-related disasters due to high climate variability are already a threat to many SA countries. Natural disasters, especially floods and droughts, affect most countries, with India, Nepal, and Sri Lanka ranked fifth, sixth, and 20th climate-risk countries in 2019 (Eckstein et al. 2021). Although ranked 100th in 2019 among the climate-risk countries, the 2022 flooding in Pakistan, which affected more than one-third of the country, shows how vulnerable the region is to climate change. Moreover, rising sea levels with climate change will profoundly impact the coastal population in Bangladesh, Maldives, Sri Lanka, and India (Harrison et al. 2021).

Many other aspects amplify the water centrality to social vulnerability. They include the large water-dependent agricultural livelihoods (FAO 2022), a large share of agricultural water withdrawals (Rasul 2016), increasing non-agricultural water demand (Sathre et al. 2022), and rapidly emerging climate change impacts with the likely increase in the variability of rainfall and temperature patterns (Almazroui et al. 2020). Climate change will exacerbate the vulnerability of many people facing recurrent extreme weather and natural disasters. The precarious economic risks, such as in Sri Lanka and Pakistan at present, and those affected by the COVID-19 pandemic could worsen the plight of the agricultural sector, people, and food security (Ahmed et al. 2021; WFP 2021). Often, the poor and disadvantaged groups are the most vulnerable. Given these multiple disaster threats, improving land and water productivity is key to enhancing the resilience of the rural population and national food security (Kumar and van Dam 2013).

The land productivity or yield (production per unit of land) in all countries in SA has increased for the past decades. The yields of main food crops have increased since 1970 (FAO 2022). Cereal yield has increased by 80% since the 1990s. However, the water scarcity due to overexploitation of groundwater resources in the west and north-west and underdevelopment of water resources in the east is a constraint for increasing land productivity. Given these constraints, increasing water productivity or more crop, value or nutrition per drop while maintaining food and nutritional security and sustainable development targets received increased attention in recent policy discourses.

Two commonly used WP indicators are physical water productivity (PWP) and economic water productivity (EWP). They assess the production and value per cubic meter of consumptive water use or crop evapotranspiration (kg/m^3 and USD/m^3) of individual crops or cropping patterns (Molden et al. 2003). The other WP indicator assesses nutritional value (calories, proteins, vitamins, fat, etc.) embedded in the crops produced, expressed as nutritional value/m^3 of consumptive water use. Instead of consumptive water use, some use applied water in the denominator. Increasing PWP is a better strategy when water and land are not constraints. However, increasing EWP could be a better strategy if climate risks are recurrent and water scarcity is imminent or already a threat (Amarasinghe et al. 2021). Given that most South Asian farmers are smallholders, other factors, such as food security, land suitability, and access to credits and markets, determine the change of focus from PWP to EWP increase. Among these factors, one major lacuna that undermines informed policy and investment decisions is the inadequate knowledge of the spatial and temporal variation of different water productivity measures.

This chapter only shows the physical and economic WP variation of the South Asian countries at the national and subnational levels (where data permits) and the changes over time. After this brief introduction, we give a Synopsis of South Asia, mainly the changes in key indicators and recent studies relevant to the WP in the SA countries. The section introduces the methodology of WP estimation and data. The Sect. 10.3 shows the spatial and temporal trends of PWP and EWP at different geographical scales. This chapter concludes with potential strategies for improving WP for South Asian countries.

10.2 Synopsis of South Asia

Poverty is still a significant development constraint in many SA countries. Over a quarter of Bangladesh, Nepal, and Pakistan's populations are below the national poverty line (Table 10.1). Over 20% of the Indian population is income-poor, but large states such as Bihar, Uttar Pradesh, and Jharkhand have income poverty of over 30%. Bihar and Jharkhand also have high multidimensional poverty (>47%). They have some of the highest multidimensional poverty in the whole world. Although India's national level of multidimensional poverty has decreased from 55% to 16% between 2005 and 2021 (UNDP and OPHI 2021), because of the large population, the absolute number of income or multidimensional poor people in the SA countries and the subnational level is very high.

Agriculture is a significant livelihood option for many SA people. Except for the Maldives, over a quarter of the population in SA has agricultural-dependent livelihoods, which in Nepal and Bhutan are over 50%. The total agricultural population in SA is close to 800 million people, more than Europe's population or the combined population of Western and Eastern Africa. Disasters due to low availability or high water supply variability, whether rainfall or irrigation, are significant constraints to human development and economic growth, affecting a large part of the agriculturally dependent population in SA (Amarasinghe et al. 2020; Amarnath et al. 2017).

As water is a critical input to production, most agricultural-dependent livelihoods are severely vulnerable to water scarcity. Physical water scarcity, defined as water availability, affects a substantial part of SA (Molden 2013), meaning that no

Table 10.1 Key social and developmental indicators of South Asian countries between 2018 and 2020

Indicators									
	Bangladesh	Buthan	India	Maldives	Nepal	Pakistan	Sri Lanka	South Asia	World
Total population (millions)	165	0.77	1380	0.54	29	221	22	1857	7761
Population density (persons/ha of land area)									
Rural population—% of total	63	58	66	60	80	63	81	66	44
Agricultural dependent population—% of total	39	56	43	9	65	38	26	43	27
Income poverty—% of population	28	10	22	5	25	30	5		
Multidimensional poverty—% of population	10	9	28	28	18	38	3	13	
% primary school enrolment—total	93	100	97	100	91	74	100	94	90
% primary school enrolment—female	97	91	98	97		61	96	91	90
Maternal mortality (per 100,000 live births)	186	193	151	54	207	146	36	170	215
Prevalence of stunting, height for age (% of children <5 years)	31	23	32	15	31	38	16	33	22
HDI									
Per capita GDP (2015 Constant USD)	1561	3072	1899	8893	1039	1479	4146	1807	10,807
Agricultural GDP-share of total (%)	12	17	17	6	22	21	8	17	4
Freshwater resource (m^3/person)	654	104,006	1075	59	7114	262	2449	1099	5690

Total water withdrawn—% of freshwater	34	0	45	16	5	357	25	47	9
Agricultural water withdrawal—% of total	88	94	90	0	98	94	87	91	72
Water productivity (Constant 2015 US$ GDP per m^3 of total freshwater withdrawal)	6	7	4	1037	3	2	7	3	20

freshwater resources exist for further development. Most countries have low per capita internal freshwater resources (<1700 m^3/person), and nearly half is already withdrawn for different uses. Pakistan has severe physical water scarcity, with total withdrawals exceeding three times its internal renewable water resources. Many regions in India also have the same predicament, where large areas and populations already faced water scarcity in 2000 (Amarasinghe et al. 2007). With a high likelihood of increasing variability of precipitation and temperature (Nepal and Shrestha 2015; Lutz et al. 2019), climate change could push more people to face moderate to severe water scarcity and water insecurity, with implications for food production and food security.

As a result of extreme weather and recurring droughts and floods, most SA countries ranked among the top ten countries with high climatic risks (Eckstein et al. 2021). In 2019, India ranked seventh highest climate-risk country, and Pakistan ranked eighth highest. Eckstein et al. ranked Sri Lanka third highest climate-risk country in 2017. Recurrent floods and droughts expose large areas (Amarnath et al. 2017), causing significant economic damage. Since 2000, the floods and drought have damaged SA countries by over 167 billion USD (EM-DAT database. https://www.emdat.be/database). The 2010 floods in the Indus River Basin affected over 20 million people, with about 2000 deaths in India and Pakistan, and destroyed 1.3 million homes, and financial losses of over 10 billion USD in financial losses, of which 2.6 billion was in Punjab. The losses from the 2022 floods, which affected one-third of Pakistan, are significantly higher.

With highly variable weather (Fig. 10.1) and severe physical and economic water scarcity, increasing water productivity could enhance South Asian resilience. The United Nations sustainable development goals (SDGs) define water productivity as the GDP per m^3 of water withdrawn, which in SA is only one-seventh of the global average. The highly industrialized countries with good services sectors contributing to higher GDP and consuming more water contribute to this world average. For example, Maldives highlights this stark difference, where its agriculture sector contributes only 6% of the GDP and accounts for less than 1% of the total water withdrawals. But, with low per capita GDP and a 90% share of agricultural water withdrawals, India, Nepal, and Pakistan have less than 4 USD/m^3 of SDG WP. For these high agricultural-dependent and water-consuming countries, it is imperative to increase the water productivity of the agricultural outputs.

10.3 WP Variations in South Asia

Three major crops—rice, wheat, and maize—account for 39% and 87% of SA's total crop and cereal harvested area (FAO 2022). Rice and wheat are essential food security crops; they supply about half the nutritional supply of 2600 kcal/day/person. Overall, cereal crops contribute 53% of the daily calorie supply. Also, all cereal crops indirectly contribute to producing animal products as feed supply (dry fodder and concentrates), especially for milk production. Milk contributes to 6% of the

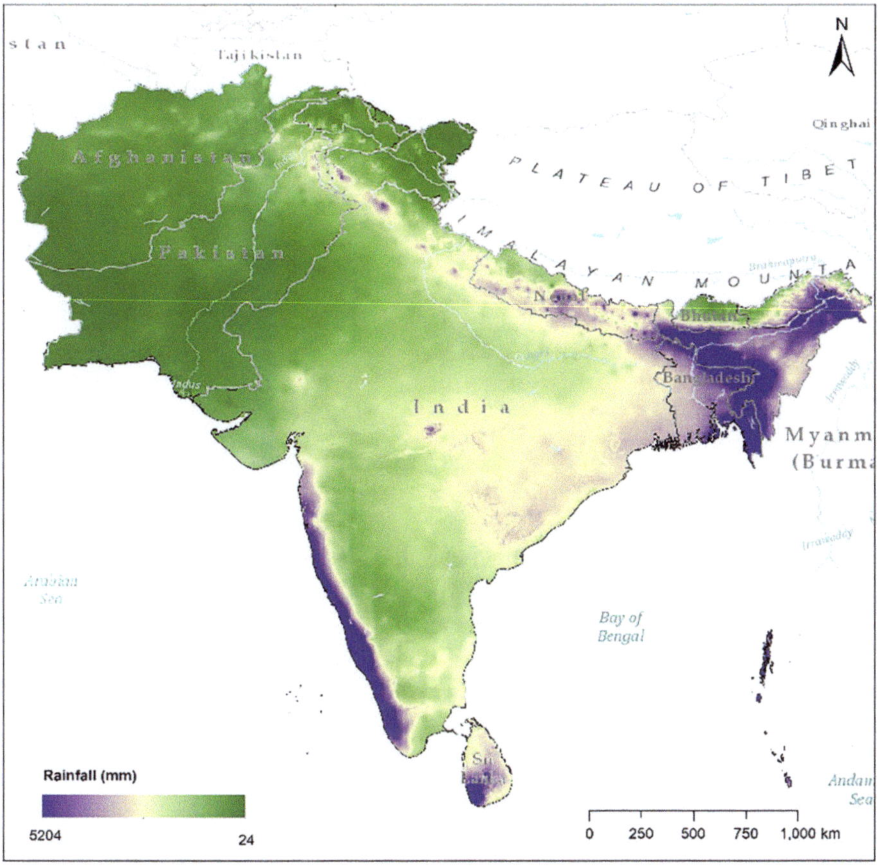

Fig. 10.1 Annual rainfall variation in South Asia

daily calorie supply. Given the increasing water scarcity, the water productivity variations of major cereals are essential for understanding SA's water and food security assessment. This paper focuses on the water productivity variation of rice, wheat, and maize.

Only a few cross-country comparisons of WP are available. Foley et al.'s (2020) meta-analysis included crop WP reporting of rice, wheat, and maize from 148 growing sites across the globe under irrigation conditions. The meta-analysis, which included Remote Sensing (RS) and non-RS estimates of 31 countries, reported eight rice and wheat WP estimates and two maize WP estimates. This analysis reported significant differences between northern and southern latitude countries. However, the SA's rice and wheat WP is moderately high, but that of maize is low compared to global standards. The global moderate WP of rice, wheat, and maize vary from 0.70 to 1.25, from 0.75 to 1.01, and from 1.25 to 1.75 kg/m^3, respectively, while the averages SA are 0.71, 0.89, and 0.91 kg/m^3. Bastiaanssen and Steduto's (2017) global assessment used RS to assess WP and reported 2.3, 2.45, and 4.9 kg/m^3 for the 99% percentiles of rice, wheat, and maize WP, indicating substantial scope for increasing the WP of many countries with locally adaptable best practices.

Many regional and national WP assessments show a substantial variation of WP within countries or river basins (Cai and Sharma 2010; Alauddin et al. 2014; Amarasinghe et al. 2014; Zwart and Bastiaanssen 2004). We examine WP's inter- and intra-country variations in these studies of major food crops in the SA.

10.4 Bangladesh

10.4.1 WP Growth and Variations

Rice is the dominant crop in Bangladesh, accounting for 80% of the gross cropped and irrigated areas. The average WP of rice in 1999–2001 was 0.44 kg/m^3 (Table 10.2). However, the rice WP varies across seasons and districts. The Boro rice (or the rabi season from November to March) has the highest WP of 0.56 kg/m^3, while that in Aus (August to November) and Aman (May to July) have relatively lower WP, 0.30 and 0.37 kg/m^3, respectively. For Boro rice, irrigation is critical, while Aus and Aman's rice receives occasional irrigation.

Rice WP increased by 64% between 1968–1980 and 2001. The variation is also significant from Non-Ganges to Ganges-dependent, no-drought-prone to drought-prone, or non-salinity to salinity-prone districts.

Productivity variation shows that Boro season rice has a clear edge in crop yield over Aus and Aman) season rice (Fig. 10.2a, b). Irrigation and temperature have a clear impact on Boro season crop yield. WP of Boro rice is still higher than Aus or Aman rice. However, higher CWU in Boro rice, much of which is from irrigation, is not necessarily associated with higher water productivity. Therefore, although access to irrigation can boost land productivity, applying the right irrigation volume can keep WP higher than the other two seasons. This is important because most

Table 10.2 Rice WP trends in Bangladesh

Seasons and districts	1968–1980[a]	1980–1990[a]	1991–2004[a]	2009–2011[b]
Kharif (Aus–Aman) rice	0.26	0.31	0.41	0.30–0.37
Rabi (Boro) rice	0.40	0.45	0.51	0.56
Total rice	0.28	0.34	0.45	0.44
Ganges dependent districts	0.33	0.31	0.36	
Non-Ganges dependent districts	0.38	0.23	0.31	
Drought-prone districts	0.21	0.34	0.44	
No-drought prone districts	0.26	0.34	0.47	
Salinity-prone districts	0.36	0.28	0.34	
Non-salinity prone districts	0.37	0.31	0.36	

Source: [a]Alauddin et al. (2014); [b]Amarasinghe et al. (2014)

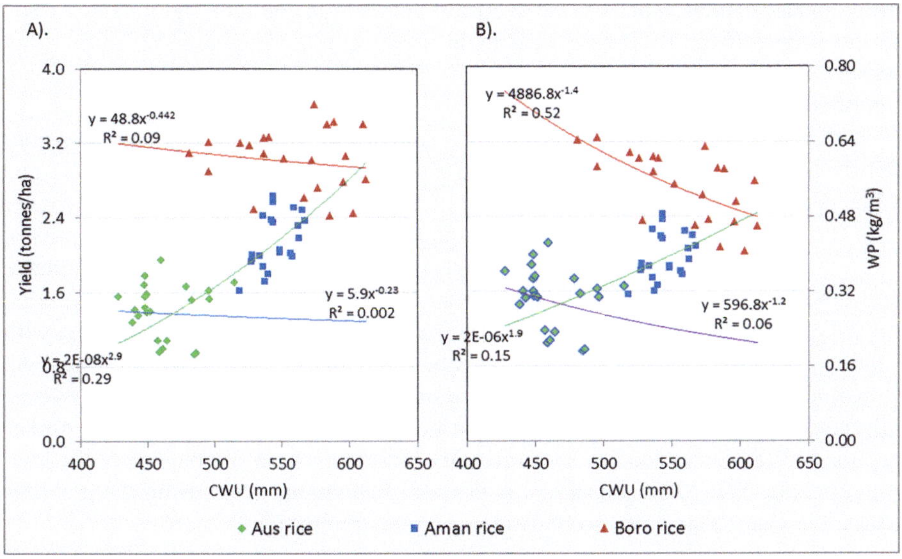

Fig. 10.2 Variation of land and water productivity across districts in 1999–2001. (**a**) Yield vs. CWU. (**b**) WP vs. CWU. (Source: Amarasinghe et al. (2014))

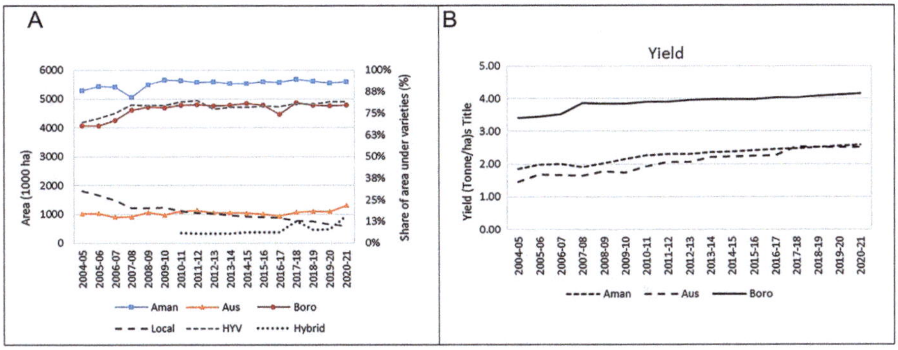

Fig. 10.3 Rice area and yield trends. (**a**) Rice area and varieties in Bangladesh. (**b**) Rice yield. (Source: Authors' estimates)

Boro rice areas use groundwater (reference). Unless used efficiently, Boro rice can contribute to significant groundwater overexploitation.

Although the Boro rice area in Bangladesh increased significantly from the 1991 to 2001 level, the growth has been levelling off in recent years (Fig. 10.3a). The primary Y-axis of Fig. 10.3a shows the rice area in the Aman, Aus, and Boro seasons. The secondary Y-axis shows the share (%) of the total rice area under local, high-yielding, and hybrid varieties. However, the yield has increased continuously (Fig. 10.3b) due mainly to the increased use of high-yielding (HYV) and hybrid varieties. Much of the Boro rice area is under groundwater irrigation. The groundwater depletion and the steep cost of abstraction could be significant reasons for levelling off the Boro rice area. Yield increase could increase WP further, but water management to reduce the excessive CWU should be a major contributor to the future growth of rice WP.

Agriculture diversification is another dimension Bangladesh could consider boosting WP, But diversification should shift the focus from PWP to EWP, i.e., value per unit of CWU ($/m^3). The EWP of most other crops is substantially higher than the rice crop (Table 10.3). In particular, the economic WP of wheat, pulses, oilseeds, potatoes, and vegetables are more than twice that of Boro rice, while maize and other cereals have substantially higher WP than Aus and Aman rice. If food security, especially rice, is still a target to reach, increasing the Boro rice area with smaller Aus and Aman rice would be a better option. If agronomically feasible, substituting a part of Aus and Aman rice crops with maize, other cereals, pulses, oilseeds, potatoes, and vegetables would be a better option for generating higher economic WP. Sugarcane has a significantly higher physical WP but one of the lowest economic WP. Since sugarcane is primarily an annual crop, using the precious irrigation water, especially groundwater, requires rethinking.

10.4.2 India

Cereals dominate India's cropping patterns, accounting for 51% of the gross cropped area (GCA) of 189 MHa, of which rice, wheat, and maize account for 23%, 17%, and 5% (GOI 2023). Other cereals and pulses account for 6% and 12% of the GCA. These crops provide 56% of the daily nutritional intake (calorie supply). Because of their importance in food security, cereals are the main focus of this section for WP analysis.

India's cereals yield and WP increased significantly during the last few decades. Rice, wheat, and maize yield increased by 44%, 38%, and 84%, respectively, between 1990 and 2020, while the physical WP increased by 45%, 33%, and 84% over the same period. Trend lines show that yields and PWPs have gradually increased over the three periods. Irrigation is a major driver of cereal yield and WP growth. Figure 10.4 shows that cereal yield and WP significantly correlate with the percentage of the irrigated cereal area of major cereal-producing states.

Irrigated rice-wheat or rice-rice is the dominant crop combination of cereals in the northern states (Table 10.4). Rice-wheat is the dominant crop combination in Punjab and Haryana, and almost all rice and wheat crop areas receive irrigation. Rice and wheat WP, and hence food grain WP in these states, are substantially high. In Uttar Pradesh, especially in the western parts, a large part of the rice and wheat areas are irrigated. Because of that, food grains in Uttar Pradesh also have high physical WP. In Bihar, most wheat areas receive irrigation but substantially less irrigation in rice areas. Thus, Bihar has a relatively lower yield and PWP. Rice is the dominant cereal crop in all seasons in Andhra Pradesh, Tamil Nadu, and West Bengal. Irrigation is a major driver behind higher WP of food grains in these states.

Table 10.3 Physical and economic water productivity of crop production in Bangladesh in 1999–2001

Water productivity (WP)	Aus rice	Aman rice	Boro rice	Wheat	Maize	Other cereals	Pulses	Oil seeds	Sugarcane	Potatoes	Vegetables
Physical WP (Kg/m^3)	0.30	0.37	0.56	1.29	1.00	0.71	0.71	0.79	3.88	3.98	1.96
Economic WP ($/m^3)	0.06	0.07	0.10	0.20	0.16	0.11	0.28	0.29	0.07	0.42	0.27

Source: Amarasinghe et al. (2014)

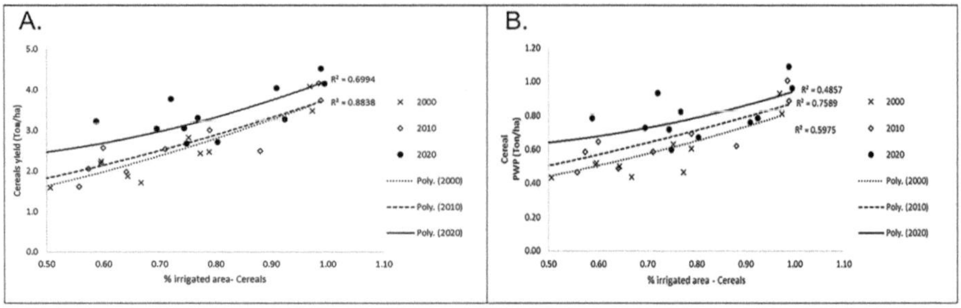

Fig. 10.4 Physical WP of food grains and irrigated crop area. (**a**) Yield and percent irrigated area of cereals. (**b**) WP and percent irrigated area of cereals. (Source: Authors' estimates)

Table 10.4 Water productivity of rice, wheat, maize, and cereals in 2018–2020

	Physical water productivity (kg/m^3)				Irrigated crop area—% of total crop area			
States	Rice	Wheat	Maize	Cereals	Rice	Wheat	Maize	Cereals
Punjab	0.81	1.87	1.00	1.20	100	99	63	99
Haryana	0.62	1.43	0.79	1.04	100	100	26	100
Tamil Nadu	0.74		3.64	0.96	94		34	72
Karnataka	0.65	0.25	1.32	0.93	76	64	30	40
Madhya Pradesh	0.53	0.90	1.09	0.83	43	95	3	71
Uttar Pradesh	0.55	1.07	0.61	0.81	86	99	45	91
Rajasthan	0.42	0.93	0.64	0.80	73	100	2	69
West Bengal	0.69	0.97	2.19	0.73	55	100	100	58
Andhra Pradesh	0.62	0.27	1.51	0.71	98	100	52	88
Bihar	0.51	0.80	1.56	0.70	69	96	72	79
Gujarat	0.49	0.69	0.62	0.60	76	90	31	73
Assam	0.61	0.42	1.14	0.57	15	100	11	22
Maharashtra	0.51	0.44	0.78	0.56	27	75	28	28
Jharkhand	0.52	0.57	0.63	0.54	4	80	2	8
Chhattisgarh	0.41	0.32	0.84	0.41	36	91	12	36
Odisha	0.40	0.39	0.76	0.41	30	100	5	30

Source: Authors' estimates
Notes: Sorted in descending order of cereals' PWP

On the other hand, rice is the major foodgrain in Chhattisgarh, Orissa, and Jharkhand, but with a relatively lower irrigated area, it has lower rice, cereal yields, and PWP.

The cereal PWP varies significantly across districts (Fig. 10.5). The PWP of many northwestern districts (in Punjab, Haryana, and Western Uttar Pradesh) and some in Bihar, Karnataka, Rajasthan, Tamilnadu, and West Bengal are more than 0.90 kg/m^3; 124 districts are in this group. A few districts in some states, such as Chhattisgarh, Bihar, Odisha, Rajasthan, and many in Maharashtra, have PWPs below 0.4 kg/m^3. The other districts have moderately high physical WP but are significantly below the benchmarks achieved by the northwestern districts.

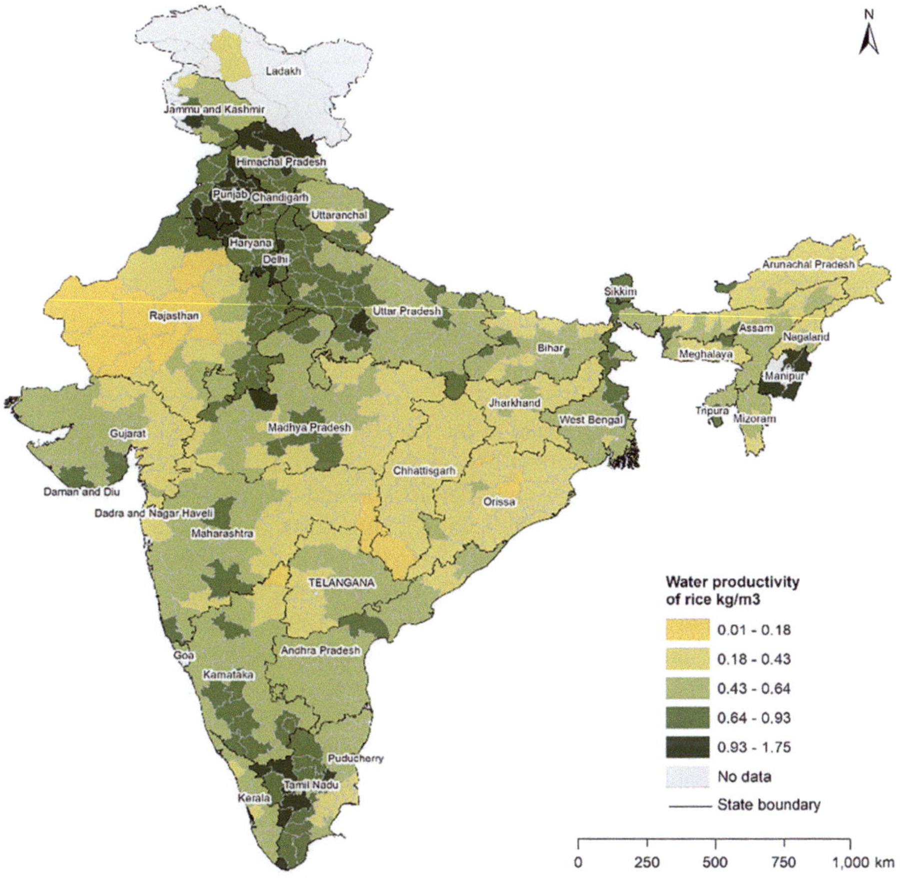

Fig. 10.5 Spatial variation of physical water productivity of cereals. (Source: Authors)

10.4.3 Pakistan

Rice, wheat, and maize are Pakistan's three major food crops. These crops occupy 50% of the total cropped area and provide half the daily calorie supply. The rice area, which accounts for 12% of the total cropped area, has increased by 76% in the last four decades (1981–2020). The rice yield and physical WP have increased by over 54% and 45% in the same period (Akbar et al. 2023). The highest WP gains are in the Sindh province, followed by Punjab, which accounts for almost the rice area. Yet the physical WP of rice is still low, with a maximum of about 0.16 kg/m^3 of applied water with a moderate variation across the country. Cai and Sharma (2010) showed a significant variation of rice WP (kg per m^3 of consumptive water use or ET) between the Pakistan part (0.69) and the Indian part (1.18) of the Indus basins, indicating the significant potential for increasing rice WP in Pakistan (Fig. 10.6).

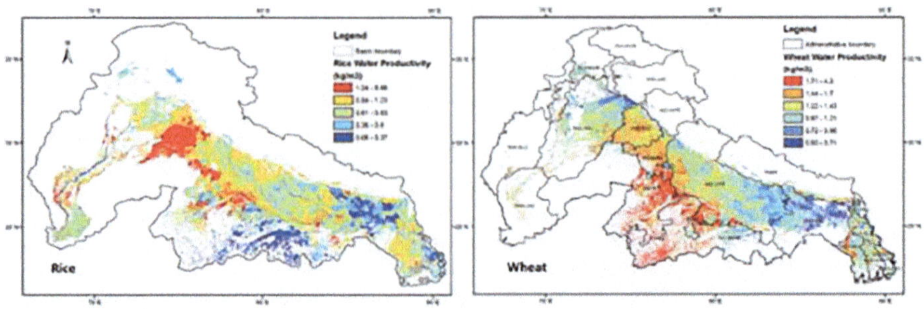

Fig. 10.6 Rice and wheat WP (kg/m^3) in Indus and Gangetic plains. (Source: Cai and Sharma (2010))

The wheat area, the largest share of the total cropped area (32%), increased significantly slower than the rice area (29% in the last four decades). However, wheat yield increased by 92% over the same period. Cai and Sharma showed that higher yields generally have higher WP, indicating a significant increase in wheat WP in the last few decades. Wheat WP in the Pakistan Punjab in the Indus is about 0.76 kg/m^3, about 26% less than the Indian Punjab state in the Indus. There are substantial intra-basin variations, too; Zwart and Bastiaanssen (2004) showed that the PWP varies between 0.6 and 1.6 Kg/m^3 in the Pinde Bhattian sub-basin with an average of 0.57 kg/m^3.

The cotton–wheat is another major cropping system in Pakistan. The total area of this system is 11.6 Mha, with Punjab province accounting for 76% of the total cotton–wheat area. Shabbir et al. (2012), in a study in Lower Bari Doab Canal (LBDC) of river Ravi, showed that there is a significant difference between applied PWP (kg/m^3 of applied water) and real PWP (kg/m^3 of CWU) of wheat in this system is 0.43 and 1.12 kg/m^3 but a substantially lower difference for cotton—0.22 and 0.26 kg/m^3.

10.4.4 Sri Lanka

Rice is the staple food of Sri Lanka, providing 42% of the total energy supply of 2524 kcal/day/person. Therefore, the WP assessment in Sri Lanka focuses on the rice crop, which occupies 90% of the crop-irrigated area.

Rice PWP in 2005 varied between 0.35 and 0.90 kg/m^3 across Divisional secretariat (DS) divisions, with an average of 0.44 kg/m^3. High rice yield in irrigated areas contributed to higher PWP in Hambantota, Polonnaruwa, Trincomalee, and Anuradhapura districts (Fig. 10.7a). The western, central, northwestern, and northern districts primarily depend on rainfall (rainfed or minor tank irrigation) and have lower WP.

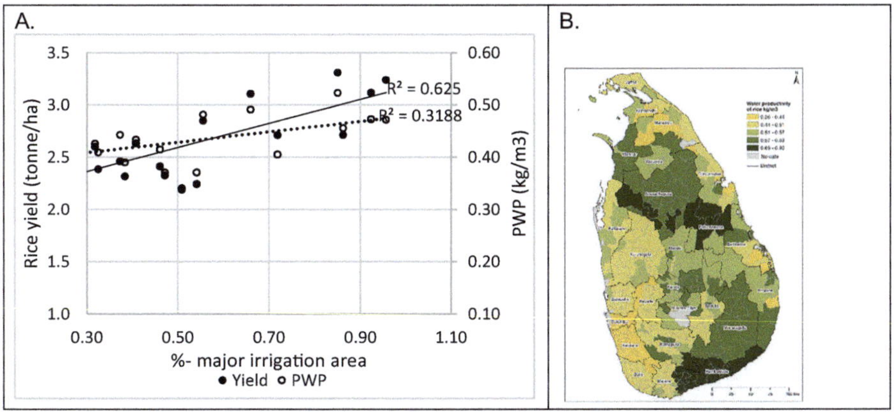

Fig. 10.7 Rice and wheat WP (kg/m^3) in Indus and Gangetic plains. (**a**) Yield and PWP variation of districts. (**b**) PWP variation across DS divisions. (Source: Authors)

Rice yield and PWP area have a high correlation (Fig. 10.7b), and higher values are associated with a higher share of major irrigation areas. This shows that there is still a substantial yield and PWP gap in many other areas where rice is irrigated. For example, the highest yield in the Main (Maha) season is 6034 kg/ha, and the median is about 3666 kg/ha. So, there is a significant potential for PWP increase in other areas with major irrigation facilities. Of the total major irrigation area of 285 ha, the four highest PWP districts—Polonnaruwa, Anuradhapura, Trincomalee, and Hambantota share only 48% of the area. A similar potential exists for the second season.

Moreover, rainfall in tank-irrigated and rainfed rice areas meets a substantial part of the CWU, but the yield is substantially lower in these regions. So, bridging the yield gap in irrigated and rainfed rice areas can increase marginal WP for consumptive water use. In fact, among the major rice-growing countries, Sri Lanka has the lowest rice yield growth in SA. India, Pakistan, Bangladesh, and Nepal have increased average rice yield from 1.91 to 3.43 (by 80%), 1.78 to 3.36 (by 88%), 2.53 to 4.89 (by 94%), and 1.89 to 3.17 (by 68%) tonnes/ha, while Sri Lanka increased its rice area from 2.92 to 4.68 (by 60%) tonnes/ha.

10.5 Recommendations and Conclusion

A major limitation of this analysis is the absence of a comparative information base on crop water productivity for South Asian countries. Indian WP data has sufficient spatial (district) and temporal (1990–2020) resolution. However, WP information available for other countries varies in spatial and temporal resolution. However, the available information shows that substantial variation in physical water productivity exists at the national and subnational levels. The WP variation exists in places where similar agro-climatic conditions prevail. This indicates there is scope for increasing crop yield and PWP by increasing the efficiency of other input use.

Many high PWP regions in SA have unsustainable groundwater use (Amarasinghe et al. 2012). Groundwater depletion from crops and milk production, i.e., crop evapotranspiration, in most parts of Punjab, Haryana, Rajasthan, Gujarat, and western Uttar Pradesh exceeds the natural recharge. The states of Karnataka, Maharashtra, Madhya Pradesh, and Tamil Nadu also have pockets of substantially high groundwater depletion. There are signs that crop yield is flattening in these regions due to water availability constraints. Given the unsustainable groundwater use for irrigation, these regions will have small marginal gains in physical WP for every unit increase in CWU and yield.

The water-scarce states such as Punjab, Haryana, and western Uttar Pradesh have many crop diversification options in India and Pakistan. The obvious option for them is to increase economic WP by diversifying crops. Amarasinghe et al. (2010) showed that by reducing rice-irrigated areas by 16%, Punjab could maintain sustainable groundwater use. Simultaneously, increasing milk and other crop production can increase economic WP to offset the income losses from rice area reduction. Deficit irrigation is also an option for water-scarce regions. In irrigated areas with high yield and consumptive use fraction, deficit irrigation to reduce consumptive water use can have large benefits. Deficit irrigation can save water and use it to increase irrigated areas in locations where land is not scarce. In areas where land is scarce, increasing the value of water should be the target.

Given the increasing weather variability, crop diversification can enhance smallholders' climate resilience. Case studies in Maharashtra show that diversifying a part of the area to horticulture crops reduces the vulnerability of farmers in drought years while they increase the income in moderate to good rainfall years (Amarasinghe et al. 2021).

Many low PWP regions have no major constraint on water availability; only access to irrigation is a major issue. The strategy for these regions is to gradually follow the paths of Uttar Pradesh, Haryana, and Punjab to increase physical WP to attain food security and then diversify the cropping patterns to enhance economic WP and income security.

Implementation of many of these interventions requires a new look at water governance mechanisms, including institutions, policies, rules, and regulations. What form of governance and how do these mechanisms increase coordination between different sectors and scales and increase farmers' access to inputs, services, and markets require further investigation?

Acknowledgments The authors would like to acknowledge the financial support from the Indian Council of Agricultural Research and CGIAR initiatives of ClimBeR: Building Systemic Resilience against Climate Variability and Extremes and Nexus Gains for collecting data and conducting analyses for this study.

References

Adger WN (1999) Social vulnerability to climate change and extremes in coastal Vietnam. World Dev 27:249–269. https://doi.org/10.1016/S0305-750X(98)00136-3

Ahmed JU, Akter S, Majumder KA (2021) Impact of COVID-19 on agricultural production and distribution in South Asia. World Food Policy 7(2):168–182

Akbar G, Hameed S, Islam Z (2023) Assessing water productivity and energy use for irrigating rice in Pakistan. Irrig Drain 72(2):478–486. https://doi.org/10.1002/ird.2772

Alauddin M, Amarasinghe UA, Sharma BR (2014) Four decades of rice water productivity in Bangladesh: a spatio-temporal analysis of district level panel data. Econ Anal Policy 44(1):51–64

Almazroui M et al (2020) Projections of precipitation and temperature over the South Asian countries in CMIP6. Earth Syst Environ 4:297–320. https://doi.org/10.1007/s41748-020-00157-7

Amarasinghe U et al (2007) India's water future to 2025–2050: business-as-usual scenario and deviations, vol 123. IWMI

Amarasinghe UA, Smakhtin V, Sharma BR, Eriyagama N (2010) Bailout with white revolution or sink deeper?: groundwater depletion and impacts in the Moga District of Punjab, India, vol 138. IWMI

Amarasinghe UA, Shah T, Smakhtin V (2012) Water-milk nexus in India: a path to a sustainable water future? Int J Agric Sustain 10(1):93–108. https://doi.org/10.1080/14735903.2012.620336

Amarasinghe UA, Sharma BR, Muthuwatta L, Khan ZH (2014) Water for food in Bangladesh: outlook to 2030. Research Report 158. International Water Management Institute, Colombo, Sri Lanka

Amarasinghe U et al (2020) How do floods and drought impact economic growth and human development at the sub-national level in India? Climate 8(11):123

Amarasinghe UA et al (2021) Improving economic water productivity to enhance resilience in canal irrigation systems: a pilot study of the Sina irrigation system in Maharashtra, India. Water Policy 23(2):447–465

Amarnath G et al (2017) Mapping multiple climate-related hazards in South Asia, vol 170. International Water Management Institute (IWMI)

Bastiaanssen WG, Steduto P (2017) The water productivity score (WPS) at global and regional level: methodology and first results from remote sensing measurements of wheat, rice and maize. Sci Total Environ 575:595–611

Cai X, Sharma B (2010) An assessment of crop water productivity in the Indus and Ganges River Basins: current status and scope for improvement, vol 140. IWMI

Eckstein D et al (2021) Global Climate Risk Index 2021. Germanwatch e.V., 2019. https://germanwatch.org/en/17303. Accessed through Resource Watch. Sept 2022. www.resourcewatch.org

FAO (2022) FAOSTAT database. Food and Agricultural Organization, Rome. Last accessed via https://www.fao.org/faostat/en/#data

Foley DJ et al (2020) A meta-analysis of global crop water productivity of three leading world crops (wheat, corn, and rice) in the irrigated areas over three decades. Int J Digit Earth 13(8):939–975. https://doi.org/10.1080/17538947.2019.1651912

Government of India (2023) Land use statistics at a glance. 2010–2011 to 2019–2020. Directorate of Statistics, Ministry of Agriculture, New Delhi

Habiba U, Abedin MA, Shaw R (2014) Defining water insecurity. In: Water insecurity: a social dilemma. Emerald Group Publishing Limited

Harrison BJ et al (2021) Future sea-level rise projections for tide gauge locations in South Asia. Environ Res Commun 3(11):115003

Kumar MD, van Dam JC (2013) Drivers of change in agricultural water productivity and its improvement at basin scale in developing economies. Water Int 38(3):312–325. https://doi.org/10.1080/02508060.2013.793572

Lutz AF, ter Maat HW, Wijngaard RR, Biemans H, Syed A, Shrestha AB, Wester P, Immerzeel WW (2019) South Asian river basins in a 1.5 C warmer world. Reg Environ Change 19:833–847

Mekonnen MM, Hoekstra AY (2016) Four billion people facing severe water scarcity. Sci Adv 2(2):e1500323

Molden D (ed) (2013) Water for food water for life: a comprehensive assessment of water management in agriculture. Routledge

Molden D et al (2003) A water-productivity framework for understanding and action. In: Water productivity in agriculture: limits and opportunities for improvement. Cabi Publishing, Wallingford, pp 1–18

Nepal S, Shrestha AB (2015) Impact of climate change on the hydrological regime of the Indus, Ganges and Brahmaputra River Basins: a review of the literature. Int J Water Resour Dev 31:201–218. https://doi.org/10.1080/07900627.2015.1030494

Paegelow M et al (2022) Water vulnerabilities mapping: a multi-criteria and multi-scale assessment in central Chile. Water Policy 24(1):159–178

Rasul G (2016) Managing the food, water, and energy nexus for achieving the Sustainable Development Goals in South Asia. Environ Dev 18:14–25

Sathre R, Antharam SM, Catena M (2022) Water security in South Asian cities: a review of challenges and opportunities. Civ Eng 3:873–894. https://doi.org/10.3390/civileng3040050

Scott CS et al (2021) Do ecosystem insecurity, and social vulnerability lead to failure of water security? Environ Dev 38. https://doi.org/10.1016/j.envdev.2020.100606

Seckler D, Barker R, Amarasinghe U (1999) Water scarcity in the twenty-first century. Int J Water Resour Dev 15(1–2):29–42

Shabbir A et al (2012) Apparent and real water productivity for cotton-wheat zone of Punjab, Pakistan. Pak J Agric Sci 49(3):357–363

Soussan J, Arriens WL (2004) Poverty and water security: understanding how water affects the poor. Asian Development Bank, Manila

UNDP and OPHI (2021) Global Multidimensional Poverty Index 2021 – unmasking disparities by ethnicity, caste and gender. United Nations Development Programme and Oxford Poverty and Human Development Initiative

WFP (2021) Food crisis in Sri Lanka likely to worsen amid poor agricultural production, price spikes and ongoing economic crisis. FAO and WFP warn. https://www.wfp.org/news/ food-crisis-sri-lanka-likely-worsen-amid-poor-agricultural-production-price-spikes-and-ongoing

World Bank (2022) Poverty and shared prosperity 2022: correcting course. World Bank, Washington, DC

Zwart SJ, Bastiaanssen WG (2004) Review of measured crop water productivity values for irrigated wheat, rice, cotton and maize. Agric Water Manag 69(2):115–133

Chapter 11
Nanotechnological Applications to Improving Agricultural Water Productivity

S. T. Hadebe

Abstract Recent scientific data suggests that nanotechnology has the potential to enhance agricultural water productivity by directly boosting yield production and crop water use. However, no studies on nanoparticles to date have quantified their impact on crop water productivity. Given the vastness of literature on nanoparticles either affecting crop water use or crop productivity, this chapter sought to theorise on a potential relationship between nanotechnology and agricultural water productivity by conjecturing on published data of nanotechnological impact on crop productivity and crop water use. Crop productivity extrapolations were made from published data on nanoparticle impacts on crop yield-related processes (germination and emergence, photosynthetic activity and biomass accumulation, root growth and water uptake potential), while crop water use was extrapolated from soil-water related processes (soil physical and hydraulic properties, plant-available water, and plant water uptake). The findings of this review chapter suggest that nanoparticle application to agricultural systems has a high potential to enhance crop water use and crop yield attributes, which implies a high potential to improve crop water productivity in agricultural systems. However, conducting a metadata analysis on existing literature and experiments using nanomaterials highlighted as potential agricultural water productivity enhancers may paint a more conclusive picture of the potential of nanomaterials to enhance agricultural water productivity and can be useful in affirming/rejecting the conjectural theory developed in this review that optimal application of nanoparticles highlighted in this chapter can improve agricultural water productivity.

Keywords Nanoparticles · Seed priming · Soil hydraulic properties · Photosynthetic enhancement · Phytotoxicity

---

S. T. Hadebe (✉)
Department of Plant Production, Soil Science and Agricultural Engineering, University of Limpopo, Sovenga, South Africa
e-mail: sandile.hadebe@ul.ac.za

T. Mabhaudhi et al. (eds.), *Enhancing Water and Food Security Through Improved Agricultural Water Productivity*,
https://doi.org/10.1007/978-981-96-1848-4_11

11.1 Water Productivity as an Integral Component of Sustainable Agricultural Systems

Agricultural systems globally suffer tremendous losses annually from degraded soils, reduced water quality, and related crop losses (Kopittke et al. 2019). Ironically, agricultural activities are a major contributor to soil degradation, water quality reduction, and an overall reduction in agricultural productivity through intensive tillage and monoculture cropping systems (Chugh et al. 2021). Agricultural activities degrade soils through soil erosion, chemical degradation (organic matter depletion, associated high nutrient losses and acidification) and biological depletion (Montgomery and Biklé 2021), reducing the productivity of agricultural soils. Poor-quality soils tend to have poor soil structure, low water holding capacity, and high water drainage, which reduce plant-available water for agricultural productivity, particularly under water-limited conditions. Several strategies have been employed to successfully mitigate water-related crop yield losses (Molden et al. 2010); however, the challenge of feeding an ever-growing global population through finite resources persists. These strategies are used to improve yields from the same amount of water used and reduce water used to produce the same crop yields. Strategies suggested to improve agricultural productivity have included agronomic and soil management (seed priming, seedling age manipulation, direct or wet seeding, crop and cultivar choice, integrating agriculture and aquaculture, soil fertility management, organic matter addition, tillage, and soil mulching practices) and engineering solutions (rainwater harvesting, deficit irrigation, proper deficit sequencing, and modernisation of irrigation system) (Ali and Talukder 2008).

Nanotechnology refers to using nanoparticles with at least one dimension in the order of 100 nm or less or as a colloidal particulate system with sizes ranging between 10 and 1000 nm (Jatav and De 2013). Nanoparticles hold great promise regarding their application in agricultural water productivity due to size-dependent qualities, high surface-to-volume ratio, and unique optical properties. Nanotechnological applications for agricultural benefit have gained popularity over the past two decades. Reported agricultural benefits have covered multiple uses such as remediation of soils and water sources, nanofertilisation, soil improvements, pest and disease control, microbial activity manipulation, abiotic stress tolerance, nanosensors for stress detection, manipulation of plant breeding techniques, and enhanced efficiency of crop metabolism for high yields (Shang et al. 2019; Pramanik et al. 2020; An et al. 2022; Nile et al. 2022; Rajput et al. 2022). Despite a litany of literature indicating the beneficial effects of nanotechnological materials in enhancing crop water uptake and crop productivity (Table 11.1), researchers have shied away from quantifying the potential impacts of nanoparticles on water productivity. This chapter attempts to bridge this gap and report on nanoparticles directly affecting soil water availability, crop water uptake, and enhanced biomass and yield production. Hence, an effort is made to extrapolate the potential impacts of such nanoparticles on crop water productivity based on their working mechanisms. The review excludes nanoparticles that may have a secondary/indirect effect on

Table 11.1 Nanoparticles can enhance agricultural water productivity by modifying crop water use and productivity

Effect	Nanomaterial	Mechanism(s) of action	References
Improved soil aggregation	Zinc oxides and iron	Induce extracellular polysaccharide secretion from soil bacteria	Tarafdar et al. (2012) and Raliya et al. (2014)
	Nanozeolite	Increases the mean weight diameter attributes of water-stable aggregates	Aminiyan et al. (2015)
Improved soil water retention and availability	Nanobentonite	Boosts water holding capacity, soil aggregate stability, exchangeable cations and plant-available water in sandy soils	Mi et al. (2020) and El Nagar and Sary (2021)
	Nanosilica	Improves soil hydraulic behaviour (field capacity, permanent wilting point, plant-available water, and water holding capacity)	AlSaeedi (2022)
Germination and emergence enhancement	Multi-walled carbon nanotubes	Regulates biosynthetic avenues by reconfiguring C18:3 enriched fatty acids involved in easing seed dormancy during germination, activating lipid metabolism in seeds, and upgrading aquaporins, ions, and water mobilisation in cell membranes	Ali et al. (2020) and Martinez-Ballesta et al. (2016)
	Iron sulphide	Generates endogenous H_2O_2 radicals in the cotyledons, thus promoting faster breakdown of the stored carbohydrates in the cotyledons; modifies dehydrogenase activity, enhancing respiration and water consumption in germinating seeds	Das et al. (2016) and Rawat et al. (2017)
	Manganese oxide	Induces phytohormone changes (abscisic acid, gibberellic acid, salicylic acid, and jasmonic acid) during germination and seedling emergence	Liu et al. (2016)
	Nanosilver	Enhance solubility of starch through regulation of α-amylase; upregulate aquaporin genes in germinating seeds; induce mild oxidative stress that triggers cell wall softening in germinating seeds	Munegumi et al. (2016) and Nile et al. (2022)
	Gold, silicon, & titanium oxide	Enhance germination percentage, radicle and plumule length, seedling vigour, seedling shoot and root length	Kumar et al. (2013), Dehkourdi et al. (2014), and Siddiqui and Al-Whaibi (2014)

(continued)

Table 11.1 (continued)

Effect	Nanomaterial	Mechanism(s) of action	References
Photosynthetic and biomass accumulation enhancement	Silicon oxide	Enhances proline activity, carbonic anhydrase activity, water capacity, carotenoid content, electron transport rate synthesis of photosynthetic pigments, stomatal conductance and transpiration rate, photosystem II activity, lignification in plant tissues, and photochemical quenching	Siddiqui and Al-Whaibi (2014), Asgari et al. (2018), Udalova et al. (2020), and Verma et al. (2021)
	Titanium oxide	Stimulates Rubisco enzyme activity, boosting chlorophyll and carotenoid light absorption, oxygen evolution, water splitting, and provision of photosystem I to photosystem II to catalyse carbon dioxide assimilation	Yang et al. (2006), Samadi et al. (2014), and Hasanpour et al. (2015)
	Aluminium oxide	Modulates reactive oxygen species production and plant redox status by increasing 2,2-diphenyl-1-picryl hydrazyl scavenging activity, total antioxidant capacity, total reducing power, total iridoids content, total saponin content, and total phenolic content	Chahardoli et al. (2020)
	Iron oxide	Boosts root growth, plant biomass accumulation, and photosynthetic pigments; modulates antioxidant enzymes and phytohormones (abscisic acid and enhanced gibberellic acid content) activity	Rui et al. (2016), Tawfik et al. (2021), and Feng et al. (2022)
	Zero-valent iron nanoparticles	Increases concentration of photosynthetic pigments (chlorophyll a/b ratio)	Majumdar et al. (2016) and Abbasi Khalaki et al. (2021)
	Zinc oxide	Boosts chlorophyll content and stimulates enzymatic activity (acid phosphatase, alkaline phosphatase, and phytase); change the concentration of carbinolic-based compounds; counters cadmium-induced stress by reducing oxidative damage	Patra et al. (2013), Reddy Pullagurala et al. (2018), and Faizan et al. (2021)
	Manganese	Boosts photosynthetic rate by mediating water splitting in the electron transport system	Pradhan et al. (2013)
	Gold	Upgrades the electron transport system by intensifying chloroplasts' electron transport rate and oxygen evolution	Das et al. (2017)
	Silver	Increase intrinsic water use efficiency of plants by moderating stomatal conductance	Guilger-Casagrande et al. (2022)
	Multi-walled carbon nanotubes	Induces aquaporin expression and promotes antioxidant defences and plant photosynthesis	Yatim et al. (2018)
	Cesium oxide	Promotes transpiration rate and stomatal conductance, increases Rubisco activity	Majumdar et al. (2016) and Cao et al. (2018)

Root growth promotion	Silicon	Enhanced root elongation and upregulates aquaporin genes, which increases plant water extraction potential	Hattori et al. (2005) and Liu et al. (2015)
	Titanium oxide	Root lengthening. Reduces the synthesis of reactive oxygen species, which lengthens the root	Thiruvengadam et al. (2015)
	Silver	Blocking of ethylene and classical stress signalling reactions and plasma membrane conductance	Rezvani et al. (2012)
	Multi-walled carbon nanotubes, titanium oxide, manganese sulphate, zinc oxide, iron, gold	Enhance root length and root proliferation	Patra et al. (2013), Deng et al. (2014), Parveen et al. (2016), Rahimi et al. (2016), Faraji and Sepehri (2019), Ye et al. (2020), and Kumar et al. (2020)

agricultural water use and crop growth, such as pests and disease control, microbial activity manipulation, abiotic stress tolerance (other than water stress), nanosensors for stress detection, and manipulation of plant breeding techniques.

11.1.1 Soil Nanotechnological Improvements with Implications to Agricultural Water Productivity

There is a limited amount of arable land on the planet, and productivity of arable land remains limited due to soil degradation and environmental factors that threaten global food security. Environmental factors such as water availability, temperature, and soil physico-chemical properties are important in agriculture production (Mi et al. 2020). Soils are mainly degraded due to soil erosion, chemical degradation, and biological depletion (Montgomery and Biklé 2021), which reduces water availability and agricultural productivity in affected soils.

11.1.2 Nanoparticles for Improved Soil Aggregation

Soil aggregation and structure are essential to ensure improved carbon management and water retention in the soil–plant–atmosphere continuum. High soil aggregate stability from applying soil organic amendments is positively associated with improved bulk density, water infiltration rate, and hydraulic conductivity in soils (Pramanik et al. 2020). Sandy and coarse-textured soils have low soil aggregate stability, reducing soil water retention and crop water uptake and reducing crop productivity. In a few studies that have assessed nanomaterial impact on soil aggregation and soil structure, zeolite, zinc oxide and iron nanoparticles were reported to improve carbon build-up and soil aggregate stability in agricultural soils through two possible mechanisms (Tarafdar et al. 2012; Raliya et al. 2014; Aminiyan et al. 2015). The first outlined mechanism occurs when zinc oxides and iron nanoparticles induce extracellular polysaccharide secretion from soil bacteria (*Bacillus subtilis*, *Aspergillus terreus*, and *Aspergillus flavus*), which significantly increases soil aggregation, soil organic carbon, and soil water retention in treated soils (Tarafdar et al. 2012; Raliya et al. 2014). Increasing the mean weight diameter of water-stable aggregates and organic carbon content in each aggregate size fraction of the mean weight diameter application constitutes the second mechanism (Aminiyan et al. 2015). High soil mean weight diameter due to nanozeolite application is attributed to the high calcium content of zeolite mineral. The presence of calcium ions facilitates the formation of cation bridges between clay crystals and organic matter from stable soil aggregates. It provides structural stability in micro-aggregates, protecting organic carbon from enzymatic and microbial degradation (Pramanik et al. 2020).

11.1.3 Nanoparticles for Improved Soil Hydraulic Properties

Soil degradation often results in poor water and nutrient holding capacity, which reduces crop productivity in the long term, even under optimal resource supply. Soil degradation-related agricultural losses can be compounded by inadequate rainfall to meet the crop water requirements in arid and semi-arid regions, where rainfall can be low, highly variable and unpredictable (Hadebe et al. 2020). Drought occurrence reduces the availability of irrigation water and further reduces the suitability of degraded lands for crop production, reducing crop yields by up to 50% per year globally (Hadebe et al. 2017). Drought events are predicted to become more frequent and extreme due to a changing climate, necessitating a shift in current soil management practices intended to improve plant water availability. Soil is a three-phase medium composed of soil grains, water, and air between the grains, which makes pore water pressure significantly influence the soil's strength, especially under dynamic loading. At a macro level, nanoparticles improve soil strength by reinforcing the soil skeleton and modifying the pore fluid. As its particle size decreases, a nanoparticle's specific surface area increases. This means the atom's surface increases, leading to larger ion exchange capacity and increased interaction with other particles. More water molecules will surround soil particles with a larger specific surface area. The existence of nanoparticles leads to an accumulation of pore water; thus, soil that contains nanoparticles usually has higher liquid and plastic limits (Huang and Wang 2016).

11.1.4 Nanoclay Applications

Sandy soils have low agricultural productivity owing to their characteristically low organic matter, cation exchange capacity and water holding capacity. Moreover, sandy soils are prone to acidification due to low buffering capacity, reducing agricultural yields in such soils (Herawati et al. 2021). Application of soil amendments to improve soil physical and chemical properties can potentially increase the agricultural productivity of such soils. Bentonite is a rock consisting predominantly of the clay mineral montmorillonite. It has excellent qualities (high cation exchange capacity and water holding capability) required to improve the productivity of coarse-textured soils (Czaban et al. 2014). Nanobentonite has been reported to significantly improve water holding capacity, soil aggregate stability, exchangeable cations and plant-available water in coarse-textured soils (Mi et al. 2020; Zhang et al. 2020; El Nagar and Sary 2021). As a result, the application of nanobentonite enhanced emergence percentage, above-ground biomass accumulation, photosynthetic rate, transpiration rate, chlorophyll pigment concentration, and leaf water use efficiency in millets (Mi et al. 2020). This suggests that increased physical qualities and textured soils increase transpiration relative to other water losses, thereby enhancing crop yields relative to available water. By implication, this increases agricultural water productivity of sandy and degraded soils. Application of nanobentonite was also observed to enhance millet grain quality parameters (grain protein, fat and fibre content) (Mi et al. 2020), further suggesting possible enhancement to nutritional water productivity of key nutritional elements. It remains difficult to predict the lasting period of nanobentonite-related soil improvements (Czaban and Siebielec 2013).

11.1.5 Nanosilica

The high surface area to volume ratio of agricultural nanotechnology materials makes incorporating nanoparticles into soils potentially boost crop-available water, water uptake and water productivity in plants. Incorporating nanosilica has been reported to improve soil hydraulic behaviour (field capacity, permanent wilting point, plant-available water, and water holding capacity), thereby improving cucumber yields and water productivity in course-structured soils (AlSaeedi 2022). Improved soil hydraulic properties were attributed to increased soluble Si^{4+} ions in the soil under nanosilica application, leading to cucumber yield and water productivity improvement of up to 178% compared to untreated soils (AlSaeedi 2022). The role of nanosilicon in improving hydraulic properties of agricultural soils has received little attention, with much of the research attention given to nanosilica application as a nanofertiliser to facilitate silicon uptake, plant growth and alleviate plant stresses (Soares et al. 2018; Mathur and Roy 2020). More research is needed to confirm the findings cited here of improved soil hydraulic performance due to nanosilica application.

11.2 Nanotechnology-Based Crop Productivity Improvements

The impact of nanoparticles on crop yields depends on their composition, concentration, size, surface charge, physico-chemical properties, and the susceptibility of the plant species utilised (Fraceto et al. 2016). Nanoparticles can have both negative and positive effects on crop productivity-related parameters such as seed germination, emergence, root elongation, cell division, metabolic functions, chlorophyll synthesis, biomass and yield accumulation (Chugh et al. 2021) depending on the application environment and conditions. By implication, nanotechnology has a high potential to increase agricultural productivity; evidence is discussed below.

11.2.1 Germination and Seedling Establishment Enhancement

Several processes, including plant tissues, hormones, and genes, induce different seed dormancy levels in crops. This can slow germination, leading to low, non-uniform, and delayed emergence. Under perceived water stress, crops' seed dormancy can be enhanced, leading to low plant stand, reduced plant vigour, and agricultural yields. Seed priming is a promising approach to improve germination and seedling development by altering the physiological state of the seed (Nile et al. 2020). The popularity of seed priming using nanoparticle solutions (nanopriming) is increasingly growing in seed science due to reported improvements in seed germination, growth, and yield of crops. Thus far, several synthetic metallic nanoparticles and carbon nanotube-based seed priming methods have been used to improve seed germination and seedling emergence (Kasote et al. 2021).

11.2.1.1 Multi-walled Carbon Nanotubes (MWCNTs)

Multi-walled carbon nanotubes (MWCNTs) are elongated cylindrical nano-objects made of sp^2 carbon. Nanotube diameter ranges from 3 to 30 nm and can be several centimetres long, which results in an aspect ratio of up to 10 million. Due to their unique physico-chemical properties, MWCNTs have received considerable scientific, commercial, and biotechnological interest (De Volder et al. 2013), particularly for agronomic applications. Contrasting effects of MWCNTs have been reported on seed germination and establishment of various crops, which implies that MWCNT seed priming can potentially have contrasting effects on crop productivity depending on application conditions. Intrinsic characteristics (shape, dimensions, electrical conductivity, stability, and limited solubility) (Scown et al. 2010), together with concentrations and crop species treated (Jackson et al. 2013), have been reported to account for the contrasting effects of seed priming. Early, uniform, and high crop emergence (Lahiani et al. 2013; Juárez-Cisneros et al. 2020) after MWCNT seed priming treatment. These observations persisted even when high vigour maize hybrids were used. On the negative side, MWCNTs have been reported to induce cell death and cause phytotoxic effects in numerous crops, including lettuce. Phytotoxic effects include germination inhibition and limitations on growth and biomass accumulation (Ikhtiari et al. 2013; Hatami 2017). During crop establishment and early vegetative growth, MWNCTs can enhance root elongation and branching (Tripathi et al. 2017), while in contrary cases, it has been reported to decrease root and shoot length and biomass accumulation (Hatami 2017). Despite contradictory reports, the consensus is that MWCNTs improve crop germination and establishment due to improved water imbibition and root uptake of seeds and seedlings if specific application guidelines are followed during seed priming (El-Sanatawy et al. 2021). We theorise that increasing crop biomass and yield due to enhanced crop stand and vigorous seedling under water stress potentially enhance agricultural systems 'crop per drop'. Multi-walled carbon nanotubes seem to regulate biosynthetic avenues by reconfiguration of C18:3 enriched fatty acids (phosphatidylethanolamine, digalactosyldiacylglycerol, phosphatidylcholine, and phosphatidic acid) involved in easing seed dormancy during germination (Ali et al. 2020). This activation of lipid metabolism in seeds upgrades aquaporins, ions, and water mobilisation in cell membranes, which results in enhanced seed germination and seedling vigour (Martinez-Ballesta et al. 2016; Nile et al. 2022).

11.2.1.2 Iron Sulphide

Metal nanoparticles have been reported to influence plant growth and productivity without completely understanding underlying molecular mechanisms. Regarding iron sulphide (FeS2), an emerging theory suggests that FeS2 nanoparticles generate endogenous H_2O_2 radicals in the cotyledons, thus promoting faster breakdown of the stored carbohydrates present in the cotyledons. This ultimately results in an enhanced number of germinating seeds and vigorous seedling growth through increased seed metabolism (Das et al. 2016; Rawat et al. 2017). Nanopriming using Fe and Zn nanoparticles reduces oxidative damage by enhancing enzymatic activity and free radical scavenging in germinating seeds (Korishettar et al. 2017). This emerging theory is further supported by observations of metallothionein genes (*MT1* and *MT4*) induction in tomato seeds, which could explain the participation of FeS_2 nanoparticles in reactive oxygen species signalling during seed germination (Anand et al. 2019). Additionally, modifications in dehydrogenase activity enhance respiration and water consumption by germinating seeds, which leads to enhanced germination and crop establishment (Nile et al. 2020). Unsurprisingly, FeS_2 has been reported to improve crop germination and seedling growth (Li et al. 2013; Kasote et al. 2019), which increases the competitive advantage for water resources, thus potentially improving the water productivity of agricultural systems.

11.2.1.3 Manganese Oxides

Manganese (Mn) is a plant growth-regulating micronutrient that sustains multiple photosynthesis metabolic roles within different plant cell compartments (Alejandro et al. 2020). In contrast, the use of Mn in seed priming is less reported. Where reported, contradicting findings on plant responses have been found (Liu et al. 2016; Ye et al. 2020; Kasote et al. 2021). For instance, positive germination responses to manganese oxide nanoparticle (MnO) seed priming have been reported on watermelons by Kasote et al. (2021), while insignificant responses were reported by Liu et al. (2016) on watermelons. Manganese oxide nanoparticle priming has also been reported to enhance seedling growth in lettuce (Liu et al. 2016) and seedling root growth in bell peppers (Ye et al. 2020). The genetic specificity of plant responses to nanometals is further emphasised in MnO nanoparticle-treated watermelon, where positive responses have only been reported in triploid watermelon lines (Kolbert et al. 2022). Enhanced germination and seedling establishment in lettuce were associated with phytohormone changes in hormones such as abscisic acid, gibberellic acid, salicylic acid, and jasmonic acid during germination and seedling establishment (Liu et al. 2016).

11.2.1.4 Silver Nanopriming

Silver nanoparticles (AgNPs) can penetrate seeds, increasing their efficacy as seed primers. After imbibition, AgNPs have been shown to enhance the solubility of sugars through the regulation of α-amylase. The active site of α-amylase starch hydrolysis activity is retained in its catalytic function (Munegumi et al. 2016). The interaction of molecules overlaid onto the coverings of silver nanocarriers intensifies enzyme conformational change, enzyme activity and stability, leading to more significant starch degradation (Nile et al. 2022). Similarly to MCWNTs, AgNPs have also been shown to upregulate aquaporin genes in germinating seeds (Munegumi et al. 2016). Another mechanism by which AgNPs enhance seed germination is by inducing mild oxidative stress that triggers cell wall softening (Nile et al. 2022). Increased catalase activity after imbibition has also been suggested as a possible mechanism by which AgNPs enhance seed germination (Munegumi et al. 2016). Overall, these mechanisms of AGNP action account for higher respiration rates, enhanced seed germination and vigour, and increases in seedling root and shoot accumulation after planting (Omelchenko et al. 2014; Siddiqui and Al-Whaibi 2014).

11.2.1.5 Other Seed Nanopriming Techniques

Multiple other nanoparticles have been reported to enhance germination in crops. High percentage germination has been reported in seeds treated with nanosilicon dioxide (SiO2) (Siddiqui and Al-Whaibi 2014; Mahakham et al. 2017) and, in some cases, gold nanoparticles (AuNPs) (Kumar et al. 2013). Furthermore, silicon has been reported to enhance seedling vigour, shoot and root length, which improves biomass accumulation during seedling establishment (Janmohammadi and Sabaghnia 2015). Anatase nanoparticles (TiO_2) have also been reported to enhance germination, radicle and plumule length (Dehkourdi et al. 2014).

11.2.2 Photosynthetic and Biomass Accumulation Enhancement

Increasing biomass yield and accumulation is an important component of increasing agricultural water productivity, which can be achieved through enhancing photosynthesis and related plant processes. Nanoparticles have a high surface area to volume ratio, which implies a high potential to modify plant morphology and physiology through modification of enzyme activities, photosynthetic activity, and nutrient metabolism (Marslin et al. 2017; Gohari et al. 2020). This can be achieved through modifications in peroxidase activity, formation of reactive oxygen species, enhanced level of antioxidant enzymes (superoxide dismutase, catalase), and leaf proteins, chlorophyll and phenolic contents (Feizi et al. 2013; Chung et al. 2018). Heightened photosynthetic activity enhances resource capture and carbon assimilation of agricultural systems, improving biomass/yield production using the existing and available resources. Numerous nanoparticles enhance crop photosynthetic activity, a selection of which are discussed below.

11.2.2.1 Silicon Nanoparticles

The efficacy and effect of silicon-based nanoparticles (SiNPs) are known to vary based on shape, size, concentration, and chemistry (composition, functional groups, and surface charges), plant species and phenological stage of application (Rastogi et al. 2017; Lu et al. 2020). Moreover, the method of SiNP application can enhance nanoparticle efficacy, where a foliar application is relatively less effective than soil-applied SiNPs (Suriyaprabha et al. 2014). The most common SiNPs applied for boosting photosynthetic activity are SiO_2 nanoparticles. Silicon nanoparticles have been reported to enhance plant photosynthetic activity by enhancing metabolic, physiological, and morphological features related to crop photosynthesis. Increased photosynthetic rate is achieved by increasing synthesis of photosynthetic pigments enhancing proline activity, carbonic anhydrase activity, water capacity, carotenoid content, electron transport rate, stomatal conductance and transpiration rate, photosystem II activity, and photochemical quenching (Siddiqui and Al-Whaibi 2014; Siddiqui et al. 2014; Mushinskiy et al. 2018; Udalova et al. 2020). Silicon also enhances crop morphology by increasing plant tissue lignification (Asgari et al. 2018), leaf area index, net assimilation rate, and relative growth rate (Suciaty et al. 2018; Mukarram et al. 2021). This Si-related boost in photosynthetic activity results in higher yields, and in the case of oilseed crops, this can sometimes lead to improved seed oil content and chemical composition (Suciaty et al. 2018; Mukarram et al. 2021).

11.2.2.2 Titanium Oxide (TiO_2) Nanoparticles

Titanium dioxide is a mainstream photo catalyst utilised to synthesise colour pigments. Titanium animates the generation of more starches, empowering development and photosynthesis rate in plants (Javed et al. 2019). Reports on the effects of titanium dioxide nanoparticles (TiO_2NPs) on photosynthesis are limited and highly contradictory where present (Dias et al. 2019). Titanium oxide nanoparticles have been reported to enhance photosynthesis by stimulating Rubisco enzyme activity, boosting chlorophyll and carotenoid light absorption, oxygen evolution, water splitting, and provision of photosystem I to photosystem II to catalyse assimilation of carbon dioxide into biological compounds in chickpea and spearmint plants (Yang et al. 2006; Samadi et al. 2014; Hasanpour et al. 2015). On the other hand, TiO_2NPs were reported to impair light-dependent and light-independent phases of photosynthesis, decrease chlorophyll content, maximal and effective efficiency of PSII, net photosynthetic rate, transpiration rate, stomatal conductance, intercellular CO_2 concentration, and starch content on wheat crops (Dias et al. 2019). The decline in photosynthetic activity ultimately limits carbon assimilation and yield development, reducing resource efficiency and thus reducing the water productivity of agricultural systems. The consensus from the limited available literature suggests that TiO_2NPs generally boost photosynthesis and antioxidant activity in crops and that adverse effects can be species and concentration specific (Tighe-Neira et al. 2020).

11.2.2.3 Aluminium Oxide (Al_2O_3)

Drought is a major plant abiotic stress that reduces crop productivity by reducing photosynthetic activity and crop yield formation. Drought stress disrupts photosynthesis and plant growth through excessive production of reactive oxygen species in compartments such as chloroplasts, mitochondrions, and peroxisomes (Djanaguiraman et al. 2018). Aluminium oxide nanoparticles (Al_2O_3NP) modulate reactive oxygen species production and plant redox status by increasing 2,2-diphenyl-1-picryl hydrazyl scavenging activity, total antioxidant capacity, total reducing power, total iridoids content, total saponin content, and total phenolic content in Al_2O_3NP-treated plants (Chahardoli et al. 2020). This is linked to maintenance or stimulation of plant physio-biochemical activities (chlorophyll a, chlorophyll b, carotenoid contents, soluble sugars, protein, amino acid, proline, lignification), and defence enzyme activity (superoxide dismutase, catalase, peroxidase, and ascorbate peroxidase) which boosts plant root length, leaf area, biomass, and yield under drought stress (Burklew et al. 2012; Latef et al. 2020; de Almeida et al. 2021). Photosynthetic boost linked to enhanced physio-biochemical and enzymatic activities is reported at lower concentrations (0.01%) of Al_2O_3NP application in Egyptian roselle (Latef et al. 2020). It is important to note that reduced root growth and decreased leaf area have been reported in crops such as maize, tobacco, cucumber, carrots, and cabbage as a result of Al_2O_3NP phytotoxicity (Yang and Watts 2005; Burklew et al. 2012; Siddiqi and Husen 2017), which suggests that aluminium oxide boosts in photosynthetic activity are dose and species dependent.

11.2.2.4 Iron Nanoparticles

Iron-based nanoparticles are usually transported to photosynthetic and reproductive organs, where they play a pivotal role in electron transport during photosynthesis and respiration, leading to enhanced photosynthesis and embryogenesis (Roschzttardtz et al. 2013; Grillet et al. 2014). Literature reports suggest that foliar application of iron oxide nanoparticles (Fe_2O_3NPs) boosts root growth, plant biomass accumulation, photosynthetic pigments (chlorophyll a, chlorophyll b, ratio of chlorophyll a to b, and carotenoids) and leaf area in crops such as peanuts, moringa, and hemp (Rui et al. 2016; Tawfik et al. 2021; Deng et al. 2022; Feng et al. 2022). Improved photosynthetic efficiency in crops has been attributed to species-specific mechanisms. For instance, high photosynthetic activity in peanuts was explained by Fe_2O_3NPs modulating antioxidant enzymes and phytohormones (diminished abscisic acid and enhanced gibberellic acid content) activity (Rui et al. 2016). Significantly high moringa crude protein, crude fibre, ash percentage, and nutrient content (N, P, K, and K/Na) were associated with increased antioxidant activity (peroxidase, polyphenol oxidase, superoxide dismutase, and nitrate reductase) in Fe_2O_3NPs-treated plants (Tawfik et al. 2021). In hemp, increased plant growth (biomass and tissue length) was attributed to increased catalase activity and reduced hydrogen peroxide content in hemp leaves, downregulation of antioxidants (e.g. tetrahydrocannabinol) and upregulation of carbohydrates and organic acids (Deng et al. 2022).

In contrast to enhanced photosynthesis in peanuts, moringa, and hemp, Fe_2O_3NPs were reported to inhibit photosynthesis and biomass production in wheat (Lu et al. 2020). The negative impacts on wheat photosynthesis were attributed to Fe_2O_3NPs modifying antioxidant enzymatic activity and malondialdehyde levels in leaves, which results in excessive generation of hydroxyl radicals that degrade chlorophyll in leaves. This suggests that mechanisms of photosynthetic stimulation/inhibition by Fe_2O_3NPs vary according to species; therefore, improvements in crop water productivity would follow similar patterns.

11.2.2.5 Zinc Nanoparticles

Zinc nanoparticles can be administered to crops as compounds that include zero-valent iron nanoparticles (nZVI) and zinc oxide (ZnO). Low (nZVI) was reported to increase photosynthetic pigments (chlorophyll a/b ratio) and stimulated biomass accumulation in rice and kidney bean plants (Majumdar et al. 2016; Abbasi Khalaki et al. 2021). Multiple mechanisms have been reported on how ZnO nanoparticles (ZnONPs) affect photosynthetic activity and related biomass accumulation enzymatic activities. Application of ZnONPs has been reported to boost chlorophyll content and stimulate enzymatic activity (acid phosphatase, alkaline phosphatase, and phytase), thereby increasing total lipids, proteins, and amino acids in mungbean (Patra et al. 2013). In cilantro, ZnONPs increased the chlorophyll content by more than half, decreased lipid peroxidation, and changed the concentration of carbinolic-based compounds (Reddy Pullagurala et al. 2018). Zinc oxide nanoparticle application was also reported to counter cadmium-induced stress by reducing oxidative damage in rice plants, which boosted biomass (root and shoot weight), photosynthetic pigments, protein, antioxidant enzymes activity and mineral nutrient contents (Faizan et al. 2018, 2021).

On the other hand, phytotoxicity has been observed due to ZnONP in crops. In spring barley, reduced photosynthetic activity was associated with reduced chloroplast sizes after ZnONP application (Rajput et al. 2021). Phytotoxicity effects of ZnONPs in *Arabidopsis* were linked to the inhibition of chlorophyll biosynthesis that reduced photosynthesis efficiency due to inhibited expression of chlorophyll synthesis genes and photosystem structure genes (Wang et al. 2016). Zinc oxide nanoparticles' phytotoxicity is attributed to their ability to induce reactive oxygen species generation, which can lead to cell death when the antioxidative capacity of the cell is exceeded (Javed et al. 2019). Above the antioxidative threshold, negative impacts on agricultural and water productivity are therefore expected, as crop growth and yields will be reduced while using unchanged amounts of water.

11.2.2.6 Other Nanoparticle Photosynthesis Enhancers

Besides the nanoparticles reported above, numerous other nanoparticles have been reported to enhance photosynthetic activity and biomass accumulation in crops, albeit to a lesser extent. Manganese nanoparticles were reported to boost the photosynthetic rate by mediating water splitting in the electron transport system (Pradhan et al. 2013). In contrast, gold nanoparticles were reported to upgrade the electron transport system by intensifying chloroplasts' electron transport rate and oxygen evolution (Das et al. 2017). Silver nanoparticles have been reported to increase plants' intrinsic water use efficiency by moderating stomatal conductance without necessarily upgrading photosynthetic machinery (Guilger-Casagrande et al. 2022). Multi-walled carbon nanotubes have also been reported to induce aquaporin expression and promote antioxidant defences and plant photosynthesis (Yatim et al. 2018). The application of cesium oxide nanoparticles has also been reported to promote transpiration rate and stomatal conductance in kidney beans (Majumdar et al. 2016) and increase the activity of Rubisco in soybean (Cao et al. 2018).

11.2.3 Root Growth Promotion

Plant roots play a vital role in water and nutrient capture, where high root volume, density, and length confer an advantage to water and nutrient capture when resource availability is low. Plant roots play a major role in the uptake of nanoparticles post-germination in plants, where nanoparticle mobility occurs from the root system through the xylem to the shoot without any downward movement (Nile et al. 2022). Some nanoparticles can also enhance root growth and function when applied to plants. In return, enhanced root growth and function improve plant anchorage, water and nutrient acquisition from the soil, thereby improving the water productivity of agricultural systems. Multiple nanoparticles have the potential to enhance root growth and function, and root growth has been positively correlated with the mitotic index, indicating that an improved root growth rate increases the frequency of plant cell division (Rico et al. 2013). As a result, improved root growth under nanoparticle application can potentially enhance biomass accumulation through increased cell division, thereby increasing marketable yields in agricultural systems. High root growth confers drought tolerance to crops due to enhanced capacity to capture soil-available water. High water capture and biomass production under water-limited conditions enhance crop water productivity.

Silicon is a quasi-essential element in plants and is considered a non-essential element to plant growth. However, evidence suggests that Si can enhance plant metabolism, vigour, and tolerance to abiotic stresses such as drought, metal toxicity and salt stress (Luyckx et al. 2017). This improvement in plant vigour and heightened tolerance to water stress further suggests that silicon can improve crop water productivity under water stress. Two mechanisms have been touted as responsible for silicon and related nanoparticles enhancing plant drought tolerance. The first is enhanced root elongation and upregulation of aquaporin genes, increasing plant water extraction capabilities (Hattori et al. 2005; Liu et al. 2015). The second is through modified guard cell turgor to limit transpiration losses under water stress (Zhu and Gong 2014) by inducing a cuticular double-layer formation under the leaf epidermis. Increased root length is linked to enhanced evapotranspiration and biomass accumulation (Syu et al. 2014), potentially improving agricultural water productivity.

Multi-walled carbon nanotubes (MWCNTs) have been reported to induce crop drought tolerance by enhancing root and shoot length under water scarcity (Rahimi et al. 2016).

Enhanced root length, proliferation, and biomass have also been reported in crops treated with TiO_2 (Khan et al. 2015; Faraji and Sepehri 2019), $MnSO_4$ (Ye et al. 2020), ZnO (Patlolla et al. 2012; Patra et al. 2013), silicon (Janmohammadi and Sabaghnia 2015), Fe (Deng et al. 2014; Alam et al. 2015; Pawar et al. 2019; Kumar et al. 2020), gold (Gunjan et al. 2014; Parveen et al. 2016), and silver (Vannini et al. 2013) nanoparticles. Root lengthening under TiO_2 treatment was attributed to reduced reactive oxygen species (Thiruvengadam et al. 2015). Similarly, root lengthening through silver nanoparticles was attributed to blocking ethylene and classical stress signalling reactions (mediated by Ca^{2+} and reactive oxygen species) and a specific effect on the plasma membrane conductance (Rezvani et al. 2012).

11.3 Environmental Toxicity and Phytotoxicity of Nanoparticles

Nanoparticles have negative effects on crop development (seed germination, root elongation, cell division, growth, metabolic functions, chlorophyll degeneration, oxidative damage, decreased biomass yield and accumulation) and the surrounding soil (heavy metal accumulation and toxicity, reduced microbial activity) environment (Nile et al. 2020; Pramanik et al. 2020; Chugh et al. 2021). The toxicity of nanoparticles is primarily influenced by solution concentration, nanoparticle type, exposure duration, and crop species. The phytotoxicity of nanoparticles can reduce agricultural productivity and, by extension, crop water productivity—issues surrounding nanomaterial phytotoxicity hamper the successful commercialisation and application of nanoparticles. Risk assessment and nano-toxicological evaluations must be conducted for existing nanoparticles and those in the development pipeline for successful commercialisation and agricultural utilisation of nanoparticles. Hasler et al. (2015) suggested that a robust life cycle assessment that thoroughly assesses nanoparticles' environmental and agricultural impact is essential to minimise potential toxic effects. This can enable evidence-based policy decision-making that reduces bioaccumulation of nanoparticles in the food chain.

11.4 Conclusion

Literature reports reviewed in this chapter show that nanoparticle application to agricultural systems can improve crop water use and yield attributes in crops, which implies a high potential to improve crop water productivity in agricultural systems. This chapter presents sufficient literature evidence (Table 11.1) to develop a conjectural theory that nanotechnological applications can enhance agricultural water productivity. However, findings from this chapter are not conclusive on the effect of nanoparticles on agricultural water productivity since literature reports have looked at crop water use components (soil physical properties and soil hydraulic properties) separately to crop yields and related parameters. This linkage between crop water use and agricultural yields is essential to understand the direct implications of nanoparticle applications to agricultural water productivity. An indirect method of establishing this linkage is conducting a metadata analysis on existing literature to paint a statistically sound picture of the potential impacts of nanoparticles on crop water productivity so that broad-scale best management recommendations can be developed for nanoscale agricultural water productivity improvements. A more direct path to assessing nanotechnological impact on agricultural productivity is increasing the number of evidence-based field trials research quantifying water productivity from various applications of nanomaterials, including the nanoparticles discussed in this chapter. Potential environmental and crop toxicity of nanoparticle applications remains a challenge to adoption and utilisation in agriculture, which could hamper the use of nanotechnology as an agricultural water improvement strategy. However, developing a life cycle assessment of nanoparticles based on crop species, solution concentration, and exposure duration before applying nanomaterials to agricultural systems could provide safety guidelines for commercialising and utilising nanotechnology to enhance agricultural water productivity. The life cycle assessment could also be beneficial in assessing the lasting benefits of nanoparticle applications in agricultural systems, an area where information is largely lacking.

References

Abbasi Khalaki M et al (2021) Influence of nano-priming on seed germination and plant growth of forage and medicinal plants. Plant Growth Regul 93(1):13–28

Alam MJ, Sultana F, Iqbal MT (2015) Potential of iron nanoparticles to increase germination and growth of wheat seedling. J Nanosci Adv Technol 1(3):14–20

Alejandro S et al (2020) Manganese in plants: from acquisition to subcellular allocation. Front Plant Sci 11:1–23

Ali MH, Talukder MSU (2008) Increasing water productivity in crop production-a synthesis. Agric Water Manag 95(11):1201–1213

Ali MH et al (2020) Carbon nanoparticles functionalized with carboxylic acid improved the germination and seedling vigour in upland boreal forest species. Nanomaterials 10(1):176

AlSaeedi AH (2022) Enhancement of soil water characteristics curve (SWCC) and water use efficiency of cucumber (*Cucumis sativus* L.) in sandy soils by using silica nanoparticles. J King Saud Univ Sci 34:101926

Aminiyan MM, Safari Sinegani AA, Sheklabadi M (2015) Aggregation stability and organic carbon fraction in a soil amended with some plant residues, nanozeolite, and natural zeolite. Int J Recycl Org Waste Agric 4:11–22

An C, Sun C, Li N (2022) Nanomaterials and nanotechnology for the delivery of agrochemicals: strategies towards sustainable agriculture. J Nanobiotechnol 20:1–19

Anand A, Kumari M, Koul TA (2019) Hydrogen peroxide signalling integrates with phytohormones during the germination of magneto primed tomato seeds. Sci Rep 9:8814

Asgari F et al (2018) Effects of silicon nanoparticles on molecular, chemical, structural and ultrastructural characteristics of oat (*Avena sativa* L.). Plant Physiol Biochem 127:152–160

Burklew CE et al (2012) Effects of aluminum oxide nanoparticles on the growth, development, and microRNA expression of tobacco (*Nicotiana tabacum*). PLoS One 7:e34783

Cao Z et al (2018) The impact of cerium oxide nanoparticles on the physiology of soybean (*Glycine max* (L.) Merr.) under different soil moisture conditions. Environ Sci Pollut Res 25(1):930–939

Chahardoli A et al (2020) Effects of engineered aluminum and nickel oxide nanoparticles on the growth and antioxidant defense systems of *Nigella arvensis* L. Sci Rep 10:1–11

Chugh G, Siddique KHM, Solaiman ZM (2021) Nanobiotechnology for agriculture: smart technology for combating nutrient deficiencies with nanotoxicity challenges. Sustainability 13:1–20

Chung IM et al (2018) Influence of silver nanoparticles on the enhancement and transcriptional changes of glucosinolates and phenolic compounds in genetically transformed root cultures of *Brassica rapa* ssp. *rapa*. Bioprocess Biosyst Eng 41:1665–1677

Czaban J, Siebielec G (2013) Effects of bentonite on sandy soil chemistry in a long-term plot experiment (II); effect on pH, CEC, and macro-and micronutrients. Pol J Environ Stud 22(6):1669–1676

Czaban J et al (2014) Long-lasting effects of bentonite on properties of a sandy soil deprived of the humus layer. Int Agrophys 28:279–289

Das CK, Srivastava G, Dubey A (2016) Nano-iron pyrite seed dressing: a sustainable intervention to reduce fertilizer consumption in vegetable (beetroot, carrot), spice (fenugreek), fodder (alfalfa), and oilseed (*Mustard, sesamum*) crops. Nanotechnol Environ Eng 1:1–12

Das P et al (2017) Green nanosynthesis and functionalization of gold nanoparticles as PTP 1B inhibitors. J Cluster Sci 28:2269–2277

de Almeida GHG et al (2021) Aluminum oxide nanoparticles affect the cell wall structure and lignin composition slightly altering the soybean growth. Plant Physiol Biochem 159:335–346

De Volder MF et al (2013) Carbon nanotubes: present and future commercial applications. Science 339(6119):535–539

Dehkourdi EH et al (2014) The effect of anatase nanoparticles (TiO_2) on pepper seed germination (*Capsicum annum* L.). Int. J Biosci 4(5):141–145

Deng Y, White JC, Xing B (2014) Interactions between engineered nanomaterials and agricultural crops: implications for food safety. J Zhejiang Univ Sci A 15:552–572

Deng C et al (2022) Effects of iron oxide nanoparticles on phenotype and metabolite changes in hemp clones (*Cannabis sativa* L.). Front Environ Sci Eng 16:134

Dias MC et al (2019) Titanium dioxide nanoparticles impaired both photochemical and non-photochemical phases of photosynthesis in wheat. Protoplasma 256:69–78

Djanaguiraman M et al (2018) Cerium oxide nanoparticles decrease drought-induced oxidative damage in sorghum leading to higher photosynthesis and grain yield. ACS Omega 3:14406–14416

El Nagar DA, Sary DH (2021) Synthesis and characterization of nano bentonite and its effect on some properties of sandy soils. Soil Tillage Res 208:104872

El-Sanatawy AM et al (2021) Maize seedling establishment, grain yield and crop water productivity response to seed priming and irrigation management in a Mediterranean arid environment. Agronomy 11:756

Faizan M et al (2018) Zinc oxide nanoparticle-mediated changes in photosynthetic efficiency and antioxidant system of tomato plants. Photosynthetica 56:678–686

Faizan M et al (2021) Zinc oxide nanoparticles alleviates the adverse effects of cadmium stress on Oryza sativa via modulation of the photosynthesis and antioxidant defense system. Ecotoxicol Environ Saf 220:112401

Faraji J, Sepehri A (2019) Ameliorative effects of TiO_2 nanoparticles and sodium nitroprusside on seed germination and seedling growth of wheat under PEG-stimulated drought stress. J Seed Sci 41(3):309–317

Feizi H et al (2013) Assessment of concentrations of nano and bulk iron oxide particles on early growth of wheat (*Triticum aestivum* L.). Annu Res Rev Biol:752–761

Feng Y et al (2022) Effects of iron oxide nanoparticles (Fe3O4) on growth, photosynthesis, antioxidant activity and distribution of mineral elements in wheat (*Triticum aestivum*) plants. Plants 11:1894

Fraceto LF et al (2016) Nanotechnology in agriculture: which innovation potential does it have? Front Environ Sci 4:1–5

Gohari G et al (2020) Titanium dioxide nanoparticles (TiO_2 NPs) promote growth and ameliorate salinity stress effects on essential oil profile and biochemical attributes of *Dracocephalum moldavica*. Sci Rep 10:912

Grillet L, Mari S, Schmidt W (2014) Iron in seeds—loading pathways and subcellular localization. Front Plant Sci 4:535

Guilger-Casagrande M et al (2022) Effects of biogenic silver and iron nanoparticles on soybean seedlings (*Glycine max*). BMC Plant Biol 22:1–9

Gunjan B, Zaidi MGH, Sandeep A (2014) Impact of gold nanoparticles on physiological and biochemical characteristics of *Brassica juncea*. J Plant Biochem Physiol 2:2

Hadebe ST, Modi AT, Mabhaudhi T (2017) Water use of sorghum (*Sorghum bicolor* L. Moench) in response to varying planting dates. Water SA 43(1):91–103

Hadebe ST, Mabhaudhi T, Modi AT (2020) Water productivity of selected sorghum genotypes under rainfed conditions. Int J Plant Prod 14:259–272

Hasanpour H, Maali-Amir R, Zeinali H (2015) Effect of TiO_2 nanoparticles on metabolic limitations to photosynthesis under cold in chickpea. Russ J Plant Physiol 62:779–787

Hasler K et al (2015) Life cycle assessment (LCA) of different fertilizer product types. Eur J Agron 69:41–51

Hatami M (2017) Toxicity assessment of multi-walled carbon nanotubes on *Cucurbita pepo* L. under well-watered and water-stressed conditions. Ecotoxicol Environ Saf 142:274–283

Hattori T et al (2005) Application of silicon enhanced drought tolerance in *Sorghum bicolor*. Physiol Plant 123:459–466

Herawati A, Mujiyo, Syamsiyah J (2021) Application of soil amendments as a strategy for water holding capacity in sandy soils. IOP Conf Ser Earth Environ Sci 724:012014

Huang Y, Wang L (2016) Experimental studies on nanomaterials for soil improvement: a review. Environ Earth Sci 75:1–10

Ikhtiari R et al (2013) Toxic effect of multi-walled carbon nanotubes on lettuce (*Lactuca Sativa*). Nano Biomed 5:18–24
Jackson P et al (2013) Bioaccumulation and ecotoxicity of carbon nanotubes. Chem Cent J 7:154
Janmohammadi M, Sabaghnia N (2015) Effect of pre-sowing seed treatments with silicon nanoparticles on germinability of sunflower (*Helianthus annuus*). Bot Lith 21(1):13–21
Jatav GK, De N (2013) Application of nanotechnology in soil-plant system. Asian J Soil Sci 8:176–184
Javed Z et al (2019) Effect of accumulation of nanoparticles in soil health- a concern on future. Front Nanosci Nanotechnol 5:1–9
Juárez-Cisneros G, Gómez-Romero M, de la Cruz HR (2020) Multi-walled carbon nanotubes produced after forest fires improve germination and development of *Eysenhardtia polystachya*. PeerJ 2020:1–13
Kasote DM et al (2019) Seed priming with iron oxide nanoparticles modulate antioxidant potential and defense-linked hormones in watermelon seedlings. ACS Sustain Chem Eng 7:5142–5151
Kasote DM et al (2021) Manganese oxide nanoparticles as safer seed priming agent to improve chlorophyll and antioxidant profiles in watermelon seedlings. Nano 11:1016
Khan M, Naqvi AH, Ahmad M (2015) Comparative study of the cytotoxic and genotoxic potentials of zinc oxide and titanium dioxide nanoparticles. Toxicol Rep 2:765–774
Kolbert Z et al (2022) Nanoforms of essential metals: from hormetic phytoeffects to agricultural potential. J Exp Bot 73:1825–1840
Kopittke PM, Menzies NW, Wang P (2019) Soil and the intensification of agriculture for global food security. Environ Int 132:105078
Korishettar P et al (2017) Influence of seed polymer coating with Zn and Fe nanoparticles on storage potential of pigeon pea seeds under ambient conditions. J Appl Nat Sci 9(1):186–191
Kumar V et al (2013) Gold nanoparticles exposure induces growth and yield enhancement in *Arabidopsis thaliana*. Sci Total Environ 461–462:462–468
Kumar VK, Muthukrishnan S, Rajalakshmi R (2020) Phytostimulatory effect of phytochemical fabricated nanosilver (AgNPs) on *Psophocarpus tetragonolobus* (L.) DC. Seed germination: an insight from antioxidative enzyme activities and genetic similarity studies. Curr Plant Biol 23:100158
Lahiani MH et al (2013) Impact of carbon nanotube exposure to seeds of valuable crops. ACS Appl Mater Interfaces 5:7965–7973
Latef AAHA et al (2020) The impact of priming with Al_2O_3 nanoparticles on growth, pigments, osmolytes, and antioxidant enzymes of Egyptian roselle (*Hibiscus sabdariffa* L.) cultivar. Agronomy 10:681
Li J et al (2013) Physiological effects of magnetic iron oxide nanoparticles towards watermelon. J Nanosci Nanotechnol 13(8):5561–5567
Liu P et al (2015) Enhanced root hydraulic conductance by aquaporin regulation accounts for silicon alleviated salt-induced osmotic stress in *Sorghum bicolor* L. Environ Exp Bot 111:42–51
Liu R, Zhang H, Lal R (2016) Effects of stabilized nanoparticles of copper, zinc, manganese, and iron oxides in low concentrations on lettuce (*Lactuca sativa*) seed germination: Nanotoxicants or nanonutrients? Water Air Soil Pollut 227:42
Lu K et al (2020) Uptake of iron oxide nanoparticles inhibits the photosynthesis of the wheat after foliar exposure. Chemosphere 259:127445
Luyckx M et al (2017) Silicon and plants: current knowledge and technological perspectives. Front Plant Sci 8:411
Mahakham W et al (2017) Nano-priming technology for enhancing germination and starch metabolism of aged rice seeds using phyto-synthesized silver nanoparticles. Sci Rep 7:8263
Majumdar S et al (2016) Soil organic matter influences cerium translocation and physiological processes in kidney bean plants exposed to cerium oxide nanoparticles. Sci Total Environ 569–570:201–211
Marslin G, Sheeba CJ, Franklin G (2017) Nanoparticles alter secondary metabolism in plants via ROS burst. Front Plant Sci 8:832

Martinez-Ballesta MC et al (2016) Multi-walled carbon nanotubes enter broccoli cells enhancing growth and water uptake of plants exposed to salinity. J Nanobiotechnol 14:42
Mathur P, Roy S (2020) Nanosilica facilitates silica uptake, growth and stress tolerance in plants. Plant Physiol Biochem 157:114–127
Mi J, Gregorich EG, Xu S (2020) Effect of bentonite as a soil amendment on field water-holding capacity, and millet photosynthesis and grain quality. Sci Rep 10:1–11
Molden D, Oweis T, Steduto P (2010) Improving agricultural water productivity: between optimism and caution. Agric Water Manag 97:528–535
Montgomery DR, Biklé A (2021) Soil health and nutrient density: beyond organic vs. conventional farming. Front Sustain Food Syst 5:1–14
Mukarram M, Khan MMA, Corpas FJ (2021) Silicon nanoparticles elicit an increase in lemongrass (*Cymbopogon flexuosus* (Steud.) wats) agronomic parameters with a higher essential oil yield. J Hazard Mater 412:125254
Munegumi T, Inutsuka M, Hayafuji Y (2016) Investigating the hydrolysis of starch using α-amylase contained in dishwashing detergent and human saliva. J Chem Educ 93(8):1401–1405
Mushinskiy AA, Aminova EV, Korotkova AM (2018) Evaluation of tolerance of tubers *Solanum tuberosum* to silica nanoparticles. Environ Sci Pollut Res Int 25:34559–34569
Nile SH et al (2020) Nanotechnologies in Food Science: Applications, Recent Trends, and Future Perspectives. Nano-Micro Lett (2020) 12: 45
Nile SH et al (2022) Nano-priming as emerging seed priming technology for sustainable agriculture—recent developments and future perspectives. J Nanobiotechnol 20:254
Omelchenko AV, Yurkova IN, Zhizhina MN (2014) Stimulating effect of silver nanoparticles on growth and development of wheat. Biol Chem 27:127–135
Parveen A, Mazhari BBZ, Rao S (2016) Impact of bio-nanogold on seed germination and seedling growth in *Pennisetum glaucum*. Enzym Microb Technol 95:107–111
Patlolla AK et al (2012) Genotoxicity of silver nanoparticles in *Vicia faba*: a pilot study on the environmental monitoring of nanoparticles. Int J Environ Res Public Health 9(5):1649–1662
Patra P et al (2013) Effect sulfur and ZnO nanoparticles on stress physiology and plant (*Vigna radiata*) nutrition. In: Giri P, Goswami D, Perumal A (eds) Advanced nanomaterials and nanotechnology. Springer proceedings in physics, vol 143. Springer, Berlin, Heidelberg
Pawar VA et al (2019) Response in chickpea (*Cicer arietinum* L.) seedling growth to seed priming with iron oxide nanoparticles. Int. J Biosci 14(3):82–91
Pradhan S et al (2013) Photochemical modulation of biosafe manganese nanoparticles on *Vigna radiata*: a detailed molecular, biochemical, and biophysical study. Environ Sci Technol 47(22):13122–13131
Pramanik P et al (2020) Application of nanotechnology in agriculture. In: Stoytcheva M, Zlatev R (eds) Environmental nanotechnology, vol 4. Springer, Berlin, Heidelberg, pp 317–348
Rahimi D et al (2016) Increasing drought resistance of *Alnus subcordata* C.A. Mey. seeds using a nano priming technique with multi-walled carbon nanotubes. J For Sci 62(6):269–278
Rajput VD, Minkina T, Fedorenko A (2021) Effects of zinc oxide nanoparticles on physiological and anatomical indices in spring barley tissues. Nanomaterials 11:1722
Rajput VD, Minkina T, Upadhyay SK (2022) Nanotechnology in the restoration of polluted soil. Nanomaterials 12:769
Raliya R et al (2014) ZnO nanoparticles induced exopolysaccharide production by *B. subtilis* strain JCT1 for arid soil application. Int J Biol Macromol 65:362–368
Rastogi A, Zivcak M, Sytar O, Kalaji HM, He X, Mbarki S, Brestic M (2017) Impact of metal and metal oxide nanoparticles on plant: a critical review. Front Chem 5
Rawat M, Nayan R, Negi B (2017) Physio-biochemical basis of iron-sulfide nanoparticle induced growth and seed yield enhancement in *B. juncea*. Plant Physiol Biochem 118:274–284
Reddy Pullagurala VL, Adisa IO, Rawat S (2018) ZnO nanoparticles increase photosynthetic pigments and decrease lipid peroxidation in soil grown cilantro (*Coriandrum sativum*). Plant Physiol Biochem 132:120–127

Rezvani N, Sorooshzadeh A, Farhadi N (2012) Effect of nano-silver on growth of saffron in flooding stress. World Acad Sci Eng Technol 6:519–524
Rico CM et al (2013) Effect of cerium oxide nanoparticles on the quality of rice (*Oryza sativa* L.) grains. J Agric Food Chem 67:11278–11285
Roschzttardtz H et al (2013) New insights into Fe localization in plant tissues. Front Plant Sci 4:350
Rui M et al (2016) Iron oxide nanoparticles as a potential iron fertilizer for peanut (*Arachis hypogaea*). Front Plant Sci 7:815
Samadi N, Yahyaabadi S, Rezayatmand Z (2014) Effect of TiO_2 and TiO_2 nanoparticle on germination, root and shoot length and photosynthetic pigments of *Mentha piperita*. Int J Plant Soil Sci 3(4):408–418
Scown TM, Van Aerle R, Tyler CR (2010) Review: do engineered nanoparticles pose a significant threat to the aquatic environment? Crit Rev Toxicol 40(7):653–670
Shang Y, Hasan K, Ahammed GJ (2019) Applications of nanotechnology in plant growth and crop protection: a review. Molecules 24:1–24
Siddiqi KS, Husen A (2017) Plant response to engineered metal oxide nanoparticles. Nanoscale Res Lett 12(1):92
Siddiqui MH, Al-Whaibi MH (2014) Role of nano-SiO2 in germination of tomato (*Lycopersicum esculentum* Mill.) seeds. Saudi. J Biol Sci 21(1):13–17
Siddiqui MH et al (2014) Nano-silicon dioxide mitigates the adverse effects of salt stress on *Cucurbita pepo* L. Environ Toxicol Chem 33:2429–2437
Soares C, Branco-Neves S, de Sousa A (2018) SiO2 nanomaterial as a tool to improve Hordeum vulgare L. tolerance to nano-NiO stress. Sci Total Environ 622–623:517–525
Suciaty T et al (2018) The effect of nano-silica fertilizer concentration and rice hull ash doses on soybean (*Glycine max* (L.) Merrill) growth and yield. IOP Conf Ser Earth Environ Sci 129:012009
Suriyaprabha R et al (2014) Foliar application of silica nanoparticles on the phytochemical responses of maize (*Zea mays* L.) and its toxicological behavior. Synth React Inorg Met Org Nano Met Chem 44:1128–1131
Syu YY et al (2014) Impacts of size and shape of silver nanoparticles on Arabidopsis plant growth and gene expression. Plant Physiol Biochem 83:57–64
Tarafdar JC, Agrawal A, Raliya R, Kumar P, Burman U, Kaul RK (2012) ZnO nanoparticles induced synthesis of polysaccharides and phosphatases by Aspergillus fungi. Adv Sci Eng Med 4:1–5
Tawfik MM et al (2021) Iron oxide nanoparticles effect on growth, physiological traits and nutritional contents of *Moringa oleifera* grown in saline environment. Bull Natl Res Cent 45:177
Thiruvengadam M, Gurunathan S, Chung IM (2015) Physiological, metabolic, and transcriptional effects of biologically-synthesized silver nanoparticles in turnip (*Brassica rapa* ssp. rapa L.). Protoplasma 252(4):1031–1046
Tighe-Neira R, Reyes-Díaz M, Nunes-Nesi A (2020) Titanium dioxide nanoparticles provoke transient increase in photosynthetic performance and differential response in antioxidant system in *Raphanus sativus* L. Sci Hortic (Amsterdam) 269:109418
Tripathi DK et al (2017) An overview on manufactured nanoparticles in plants: uptake, translocation, accumulation and phytotoxicity. Plant Physiol Biochem 110:2–12
Udalova ZV et al (2020) Effects of silicon nanoparticles on photosynthetic pigments and biogenic elements in tomato plants infected with root-knot nematode *Meloidogyne incognita*. Biochem Biophys Mol Biol 495:329–333
Vannini C et al (2013) Bracale M. Morphological and proteomic responses of *Eruca sativa* exposed to silver nanoparticles or silver nitrate. PLoS One 8(7):e68752
Verma KK et al (2021) Silicon induced drought tolerance in crop plants: physiological adaptation strategies. Silicon 14(6):2473–2487
Wang X, Yang X, Chen S (2016) Zinc oxide nanoparticles affect biomass accumulation and photosynthesis in Arabidopsis. Front Plant Sci 6:1–9

Yang L, Watts DJ (2005) Particle surface characteristics may play an important role in phytotoxicity of alumina nanoparticles. Toxicol Lett 158:122–132
Yang F et al (2006) Influences of nanoanatase TiO_2 on the nitrogen metabolism of growing spinach. Biol Trace Elem Res 110(2):179–190
Yatim NM et al (2018) Effect of functionalised and non-functionalised carbon nanotubes-urea fertilizer on the growth of paddy. Trop Life Sci Res 29(1):17–35
Ye Y et al (2020) Manganese nanoparticles control salinity-modulated molecular responses in Capsicum annuum L. through priming: a sustainable approach for agriculture. ACS Sustain Chem Eng 8:1427–1436
Zhang H, Chen W, Zhao B (2020) Sandy soils amended with bentonite induced changes in soil microbiota and fungistasis in maize fields. Appl Soil Ecol 146:103378
Zhu Y, Gong H (2014) Beneficial effects of silicon on salt and drought tolerance in plants. Agron Sustain Dev 34:455–472

Chapter 12
Scaling Resource Recovery and Reuse (RRR) Innovations in Low- and Middle-Income Countries for Climate Change Adaptation and Mitigation

George Danso, O. Cofie, Andrew Emmanuel Okem, and Pay Drechsel

Abstract Wastes, including biomass and wastewater, have valuable resource recovery and reuse (RRR) potential. However, previous work has paid limited attention to the climate change adaptation and mitigation potential of RRR. This chapter reviews the linkages between climate change and business models of RRR innovations in building sustainable food systems. The review demonstrates that RRR can help societies adapt to climate change mitigation strategies by providing an additional value and sustainable source of nutrients—food, water, and energy. Water reuse as a mitigation strategy has demonstrated resource recovery more than technical challenges. Engineered but simple treatment systems with great cost-effectiveness and higher change of cost recovery, if integrated into an RRR concept, can help mitigate climate change. Effective adaptive measures require nature-based solutions to treat wastewater. The International Water Management Institute (IWMI) documented innovative business models with a focus on circular economy principles and sustainable food systems and has the potential to mitigate climate change impacts. Policies on reducing GHG emissions and achieving a more circular economy with wastewater use options that can mitigate negative climate change impacts are discussed in this chapter.

Keywords Water reuse · Resilience and adaptation · Circular economy · Transformative approaches · Nutrient generation

G. Danso (✉)
Trackpact, Edmonton, AB, Canada
e-mail: gdanso@ualberta.ca

O. Cofie · A. E. Okem · P. Drechsel
International Water Management Institute, Accra, Ghana

T. Mabhaudhi et al. (eds.), *Enhancing Water and Food Security Through Improved Agricultural Water Productivity*,
https://doi.org/10.1007/978-981-96-1848-4_12

12.1 Introduction

The global food supply system is affected by various factors, including the impacts of climate change, the COVID-19 pandemic, and the ongoing war in Ukraine. These factors have exposed the global food systems' fragility. Climate change remains one of the most significant threats to the global food systems. Millions have been exposed to acute food insecurity through increasing weather and climate extremes (FAO 2020). Agriculture accounted for most freshwater withdrawal between 1960 and 2014, experiencing a 100% increase in the last century. Water extracted for irrigation is projected to increase to 2.9 thousand km^3 by 2050 (Ungureanu et al. 2020). Changing rainfall patterns and melting glaciers and snow affect the water available for agriculture (Pörtner et al. 2022). In most developing regions where most production systems depend on rain-fed and snow-melt water systems, decreasing water availability has significant implications for food security. According to the World Bank (2021), climate change could reduce agricultural productivity by up to 30% by 2050. Besides its impacts on the production system, food prices are projected to increase because of the impacts of climate change (Mbow et al. 2019). Reduced agricultural productivity and increased food prices will increase the number of people at risk of hunger and malnutrition.

COVID-19 demonstrates that the global food supply change is vulnerable to pandemics, putting millions of people at risk of hunger and malnutrition (FAO 2020). COVID-19 has affected all aspects of the global food supply chain (production, processing, distribution, and retail). The pandemic affected the food production system because of labor shortages resulting from illness, quarantine, and lockdown restrictions. Lockdown measures delayed the imports and exports of goods, further affecting the food system. Disruptions to the supply chain led to shortages of inputs and food distribution. Where food is available, the loss of jobs and livelihoods and the resultant loss of income affected people's ability to purchase food. The impact of the pandemic on the global food system has been particularly severe in low- and middle-income countries (FAO 2020).

The war in Ukraine highlights another dimension of the fragility of the global food system. The war further worsens the global food system, which is already severely affected by climate change and COVID-19. Ukraine is a major exporter of wheat, corn, and sunflower oil. The war has constrained the country's ability to export these products. The production system has also been affected by conscription into the army, displaced population, and labor shortages. The FAO (2022) estimates that between 20% and 30% of winter crops will likely remain unharvested because of the war. Reduced agricultural productivity because of the war has increased global food prices, affecting millions of people.

Against the backdrop of the impacts of climate change, a global pandemic, and the ongoing conflict in Ukraine, the need to develop resilient and sustainable food systems that can withstand shocks and ensure food security for all remains urgent. Transformative approaches such as the circular economy and sustainable food systems emphasize cross-sectoral sustainability and demonstrate strategies for building resilience against current and future shocks. This paper focuses on the linkages between climate change impacts and business models of RRR innovations in building sustainable food systems. The intent is to show how RRR innovations support climate change mitigations and water scarcity.

12.2 RRR Potential of Wastewater

Water reuse is a reliable alternative to conventional water resources for several uses if treated and/or used safely. Wastewater reuse is particularly important in regions characterized by water scarcity. Ungureanu et al. (2020) state that wastewater accounts for about 80% of domestic water use. On the global front, it discharges an estimated 400 billion m^3 per year, leading to 5500 m^3 of water per year of pollution.

Wastewater reuse can reduce water stress and food insecurity. In Jordan, it is estimated that wastewater reuse can result in about a 48% reduction in the demand-supply gap (Ministry of Environment 2020). In Israel, effluents constitute a major source of water for the agriculture sector, and over 85% of effluents are reused (United Nations Economic Commission for Europe 2022). Wastewater reuse also offers significant economic potential. In many circumstances and depending on the type and location, wastewater practitioners generate more economic benefits per hectare than farmers not using wastewater. Although there is a growing trend in using treated and untreated wastewater for irrigation, irrigation with wastewater only accounts for 1% of total global water consumption for agriculture (Ungureanu et al. 2020).

A range of RRR business models documented by IWMI can serve as dependable water sources, nutrients, and energy for domestic and industrial uses, contributing to more sustainable resource utilization and sound management (Otoo and Drechsel 2018). One way of transferring nutrients back to agriculture is to sell treated water to farmers, reducing the amount of freshwater use and supporting nutrient reuse in the sector. This approach will also reduce the use of inorganic fertilizers. Finally, this approach, especially when recycling solid and liquid waste with modern technologies, may reduce the severity of pollution (Lazurko 2018; Cofie 2003). IWMI and partner projects in Ghana and Sri Lanka worked on co-composting, and composting provided scientific evidence of its benefits.

Hanjra et al. (2015) outlined that energy recovery from biogas supports climate change mitigation and adaptation strategies. As documented in Otoo and Drechsel (2018), energy can be recovered from agricultural, municipal, livestock, and kitchen waste. These reuse options can reduce overall water consumption and treatment needs, resulting in cost savings and contributing to food security. Regarding the environment, reusing treated water allows for the conservation and allocation of freshwater and can enhance the restoration of streams, wetlands, and ponds. Treatments of wastewater with reuse options have direct climate benefits in reducing GHG emissions. When the system is well designed with efficiency options, it allows for better sludge management solutions, leading to GHG mitigations (Saporiti and Robins 2021; Rodriguez et al. 2020).

The total benefits of wastewater use have not been captured as the sector focuses on regulation rather than economic or business principles. However, this is changing, and policymakers, researchers, and citizens are discussing how wastewater can achieve food security. This is a policy topic because wastewater is one of the resources that directly relates to population growth and provides an opportunity for converting it into a valuable resource for the economy.

Wastewater reuse provides a sustainable opportunity for responding to population growth, urbanization, and water scarcity (Rodriguez et al. 2020). Since 2012, the International Water Management Institute (IWMI) has implemented RRR programs across the Global South and analyzed existing and successful enterprises and public-private partnerships (PPPs) for their business models. Over 70 models have emerged, mostly wastewater and solid waste management at suburb or city scales for recovering water, nutrients, biosolids, and energy. Entrepreneurs, policymakers, and stakeholders can derive economic, social, and environmental benefits from these models. Also, there are engineered but simple treatment systems with great cost-effectiveness and higher change of cost recovery if integrated into an RRR concept, which can help mitigate climate change.

Previous studies reported that treating wastewater reduces GHG emissions by about one-third (Warming 2020; Madsen 2022). Recent technologies and improved infrastructure can eliminate other portions of GHG emissions and further reduce them. Strategies such as reducing operational costs and applying innovative cost-sharing models could augment the energy generated from wastewater treatment plants. As reported by Drechsel and Hanjra (2018), the energy generated from the As Samra plant could support the entire energy generated in Jordan, and this might be able to limit the use of fresh water for non-essential services, as it can generate 95% of energy needs. Wastewater is a major resource for a country and has many opportunities in resource recovery systems. When these resources are recovered and safely used for diverse services, it could reduce the impact of climate change, as countries will tend to use less fresh water for non-essential services.

A limitation to that work is that climate change impacts were not the focus for most of the viable business documented—this is a key gap, and further research is needed. Climate change as a catalyst for RRR Innovations Addressing climate change entails "reducing and stabilizing the levels of heat-trapping greenhouse gases in the atmosphere ("mitigation") and/or adapting to the climate change already in the pipeline ("adaptation")" (NASA 2021). Policymakers, researchers, and key stakeholders know climate change is the most significant challenge of the twenty-first century, with the potential impacts on food security, health, water, and sanitation services. Climate change generates additional risks to water-related infrastructure, requiring an ever-increasing need for adaptation measures. Wastewater use for multiple benefits and a climate change mitigation and adaptation strategy has been considered essential in many global strategies, including the Sustainable Water Management Scheme (Marlow et al. 2013), IWMI policy papers, the WHO guidelines, and the World Bank Policy papers. The benefits of wastewater reuse as an adaptive measure to climate change impacts have been highlighted as part of achieving the SDGs (Obaideen et al. 2022). Most of the RRR innovations documented by IWMI and summarized in Otoo and Drechsel (2018) fall under the green infrastructure measures.

Access to adequate water supplies is central to a sustainable future. Climate change is projected to exacerbate water scarcity in several countries, especially low- and middle-income countries (Pörtner et al. 2022). Water reuse is an effective adaptation and mitigation measure. Reuse offers multiple benefits, such as recovering water, energy, nutrients, and other resources and reducing greenhouse gas (GHG) emissions. Water reuse practices worldwide have demonstrated that resource recovery is far beyond just a matter of technology. With the policy discourse of mitigating climate change, efforts in reducing natural resource use stress and strategies to enhance circular economy principles will benefit the ecosystems and support the Sustainable Development Goals (SDGs) 11, 12, and 13 and other interlinked SDGs.

Water and sanitation services generally contribute to GHG due to the energy required to power wastewater treatment systems. Reusing treated wastewater in agriculture also has a significant GHG reduction potential. Recent studies have demonstrated that good wastewater management reduces GHG. For instance, Santos et al. (2015) found that sewage treatment was the primary source of emissions reduction from a wastewater system in Bahia state in Brazil. Freidrich et al. (2009) showed that onsite sanitation systems produced less GHG emissions in South Africa than wastewater treatment systems due to lower energy requirements.

These studies also used life cycle assessment tools and showed that sanitation options focusing on recycling to meet increasing demand had less carbon footprint compared to established industry benchmark numbers. A study in Spain found that using treated wastewater for irrigation instead of fresh water could reduce GHG emissions by about 75% (Chen et al. 2018). Water reuse can contribute to helping cities adapt to climate change by providing an additional and sustainable source of fresh water. For effective adaptive measures, the recommendation has been to use simple engineered treatment systems, like pond-based systems, that have great cost-effectiveness and a higher chance of cost recovery if integrated into an RRR concept. In the following subsections, we show how nutrient recovery, energy generation, and water swaps in the context of waste management contribute to climate change action.

12.3 Nutrients

Many cities continue to be sinks for food waste without a consistent policy on recycling options. While this is a challenge, there is the issue where many farmers in rural and peri-urban areas are looking for good soils to grow crops to feed the growing population (Otoo and Drechsel 2018). These are not the only two variables playing an issue here; the cost of poor waste management threatens sustainable urban growth. Policies that seek to reduce food waste by 50% of the current levels would enhance waste management and reduce GHG by up to 28% (WEF 2016).

In this space, the IWMI program on RRR has conducted extensive research on how nutrients can be recovered from solid and liquid waste in many developing countries, with key implementation projects in Ghana and Sri Lanka.[1] For instance, IWMI and partners have built a large new purpose fecal sludge recycling plant in Tema, Ghana, to produce *Fortifer.*[2] Tema Metropolitan Assembly and Jekora Ventures Ltd. established a public-private partnership. They built a plant with an annual treatment capacity of 12,500 cubic meters, almost the same as a plant with a servicing capacity of about 100,000 people. The plant has the additional capacity to treat organic food waste (i.e., 700 MT), and 500 MT will be the total capacity of the plant, which will generate one of the finest compost in West Africa and provide the benefits of reducing waste that will be at the landfill sites.

Urine separation is another source of nutrients necessary to improve soils for urban and peri-urban farmers, which provides a feasible means of recovery that significantly reduces nutrient loads in treatment systems. This approach ensures that maximum values are recovered from wastewater. Most nitrogen (80%) and phosphorus (50%) in domestic wastewater are in the urine (Jönsson 2001). An example is a multi-partner project implemented by GIZ, ONEA, Water and Sanitation for Africa and funded by ECOSAN-EU in Ouagadougou, Burkina Faso. The project supplied 1000 households with affordable urine-diverting dry toilets and provided collection and treatment services, including reuse. Project outcomes include but are not limited to human health protection, improved food security measures and the promotion of small-scale businesses (Otoo and Drechsel 2018).

There is a recurrent use of phosphorus in agriculture, mostly in Morocco, China, South Africa, and the United States (Childers et al. 2011). It is projected that the production of phosphorus will decline in 2033. However, its use will still be on the rise. Therefore, recovering these finite resources from wastewater is critical for the global food systems. A practical example of how P is extracted and reused in the agricultural sector has been documented for Ouagadougou and the Ostara plant in Canada (Otoo and Drechsel 2018).

Another energy-intensive resource used in agriculture is nitrogen, which generates GHG emissions. Compared to the fertilizer market, the annual capacity of the wastewater treatment industry is between 100,000 and 210,000 MT of phosphorus pentoxide and 220,000 MT of Nitrogen (Latimer et al. 2016). Still, this is a very small market compared to the global market, so the trend is expected to continue, and the sector will grow over time. The situation is different when we compare sector to sector, as it is clear that in niche markets such as the fertilizer for ornamental plants, we see that about 30% and over 100% are covered. Since 2001, the Dutch GMB Bioenergy company has been running a urine treatment business for music

---

[1] https://www.iwmi.cgiar.org/success-stories/transforming-human-waste-into-an-economic-opportunity-in-ghana/

[2] See http://www.iwmi.cgiar.org/Publications/wle/fortifier/the-fortifer-production-plant.pdf

festivals and treats about 1300 m^3 each year. The SaNPhos treatment process ensures that each cubic meter of urine generates 3–4 kg of solid fertilizer and about 60 kg of liquid fertilizer.

In Europe, Further research is required to provide a holistic understanding of how RRR innovations reduce nitrogen in treatment plants.

12.4 Energy

In this regard, business models documented by IWMI, such as the wastewater treatment plants in Egypt, Tunisia, Jordan, and Morocco (Drechsel and Hanjra 2018; Danso et al. 2018; Drechsel et al. 2018), have energy potentials that can be delivered to the national grid. Also, good wastewater treatment plants can potentially be a net contributor to energy, thus making systems energy-positive. This mainly comes from biogas generators with treatment from septic tanks. However, micro hydro systems within pipes can also generate electricity. To reduce greenhouse gases, the design of wastewater treatment plants should not only consider the treatment process but also cover innovations in equipment, process control, and energy and resource recovery.

While past efforts have been on treating effluent to specific standards, recent developments have focused on the energy balance of treatment systems. Although this discussion is more relevant in developed nations, it is still crucial for developing countries with more untreated water to develop strategies to address this issue as well. Simple treatment systems such as ponds and die-offs are more encouraged for climate change mitigation. As documented in Otoo and Drechsel (2018), options such as facultative ponds, duckweed-based ponds, aquaculture, forestry, and agriculture (i.e., Ghana, Bangladesh, and India), groundwater research and water swap (e.g., in Spain), and groundwater recharge (in India) support reuse to mitigate climate and water scarcity. In this system, the goal is to maximize reuse potentials while reducing energy demands.

Energy use and generation is a significant policy topic in the discourse of water and waste treatment systems and reuses to mitigate the negative impact of climate change. The focus of the discussion revolves around the utilization of energy during production, which can be derived from wastewater treatment or through composting processes, allowing for potential recovery. The energy generated from the plant is expected to be greater than the plant's energy needs, or the carbon credit option will be beneficial. Thus, the current discussions focused on energy neutrality more than energy efficiency.

As many SDGs deal with several issues, seven are relevant regarding climate change and energy uses. This number is dedicated to affordable and sustainable energy use in all countries. Target 7.2 is important as it deals with the global need to increase the share of renewable energy, emphasizing power generated from solid and liquid waste. Policymakers and donors call for significant investment to improve energy security to achieve SDG 7 indicators. For instance, they are asking for current investment to increase three times from US$400 to 1.25 trillion. This should be achieved by 2023, with all other things equal. This is crucial as RRR waste-to-energy business models support achieving SDG 7 and contribute to SDG 12.5 in reducing waste generation through prevention, reduction, recycling, and reuse (Rao and Gebrezgabher 2018).

The IWMI RRR program documented viable energy recovery business models and cases from different waste streams to produce energy products in solid (briquette), liquid (bio-fuel/ethanol), and gaseous (producer gas and biogas) forms (Otoo and Drechsel 2018; Gebrezgabher and Musisi 2018). In 2018, Otoo and Drechsel outlined the benefits of waste-to-energy business models and showed various energy sources. Most common examples came from briquettes from either agro-waste or municipal solid waste, biogas from fecal sludge and power from manure, agro-waste, or municipal solid waste. These authors concluded that when these business models are combined with efficient systems, households can reduce the energy used in cooking and heating, thus further reducing health-demanding pollutants—carbon monoxide. Another research from Waste COOCEN showed that the use of 2000 tons of firewood could be avoided every year, and this could reduce deforestation. This amount also represents about 9 ha of forest plantation. Additionally, this project has led to a reduction of 300 tons of CO_2 emissions per year (Adam-Bradford and Gebrezgabher 2018). In India, IWMI researchers reported in Otoo and Drechsel (2018) that about 105,000 households received clean cooking fuel daily. These averted 202,343 hectares saved in deforestation and about 127,650 tons of CO_2 reduced yearly.

In Uganda, an agriculturally based country, there is a large supply of biomass, which is often not used or recovered but rather buried in the fields, causing pollution in many regions. When these untapped resources are converted to briquettes, households have access to energy-efficient cooking tools such as stoves and save money. By using briquette, there is the opportunity to reduce rapid wildfires and methane due to biomass decomposition (Gebrezgabher and Musisi 2018).

To attain the Clean Development Mechanisms project, the 3S program in Brazil provided 3500 pig producers with bio-digesters designed to reduce GHG emissions. In Sukarne, a project was set up to recover methane from animal waste generated from five facilities and further generate biogas and compost while reducing GHG emissions significantly. Many other composting plants and GHG emissions reduction projects can be found in Gebauer and Gebrezgabher (2018) and Otoo and Drechsel (2018).

All these examples and cases show that concerted efforts to reduce open dumping and waste burning could decrease GHG emissions and exposure to untreated waste. To achieve optimal gains for the societies, an efficient strategy will be to convert MSW and FS to compost, as this will improve soil health and crop productivity. Overall, this approach reduces waste management costs, GHG emissions and the use of inorganic fertilizer.

12.5 Water Swaps

At the city or country level, multipronged approaches seem to be the safest bet to address climate-induced water shortage (Drechsel et al. 2022). A prudent strategy of water swaps can offer multiple benefits, including moving water among various sectors in a given city. For instance, water can move from agriculture and other sectors within a city, and this approach offers the option of moving water from low-value to high-value uses. Many use cases can be found in Mexico, Spain, and Iran. In Durango (Mexico), the city treats nearly 100% of wastewater and reuses water for irrigation and urban purposes such as parks and gardens. Further uses can be found in the industrial and mining sector (Comisión Nacional del Agua 2015). The project was initiated due to water scarcity challenges and a prolonged drought in the region, with a deficit of 35 Mm^3 per year.

The city initiated a wastewater-to-freshwater exchange with farmers in one of the irrigation districts—District 052. The district has about 13,455 hectares and is coordinated by several farmers (estimated 3000) using 140 Mm^3 of water for agricultural purposes, including irrigation (Subdirección General de Infraestructura Hidroagrícola 2017). About 3000 farmers are within this district and cultivate 13,455 hectares managed by over 3000 farmers with an average water usage of 137 Mm^3 for irrigation. The state congress supported this deal. The city provided 34 Mm^3 of treated wastewater to farmers in exchange for 17 Mm^3 concession titles. The IWMI documented similar models in Spain and Iran (Otoo and Drechsel 2018). However, these experiences show that farmers who abide by the rules cannot release quality water without following the rules implemented by the authorities or strong compensation. A key recommendation is to conduct an economic analysis to decide on the parameters in the negotiated contracts, which should be binding for all parties (Drechsel et al. 2022).

12.6 Policy Gaps

Available evidence indicates that climate change negatively affects the planet and is urgently needed for adaptation and mitigation. A major challenge with climate change impact is a reduction of GHG emissions—CO_2, CH_4, N_2O, hydrofluorocarbons, and ozone in the lower atmosphere. Wastewater and solid waste systems generate these gases, and strategies to reduce any of these gases will improve the environment and reduce global warming. Traditional treatment systems are inadequate to mitigate these gases and improve the environment. The IWMI has conducted research through the RRR program and identified viable business models that maximize profit while minimizing the environmental impact, including GHG reduction (Table 12.1). For instance, a wastewater treatment plant with a reuse component has the potential to generate energy for the plant and reduce GHG. The challenge for this business model is obtaining carbon neutrality with biogas production.

Table 12.1 RRR business models and climate change mitigation and adaptation strategies

Strategies	Technology-business model	GHG impacts	Sources
Mitigation	Composting[a]	–	Luske (2010) and Lahmouri et al. (2019)
Mitigation	Co-composting	–	Otoo and Drechsel (2018)
Adaption	Fecal sludge management	+	Mills et al. (2020)
Mitigation	Wastewater treatment plant	+	Parravicini et al. (2016)
Mitigation and adaption	Wastewater treatment with reuse[b]	–	Yapıcıoğlu (2019) and Biswas and Yek (2016)
Mitigation	Constructed wetlands	+	Zhang et al. (2017) and Rosli et al. (2017)
Mitigation	Biogas generation	–	Paolini et al. (2018)
Mitigation	Briquette	–	Otoo and Drechsel (2018) and Gebrezgabher and Musisi (2018)
Adaption	Aquaculture	–	Amoah et al. (2018)
Mitigation	Facultative ponds	–	Amoah et al. (2018)
Mitigation and adaption	Food waste reduction	?	Crippa et al. (2021)
Mitigation and adaption	Decentralized wastewater for irrigation with	–	Lahmouri et al. (2019)
Mitigation	Phosphorus recovery	+[1]	Lahmouri et al. (2019), Amoah et al. (2018), and Latimer et al. (2016)
Adaption and mitigation	Water swaps	+	Drechsel et al. (2022)

– positive contributions, + negative contributions

[a]In producing one ton of compost, 129.35 kg CO_2 was emitted. Most of the emissions occurred during the composting process. However, it must be noted that methane and nitrous oxide emissions are very dependent on managing the compost site and waste types used; they can be lower but also much higher

[b]CO_2 is released from these plants because of the treatment process and electricity consumption. Due to biogas cogeneration, CH_4 is not emitted into the atmosphere from these plants.

Other business models such as aquaculture, use of duckweed, groundwater recharge, and water swapping have great potential because they use less energy and provide substantial benefits to improve the environment and freshwater and reduce negative climate change impacts. One critical factor is scaling these RRR innovations in developing countries to reach the derived target impact at the community, city, and country levels. Another factor that needs full consideration is the nitrogen level in the treatment systems. Nitrous oxide (N_2O) is a significant GHG and has a global warming potential of approximately 300 times that of carbon dioxide in equivalent terms (NGER 2016). Future research should explore the synergy between water reuse and the effect of N_2O on the ecosystem.

12.7 Conclusion

This chapter aims to demonstrate the potential for scaling RRR innovations in low- and middle-income countries and the contributions of RRR to climate change adaptation and mitigation. This chapter shows that water reuse is an effective adaptive measure to reduce the impact of climate change on natural resources. Engineered but simple treatment systems, which require less energy, are encouraged. Resource recovery and reuse support cities in adjusting to climate change mitigation options by providing additional values and generating sustainable sources of nutrients such as food, water, and energy. Water reuse options are essential for green infrastructure and economies, and strategic policies should be directed toward scaling these RRR business models.

RRR business models that balance water use, nutrient recycling for agriculture, and energy neutrality are encouraged, as they have the potential to mitigate climate change. Water reuse practices worldwide have proved that resource recovery and reuse are technical challenges. Investments are needed to promote and scale these businesses in the developing world, where financing options are limited.

Phosphorus and nitrogen are crucial in agriculture and resource recovery. RRR business models focus on water, nutrients, and energy but less on phosphorus and nitrogen recovery. IWMI should expand its RRR program to explore the linkage between RRR business models and nitrogen generation for agriculture and energy demands.

Different treatment systems provide various levels of GHG emissions. Future discussions on treatment selections should consider calculating greenhouse gas emissions, and further research is needed to quantify total GHG from the available solid waste and wastewater treatment options.

Water swap is a practical approach to address climate-induced changes in water shortages. However, a pragmatic approach with a detailed cost–benefit analysis showing mutual gains for the parties in exchange for water without unnecessary restrictions may be helpful.

References

Adam-Bradford A, Gebrezgabher S (2018) Briquettes from municipal solid waste (COOCEN, Kigali, Rwanda). In: Otoo M, Drechsel P (eds) Resource recovery from waste. Business models for energy, nutrient and water reuse in low- and middle-income countries. Routledge-Earthscan, Oxon, pp 61–71

Amoah P et al (2018) A public-private partnership linking wastewater treatment and aquaculture (Ghana) – case study. In: Otoo M, Drechsel P (eds) Resource recovery from waste: business models for energy, nutrient and water reuse in low- and middle-income countries. Oxon, Routledge-Earthscan, pp 617–630

Biswas WK, Yek P (2016) Improving the carbon footprint of water treatment with renewable energy: a Western Australian case study. Sustain Energy Res 3:14. https://doi.org/10.1186/s40807-016-0036-2

Chen Z et al (2018) Sustainability evaluation and implication of a large scale membrane bioreactor plant. Bioresour. Technol 269, pp 246–254, https://doi.org/10.1016/j.biortech.2018.08.107

Childers DL et al (2011) Sustainability challenges of phosphorus and food: solutions from closing the human phosphorus cycle. Bioscience 61:117–124

Cofie O (2003) Co-composting of faecal sludge and solid waste for urban and peri-urban agriculture in Kumasi, Ghana. Programme 'Gestion durable des déchets et de l'assainissement urbain'. Action A09. IWMI, SANDEC, KNUST and KMA

Comisión Nacional del Agua (2015) Actualización de la disponibilidad media anual de agua en el acuífero Valle del Guadiana (1003). Ciudad de Mexico, Mexico: Estado de Durango. CONAGUA. https://www.gob.mx/cms/uploads/attachment/file/103634/DR_1003.pdf

Crippa M et al (2021) Food systems are responsible for a third of global anthropogenic GHG emissions. Nat Food 2:198–209

Danso GK, Hanjra M, Drechsel P (2018) Suburban wastewater treatment designed for reuse and replication (Morocco). In: Otoo M, Drechsel P (eds) Resource recovery from waste. Business models for energy, nutrient and water reuse in low- and middle-income countries. Routledge-Earthscan, Oxon, pp 569–583

Drechsel P, Hanjra A (2018) Wastewater and biosolids for fruit trees (Tunisia). In: Otoo M, Drechsel P (eds) Resource recovery from waste. Business models for energy, nutrient and water reuse in low- and middle-income countries. Routledge-Earthscan, Oxon, pp 569–583

Drechsel P, Danso GK, Hanjra M (2018) Viability gap funding (As Samra, Jordan) – case study. In: Otoo M, Drechsel P (eds) Resource recovery from waste. Business models for energy, nutrient and water reuse in low- and middle-income countries. Routledge-Earthscan, Oxon, pp 642–665

Drechsel P, Qadir M, Baumann J (2022) Water reuse to free up freshwater for higher-value use and increase climate resilience and water productivity. Irrig Drain Spec Ed 71(S1):100–109. https://doi.org/10.1002/ird.2694

FAO (2020) The impact of COVID-19 on food security and nutrition. Retrieved from http://www.fao.org/2019-ncov/q-and-a/impact-on-food-and-agriculture/en/

FAO (2022) Ukraine: note on the impact of the war on food security in Ukraine. FAO, Rome. https://www.fao.org/3/cc1025en/cc1025en.pdf

Freidrich E, Pillay S, Buckley CA (2009) Carbon footprint analysis for increasing water supply and sanitation in South Africa: a case study. J Clean Prod 17(1):1–12

Gebauer H, Gebrezgabher S (2018) Power from swine manure for industry's internal use (Sadia, Concordia, Brazil). In: Otoo M, Drechsel P (eds) Resource recovery from waste. Business models for energy, nutrient and water reuse in low- and middle-income countries. Routledge-Earthscan, Oxon, pp 162–171

Gebrezgabher S, Musisi A (2018) Briquettes from agro-waste. In: Otoo M, Drechsel P (eds) Resource recovery from waste. Business models for energy, nutrient and water reuse in low- and middle-income countries. Routledge-Earthscan, Oxon, pp 41–51

Hanjra MA et al (2015) Assessing the finance and economics of resource recovery and reuse solutions across scales. In: Drechsel P, Manzoor Q, Wichelns D (eds) Wastewater: an economic asset in an urbanizing world. Springer, Dordrecht, pp 113–136

Jönsson H (2001) Source separation of human urine separation efficiency and effects on water emissions, crop yield, energy usage and reliability. In: Proceedings of the first international conference on ecological sanitation, Nanning, China, 5–8 November 2001

Lahmouri M, Drewes JE, Gondhalekar D (2019) Analysis of greenhouse gas emissions in centralized and decentralized water reclamation with resource recovery strategies in Leh Town, Ladakh, India, and potential for their reduction in context of the water-energy-food nexus. Water 11:906. https://doi.org/10.3390/w11050906

Latimer R et al (2016) Towards a renewable future: assessing resource recovery as a viable treatment alternative state of the science and market assessment. IWA, London

Lazurko A (2018) Assessing the value of resource recovery and reuse. Social, environmental and economic costs and benefits for value creation and human well-being. Resource recovery and reuse series 13. IWMI/WLE Publication

Luske B (2010) Reduced GHG emissions due to compost production and compost use in Egypt, comparing two scenarios. Louis Bolk Institute

Madsen JA (2022) The state of knowledge on GHG emissions in the wastewater sector. International Water Association. https://iwa-network.org/the-state-of-knowledge-on-ghg-emissions-in-the-wastewater-sector/

Marlow DR et al (2013) Towards sustainable urban water management: acritical reassessment. Water Res 47:7150–7161

Mbow C et al (2019) Food security. In: Shukla PR, Skea J, Buendia EC, Masson-Delmotte V, Pörtner H-O, Roberts DC, Zhai P, Slade R, Connors S, van Diemen R, Ferrat M, Haughey E, Luz S, Neogi S, Pathak M, Petzold J, Portugal Pereira J, Vyas P, Huntley E, Kissick K, Belkacemi M, Malley J (eds) Climate change and land: an IPCC special report on climate change, desertification, land degradation, sustainable land management, food security, and greenhouse gas fluxes in terrestrial ecosystems. https://doi.org/10.1017/9781009157988.007

Mills F et al (2020) Costs, climate and contamination: three drivers for citywide sanitation investment decisions. Front Environ Sci 8. https://doi.org/10.3389/fenvs.2020.00130

Ministry of Environment (2020) MoENV. Water Sector Green Growth national Action Plan 2021-2025, 2020. Amman, The Hashemite Kingdom of Jordan.

NASA (2021), Global climate change. Vital signs of the planet. Responding to climate change. https://climate.nasa.gov/solutions/adaptation_mitigation/#:~:text=Responding%20to%20climate%20change%20involves%20two%20possible%20approaches:%20reducing%20and,pipeline%20

NGER (2016) National Greenhouse and Energy Reporting scheme measurement – technical guidelines for the estimation of emissions by facilities in Australia (August 2016). Department of the Environment and Energy, Australian Government, Canberra. www.environment.gov.au

Obaideen K et al (2022) The role of wastewater treatment in achieving sustainable development goals (SDGs) and sustainability guidelines. Energy Nexus 7:100112. https://doi.org/10.1016/J.NEXUS.2022.100112

Otoo M, Drechsel P (eds) (2018) Resource recovery from waste: business models for energy, nutrient and water reuse in low- and middle-income countries. Routledge-Earthscan, Oxon, p 816

Paolini V et al (2018) Environmental impact of biogas: a short review of current knowledge. J Environ Sci Health 53(10):899

Parravicini V, Svardal K, Krampe J (2016) Greenhouse gas emissions from wastewater treatment plants. Energy Procedia 97:246–253

Pörtner HO et al (2022) IPCC, 2022: summary for policymakers. In: Pörtner H-O, Roberts DC, Tignor M, Poloczanska ES, Mintenbeck K, Alegría A, Craig M, Langsdorf S, Löschke S, Möller V, Okem A, Rama B (eds) Climate change 2022: impacts, adaptation and vulnerability. Contribution of Working Group II to the Sixth Assessment Report of the Intergovernmental Panel on Climate Change. Cambridge University Press. https://doi.org/10.1017/9781009325844.001

Rao K, Gebrezgabher S (2018) Briquettes from agro-waste. In: Otoo M, Drechsel P (eds) Resource recovery from waste. Business models for energy, nutrient and water reuse in low- and middle-income countries. Routledge-Earthscan, Oxon, pp 52–71

Rodriguez DJ et al (2020) From waste to resource: shifting paradigms for smarter wastewater interventions in Latin America and the Caribbean. World Bank, Washington, DC

Rosli FA et al (2017) The use of constructed wetlands in sequestrating carbon: an overview. Nat Environ Pollut Technol 16(3):813–819

Santos JO et al (2015) Greenhouse gas inventory of a state water and wastewater utility in Northeast Brazil. J Clean Prod 104:168–176

Saporiti E, Robins N (2021) Scaling up water reuse: why recycling our wastewater makes sense. https://www.weforum.org/agenda/2021/09/scaling-up-water-reuse-recycling-wastewater/

Subdirección General de Infraestructura Hidroagrícola (SGIH) (2017) Estadísticas agrícolas de los Distritos de Riego Año Agrícola 2015–2016. Ciudad de Mexico, Mexico: Subdirección General de Infraestructura Hidroagrícola, Comisión Nacional del Agua.
Ungureanu N, Vlăduţ V, Voicu G (2020) Water scarcity and wastewater reuse in crop irrigation. Sustainability 12:9055. https://doi.org/10.3390/su12219055
United Nations Economic Commission for Europe (2022) Summary report of Israel under the Protocol on Water and Health 5th Reporting cycle. https://unece.org/sites/default/files/2022-04/Israel_summary_report_5th_cycle_18Apr22_ENG.pdf
Warming M (2020) How can more water treatment cut CO2 emissions? Climate change. IWA publication. https://iwa-network.org/how-can-more-water-treatment-cut-co2-emissions/
WEF (2016) The global risks report 2016, 11th edn. World Economic Forum, Geneva
World Bank (2021) Climate change and agriculture. Retrieved from https://www.worldbank.org/en/topic/climatechange/brief/global-challenges-agriculture
Yapıcıoğlu P (2019) Seasonal water footprint assessment for a paint industry wastewater treatment plant. Sakarya Univ J Sci 23(2):175–183. https://doi.org/10.16984/saufenbilder.411137
Zhang Z et al (2017) Emerging role of wetland methane emissions in driving 21st century climate change. PNAS Vol.114 No.36. https://doi.org/10.1073/pnas.1618765114

Chapter 13
A Systematic Scoping Review of Irrigation Development and Agricultural Water Management in South Africa

Tafadzwanashe Mabhaudhi, T. L. Dirwai, C. Taguta, E. K. Kanda, L. Nhamo, and O. Cofie

Abstract An estimated 80% of Africa's population depends on agriculture; hence, efficient agricultural water management (AWM) is crucial for enhancing agricultural and water productivity and building climate resilience across different farming scales to ensure food security. Approximately 90% of sub-Saharan Africa's agriculture is rain-fed, and the bulk of the farmers are classified as smallholder farmers, constituting 70% of the continent's population. Climate change (CC) and climate variability (CV) have increasingly imposed yield penalties on Africa's irrigated and

T. Mabhaudhi
Centre on Climate Change and Planetary Health, London School of Hygiene and Tropical Medicine, London, UK

Centre for Transformative Agricultural and Food Systems, University of KwaZulu-Natal, Carbis Road, Scottsville, Pietermaritzburg, KwaZulu-Natal, South Africa

T. L. Dirwai (✉)
International Water Management Institute, Pretoria, South Africa

Centre for Transformative Agricultural and Food Systems, University of KwaZulu-Natal, Pietermaritzburg, South Africa
e-mail: T.Dirwai@cgiar.org

C. Taguta
Centre for Transformative Agricultural and Food Systems, University of KwaZulu-Natal, Pietermaritzburg, South Africa

E. K. Kanda
Department of Civil and Structural Engineering, Masinde Muliro University of Science and Technology, Kakamega, Kenya

L. Nhamo
Centre for Transformative Agricultural and Food Systems, University of KwaZulu-Natal, Pietermaritzburg, South Africa

Water Research Commission of South Africa, Pretoria, South Africa

O. Cofie
International Water Management Institute, Accra, Ghana

T. Mabhaudhi et al. (eds.), *Enhancing Water and Food Security Through Improved Agricultural Water Productivity*,
https://doi.org/10.1007/978-981-96-1848-4_13

non-irrigated farming sectors. This has put significant pressure on the African farmer to produce more on less water. As such, the African Union (AU), through the AU-Irrigation Development and Agricultural Water Management (AU-IDAWM) framework, proposed four IDAWM pathways as potential countermeasures to yield penalties. Despite inroads made to equip the African farmer with extension services, a dearth of information related to performance challenges faced in the continent's irrigation fraternity exists. This information is vital as a feedback loop to identify opportunities, challenges, and potential investment gaps to boost irrigation development in the continent. Therefore, using a South African case study, this research sought to assess the existing knowledge, gaps, challenges, and opportunities related to irrigation development and agricultural water management. PRISMA-P protocols guided the study. The SPIDER (Sample, Phenomenon of Interest, Design, Evaluation, and Research type) framework informed the eligibility criteria, which we used to formulate the inclusion–exclusion criteria. The different AU-IDAWM exhibited varied developments in infrastructural and governance structures. The AWM practices exhibited overlaps than variances under the rain-fed, FLID, and modernisation pathways.

Keywords Bright spots · Irrigation · Water governance · Water productivity · Population density

13.1 Introduction

Agricultural production under climate change (CC) and climate variability (CV) are fraught with challenges. Burgeoning global populations have put pressure on natural resources and target planetary boundaries (PBs) to produce more on less water resources. For example, population growth presents a dual challenge because large tracts of land are required for agricultural purposes, and, subsequently, irrigation to augment rain-fed systems and more land for human occupation will be needed, hence putting pressure on the land resource (DEA 2018). Climate change has significantly affected humanitarian crises such as food insecurity and displacement (FSIN 2019). Global food crises have been perpetuated by over-reliance on rain-fed food production and inadequate irrigation and agricultural water management (AWM) malpractices. AWM is a holistic approach or practice that aims at (1) increasing total available water in the plant's root zone, (2) improving soil water holding capacity through mulching practices, (3) water harvesting, (4) irrigation, and (5) employing strategies for improved soil drainage (AU 2020).

Climate change (CC) and variability have been the primary drivers of water scarcity and acute food insecurity. Acute food insecurity is predominant in Africa; the FSIN (2019) reported that more than 50% of the 113 million food insecure population was in Africa. Furthermore, poor AWM strategies at local and national levels exacerbate poor crop performance in Africa. This is evidenced by a 53% margin of poor implementation of AWM-related strategies such as integrated water resources management (IWRM) in the continent (WMO 2021). To counter the acute food shortages, Africa has seen an expansion in irrigation and lands under AWM, which stand at 18.6 Mha (AU 2020; FAO 2016). However, critics argue this is an underestimation because the hectarage under farmer-led irrigation development (FLID) is unknown. Considering the outlook on African agricultural production systems, one can hypothesise that irrigation development and bespoke AWM strategies present an opportunity to sustain food production with minimal yield penalties.

South Africa is a relatively arid country with an average of 500 mm per annum, a minimum of which sustains dryland cropping (Fanadzo and Ncube 2018). Agricultural production is a strategic industry of importance as it aims to eliminate poverty. South Africa is regarded as a middle-income country with high levels of income distribution; as such, the government has applied deliberate efforts to counter the narrative through investment in the agricultural sector. For example, during the early 1980s and early 1990s, the government supported farmers with debt consolidation subsidies of R344 million, crop production loans of R470 million, and drought relief of R120 million. It acted as a guarantor of consolidated debt of R900 million (Kirsten and Vink 2003). In addition, the revitalisation of smallholder irrigation schemes (RESIS) programme promoted the rehabilitation of smallholder irrigation schemes for improved performance through the installation of new efficient infield irrigation systems and provision of the necessary services such as water, electricity, and access roads (Maepa et al. 2014). This information attests to the government's dual effort to tackle poverty and improve agricultural production and AWM.

Irrigation is a crucial adaptation strategy for coping with CC and C; hence the AU devised the AU-IDAWM strategy to drive irrigation adoption and identify key investment points in the African irrigation fraternity. The AU-IDAWM framework is anchored on four pillars identifying strategic development pathways for enhanced spatial-AWM practices across scales (AU 2020). The four pathways are as follows:

- Pathway 1: Improved water control and watershed management in rain-fed farming
- Pathway 2: Farmer-led irrigation development (FLID)
- Pathway 3: Irrigation scheme development and modernisation
- Pathway 4: Unconventional water use for irrigation

The four pathways were developed to respond to the varied farming scales in Africa. For example, pathway 1 targets rain-fed farming, which is predominant compared to irrigated agriculture in Africa. Rain-fed farming is mainly practised by most smallholders in rural areas. Pathway 2 addresses FLID issues that have dominated the research space for nearly two decades. A smallholder population depends on common pool resources (CPR), such as water, to irrigate their fields. The FLID setting resembles a typical irrigation scheme with a water user association (WUA) that manages water politics amongst the farmers. Pathway 3 emphasises modernisation versus building new schemes because of the high establishment costs involved in constructing new schemes. In contrast, pathway 4 emphasises utilising waste water for irrigation to reduce freshwater demands (AU 2020).

To fully assess the status of irrigation development and AWM in South Africa, it is important to juxtapose the current irrigation development and AWM status against the AU irrigation development pathways. Hence, this systematic and scoping review sought to assess existing knowledge, gaps, challenges, and opportunities related to irrigation development and agricultural water management in South Africa. This chapter is organised as follows; the next section describes the methodology employed for the literature review; the results and discussion are presented.

13.1.1 Methods

The case study followed the PRISMA-P systematic review approach Arksey and O'Malley (2005) described. The approach offered a holistic methodology for scoping, gathering, screening, and reporting literature (Dirwai et al. 2021a, b).

13.1.2 Study Area

South Africa is considered a semi-arid country characterised by low annual precipitation. Agriculture contributes significantly to the economy (Fig. 13.1). The sector employs 860,000 people and comprises commercial and subsistence farmers. The commonly grown crops are maize, wheat, and sugarcane (World Bank 2021). Approximately 80% of the country is semi-arid to arid; only 18% is classified as dry or sub-humid (FAO 2008; Moswetsi et al. 2017).

13.1.3 Eligibility Criteria

The SPIDER (Sample, Phenomenon of Interest, Design, Evaluation, and Research type) framework informed the eligibility criteria (Table 13.1), which we used to formulate the inclusion–exclusion criteria for the systematic review (Table 13.2). The SPIDER criteria are preferred over other search criteria because they yield high search hits (Methley et al. 2014). Evidence search was conducted in the following databases: Web of Science, Scopus, SciELO, and Google Scholar. Database selection was based on the overarching nature in terms of information archiving. We did not emphasise the publication date. The search strategy utilised the Boolean operators (OR & AND). The search queries combined the Boolean operators to form expanded search terms or queries. The search strategy was a trial-and-error process. Different search terms were matched to yield many articles for literature extraction.

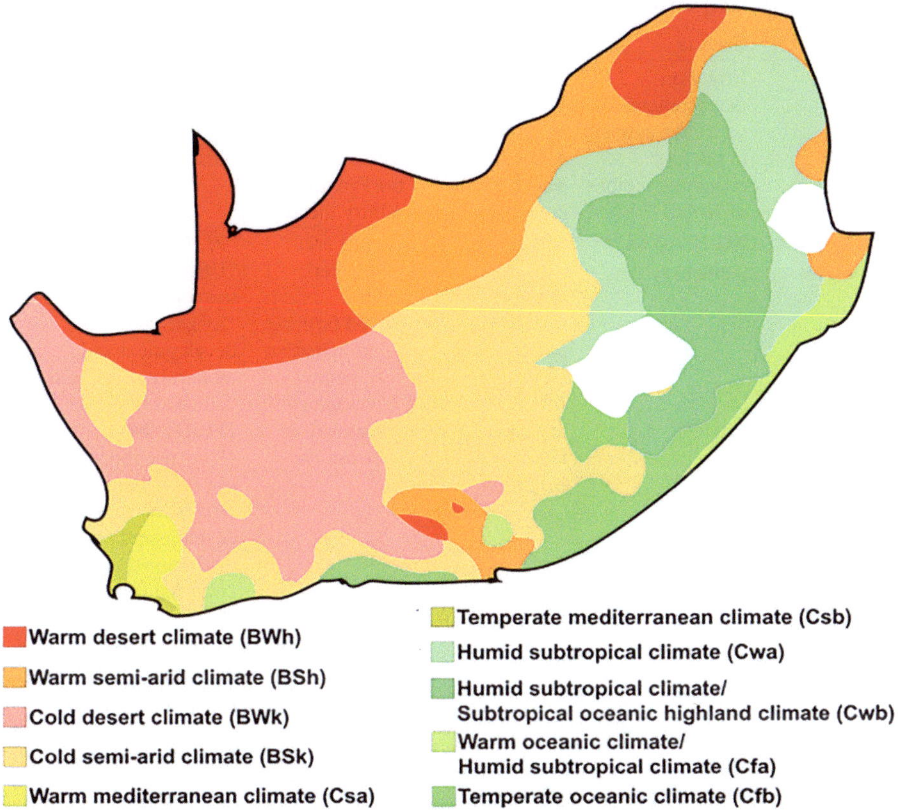

Fig. 13.1 South African climates according to Koppen classification. (Source: Anon (2022a))

13.1.4 Selection Process and Assessing Risk of Bias

The authors employed a manual screening process. The screening process ensured that we retained many articles that facilitated a critical weight regarding information captured in the Excel database. The screening is summarised in Table 13.2.

13.1.5 Strength of Evidence

We applied the Koutsos et al. (2019) strength of evidence approach to assess the strength of evidence for each screened article. The strength of the evidence is summarised in Table 13.3. The grading is defined as follows: substantiated and documented research is given a high score, whereas an average score is assigned to partially substantiated studies and studies with conditional conclusions. Opinion

Table 13.1 The adapted SPIDER structure for formulating the inclusion and exclusion criteria

	Criteria	Definitions	Example search terms
Sample	Irrigation development pathways adopted in the four economic regions of Africa	The target economic regions were East Africa, Southern Africa, West and Central Africa, and MENA. The target country was South Africa and the Provinces contained in the country thereof.	"Southern Africa*" ***OR*** "South Africa*" ***OR*** "province in South Africa*"
Phenomenon of interest	AU-IDAWM pathways	The pathways of interest are: pathway 1: Improved water control and watershed management in rain-fed farming; pathway 2: Farmer-led Irrigation Development (FLID); pathway 3: Irrigation scheme development and modernisation; pathway 4: Unconventional water use for irrigation	"FLID*" ***OR*** "Unconventional water use for irrigation *" ***OR*** "Irrigation scheme development and modernization*" OR "rain-fed farming*" {FLID} ***OR*** {Unconventional water use for irrigation} ***OR*** {Irrigation scheme development and modernization} ***OR*** {rain-fed farming}
Design	Literature extraction on different studies on AWM practices in Africa	Literature extraction	"Questionnaire*" ***OR*** "survey*" ***OR*** "interview*" ***OR*** "focus group*" ***OR*** "case study*" ***OR*** "observations*"
Evaluation	Comparative analysis of performance and adoption rate of each AU-IDAWM pathway	Case-by-case comparison of the four AU-IDAWM pathways in the respective countries	"Attitudes*" ***OR*** "perception*" ***OR*** "imprecision*" ***OR*** "conflicting evidence*" ***OR*** "*"
Research type	Mixed methods	Qualitative and quantitative	"Qualitative*" ***OR*** "mixed method*"

papers are assigned a low score, and non-evidence-based research is ungraded (Dirwai et al. 2021a, b; Koutsos et al. 2019).

13.2 Results

A two-step quality assessment was done: (1) literature screening and (2) article grading according to the strength of evidence. The search results from different databases yielded different results. The Scopus database had the highest number of articles (n = 159) and Web of Science yield ($n = 5$). The study considered the two databases because of their comprehensive and overarching nature. Data on pathway

Table 13.2 Inclusion–exclusion criteria

Inclusion	Exclusion
Articles published in English	Articles from predatory journals
Original research in peer-reviewed journals	Articles not published in English
Conference proceedings	Full articles that could not be retrieved
MSc and PhD theses Government gazettes	Articles with insufficient and irrelevant results, discussions and conclusions
Article profiling smallholder irrigation research globally	
Books and book chapters	

Note: Predatory journals are those listed in the Beall (2015) list of potential predatory publishers

Table 13.3 Strength of evidence grading

Strength of evidence	Type of research	Study design	Study type	Evidence
Strong (I)	Applied research, adaptive/farm-level research, strategic research, systematic reviews, and meta-analyses	Experiments, field trials, systematic reviews	Experimental/structured reviews	Substantiated (+++)
Moderate (II)	Case studies, reviews, modelling/simulations	Case studies, narrative reviews, simulations	Observational	Partially substantiated (++)
Low (III)	Opinion papers, conference papers, workshop papers	Qualitative research, opinion papers, reports of expert committees	Descriptive	Unsubstantiated, qualitative analysis and opinions (+)
Very low (IV)	N/A	N/A	N/A	N/A

4 (unconventional water use) was drawn from grey literature. The PRISMA flow chart (Fig. 13.2) depicts the summary of studies used for literature synthesis, and the strength of the evidence assessment summary is presented in Table 13.8 (see Appendix I).

13.3 Discussion

The South African smallholder irrigation development landscape can be described and defined over four eras, namely: (1) the peasant and mission diversion era (nineteenth century), which focussed on developing river diversion technologies by individuals and small farming groups; the technologies went obsolete at the dawn of the

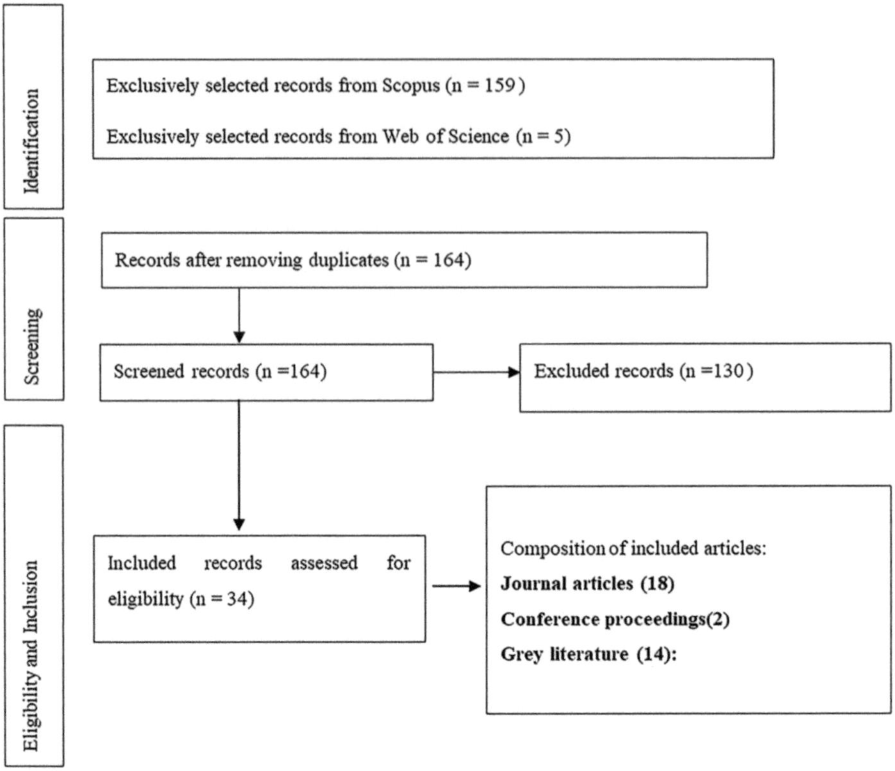

Fig. 13.2 PRISMA flow chart

twentieth century, (2) the smallholder canal scheme era (1930–1960), which primarily focussed on developing and providing livelihoods in "Bantustans". A total of 122 irrigation schemes which covered 11,406 ha, comprising 7538 plots ranging from 1.28 to 1.71 ha, were developed, (3) the independent homeland era (1970–1990), which focussed on irrigation expansion in which an additional 64 irrigation schemes were constructed across the country, and (4) the irrigation management transfer and revitalisation era (1990–to date), which has thus far engaged the revitalisation of irrigation schemes in the former homelands and commercial farming areas (DAFF 2012; Van Averbeke and Mohamed 2006). This timeline provides a picture of the socio-economic dynamics involved in irrigation development. We hypothesise that the dynamics directly or indirectly influence the AWM approaches adopted across scales. The following sections detail the extent of irrigation development and the subsequent AWM practices implemented in each AU-IDAWM pathway.

13.3.1 Pathway 1: Improved Water Control and Watershed Management in Rain-Fed Farming

Climate variability has hindered production under rain-fed or dryland farming. The effects are compounded by poor access to agronomic and training resources. The resource-poor dryland farmers lack access to production inputs such as fertilisers, knowledge, and subsequent information dissemination avenues. A significant population of smallholder farmers practice conventional tillage, which promotes the accelerated formation of hard pans that impede soil infiltration (Moswetsi et al. 2017). The hard pans also increase the incidence of run-off, which the farmers, more often than not, are ill-equipped to harvest. Rainwater harvesting is a technique practised amongst smallholder rural communities. The communities utilise potholing or basins for retaining rainfall. A summary of the AWM approaches employed under this pathway is outlined in Table 13.4.

13.3.2 Pathway 2: Farmer-Led Irrigation Development (FLID)

South Africa has a multi-layered irrigated landscape. The adoption of irrigation technologies varies across scales with different operational statuses (Table 13.5), and it is imperative to have a broad-based effort that delineates and aligns each farming category with the envisioned African Union (AU) irrigation development and agricultural water management strategy (AU-IDAWM). Water use remains rhetoric since there are no measures to ensure and monitor individual water use; thus, there tends to be over-application or over-irrigating, resulting in unsustainable water use.

Table 13.4 AWM typologies and crop management strategies under rain-fed or dryland farming

AWM practice	Description
Conservation agriculture (CA)	The practice involves minimum soil disturbance (MSD). MSD prevents the creation of hard pans that promote run-off. This definition is used in the South African context.
In situ rainwater harvesting	The technology involves manually digging basins meant to collect rainwater and run-off. The basins are dug during the dry seasons to spread labour requirements.
Ridging	Although it primarily prevents soil erosion, it subsequently increases contact time between the water and the soil interface. This increases the plant's available water.
Mulching	This technology utilises crop residue retention to increase ground cover for minimal bare soil evaporation.
Rotation and intercropping	Rotation implies continuous sequencing of crops in a defined pattern. The dryland farmers practice rotation between cereals (maize) and legumes to prevent pests and disease and diversify their diets. This practice subsequently improves soil fertility and improves soil water holding capacity. In addition, intercropping provides soil cover, which minimises bare soil evaporation.

Table 13.5 Spatial variation in functional and non-functional irrigation developments within the smallholder irrigation sector of South Africa

	Number of operational schemes by an irrigation system				Number of non-operational schemes by an irrigation system				
Province	Gravity-fed	Pumped surface	Overhead	Micro	Gravity-fed	Pumped surface	Overhead	Micro	Total
Limpopo	49	09	30	13	12	05	41	11	170
Mpumalanga	03	00	04	00	01	00	11	00	09
North West	00	02	00	00	00	00	00	00	02
KwaZulu-Natal	05	00	30	00	00	00	00	00	35
Free State	00	01	00	00	01	00	00	00	02
Northern Cape	00	02	00	00	00	01	00	00	03
Eastern Cape	04	00	46	01	00	00	16	00	67
Western Cape	06	00	1	00	00	00	01	00	08
Total	67	14	111	14	14	06	59	00	296

Adopted from Van Averbeke et al. (2011)

The FLID pathway is characterised by diverse farming systems, including perennial cereal production under conventional tillage and conservation agriculture (CA). Apart from practising CA or minimum soil disturbance, the irrigators, through water user associations (WUA), collect water fees for the operation and maintenance (O&M) of water conveyance structures. The O&M primarily focusses on removing silt in canals and mending broken canals to prevent water losses through leakages. Dirwai et al. (2019) performed an infrastructure condition assessment on the Tugela Ferry and Mooi River irrigation schemes, and their study revealed that a significant portion of the irrigation schemes had dilapidated water control and conveyance infrastructure, which consequently affected water adequacy.

There exists a disconnect between national government and provincial government policy coordination. Under such circumstances, irrigation development and AWM practices are implemented without a clear guiding framework that facilitates monitoring and evaluation procedures. Provincial governments formulate different policy pathways for irrigation development, that is, hardware/infrastructure/technology and the subsequent governance mechanisms; for example, the Limpopo provincial government put in place the revitalisation of smallholder irrigation schemes recharge (RESIS Recharge) programme (Denison and Manona 2007) which targeted re-tooling irrigation scheme without necessarily modernising them. The programme also targeted creating public-private partnerships (PPPs) (Anon 2022b). The Eastern Cape government embarked on a green revolution (Denison and Manona 2007), which sought to promote rural development and agrarian transformation (Blaai-Mdolo 2009). The two examples highlight the differing pathways embarked on to achieve the same goal using different methods. This approach often leads to policy discordancy and inhibits the desired development.

13.3.3 Pathway 3: Irrigation Scheme Development and Modernisation

The irrigation scheme development and modernisation pathway has overlapped with the FLID. The smallholder and commercial farmers use similar technologies and AWM practices, albeit at different scales of production. The Growth, employment and Redistribution (GEAR) policy was one of the significant developmental policies that aimed to promote irrigation scheme modernisation for poverty eradication. The revised GEAR programme revitalised smallholder irrigation schemes (RESIS). RESIS was a dualistic programme aimed to revitalise and modernise South Africa's irrigation schemes (Denison and Manona 2007). The Water Care programme was the RESIS driver through which struggling schemes were identified for revitalisation and modernisation. The programme targeted hydraulic or water control infrastructure and capacity building on leadership and management (Denison and Manona 2007; Dirwai 2019).

Significant upgrades have putatively improved water usage in certain irrigation schemes in South Africa. For example, some sections in the Tugela Ferry irrigation scheme in Msinga, KwaZulu-Natal, were revitalised, that is, canal repairs, whilst other sections were upgraded to surge systems (Fig. 13.3a) whereby water is availed through hydrants and in-field flooding happens through water-hoses. The hydrants have a known discharge capacity, allowing farmers to calculate application times. The provincial department of agriculture also embarked on providing diesel irrigation pumps (Fig. 13.3b) that abstract water from the uThukela River to the fields. The Mooi River irrigation scheme in Msinga district in KwaZulu-Natal is also set to have a surge system (Dirwai 2019).

South Africa has an estimated 60,000 commercial farmers occupying 102 million ha of land (Moswetsi et al. 2017). The number is decreasing due to low productivity and droughts. This has subsequently created a surge in the smallholder population (WWF-SA 2012) cited by (Moswetsi et al. 2017). The commercial sector has access to bank loans that they utilise for irrigation development and modernisation.

13.3.3.1 AWM Typologies

The pathway employs different AWM prcatices depending on scale. The AWM practices are summarised in Table 13.6.

A study by Nyam et al. (2020) quantified the social, technological, and biophysical factors that drive sustainable water management. Their findings highlighted the low adoption rate (40%) for precision agriculture at different scales. The low adoption rates of precision agriculture are attributed to the concept becoming rhetoric and its delayed benefits. Rainwater harvesting and irrigation technology had a high adoption rate of 80% (Nyam et al. 2020). Other economic factors that improved AWM practices adoption were profitability and government subsidy. Technology and or practice adoption is a function of rewards; irrigators are likely to adopt a practice if it carries an incentive.

Fig. 13.3 (**a**) Hydrant and hose combination used for flood irrigation and (**b**) Diesel pump abstracting water directly from the uThukela River for field flooding. (Source: Dirwai et al. (2019))

Table 13.6 AWM typologies under Pathway 3: Irrigation Scheme Development and Modernisation

Scale	AWM practice	Description
Smallholder	Ridging	Although it primarily prevents soil erosion, it subsequently increases contact time between the water and the soil interface. This increases the plant's available water
	Mulching	This technology utilises crop residue retention to increase ground cover for minimal bare soil evaporation
	Rotation and intercropping	Rotation implies continuous sequencing of crops in a defined pattern. The dryland farmers practice rotation between cereals (maize) and legumes to prevent pests and disease and to diversify their diets. This practice subsequently improves soil fertility and improves soil water holding capacity. In addition, intercropping provides soil cover, which minimises bare soil evaporation
Commercial	Mulching	Build up soil organic matter to reduce evaporative water loss and maximise the soil's water-holding capacity
	Technology shift: Micro and sub-surface irrigation	Use more efficient irrigation systems, such as drip irrigation
	Drones	UAVs for mapping soil moisture stress for targeted irrigation
	Operation and maintenance (O&M)	Maintain irrigation systems regularly
	Water budgeting	Recording actual water use to compare against registered use
	Drought-tolerant cultivars	Using drought-resistant varieties

13.3.4 Pathway 4: Unconventional Water Use for Irrigation

Wastewater irrigation reduces freshwater demands for crop production. Wastewater variants such as treated (reclaimed water) or non-treated (raw wastewater) can be used to directly irrigate crops or can be diluted before irrigation (Jiménez 2006). An estimated 67% of the global population will face moderate to high water stress by 2025. Currently, 40% of the global population is in high-water stress zones that face yield penalties due to water shortages (Lazarova et al. 2001; Zhang and Shen 2019). Urban areas are key producers of wastewater, and the volume of wastewater generation is set to increase with increasing urbanisation. In addition, oil, mining, manufacturing, and logging industries produce wastewater. The rapid urbanisation and the scale of industrialisation present opportunities to harvest and harness technologies for leveraging this resource for agricultural production.

South Africa has provided low-cost on-site Decentralised Wastewater Treatment Systems (DEWATS) in informal settlements (Tesfamariam et al. 2018). DEWATS is a four-modular wastewater treatment facility that consists of an anaerobic baffled reactor (ABR) and anaerobic filter (AF) that degrade black water and greywater to produce biogas and treated wastewater with a low chemical oxygen demand. The wastewater is passed through horizontal and vertical flow-constructed wetlands for further treatment. The process removes solids and deactivates pathogens to a certain degree (Kadlec and Wallace 2008; Singh et al. 2009; Tesfamariam et al. 2018). There is no clear policy regarding wastewater usage for irrigation; however, a combination and cascading of policies, for example, the Water Services Act 108 (1997), produced the white paper on Basic Household Sanitation (2001) and the Strategic Framework for Water Services (2003), which updates the White Paper on Water Supply and Sanitation Policy (1994). The continuous policy revision led to the development of the National Environmental Management Act 107 (NEMA), from which the Sectoral Environmental Management Acts and numerous more specific pieces of legislation, such as the Integrated Coastal Management Act and the Waste Management Act (South Africa 1998) were derived (Cross and Buckely 2016). In short, these policies and acts give local governments and their corresponding municipalities the freedom and flexibility to develop policies, procedures, and strategies in line with national government policies. Sea water desalination is considered an option to provide freshwater for irrigation. The Western Cape (WC), KwaZulu-Natal (KZN), and Eastern Cape (EC) provincial governments are strongly considering the establishment of desalination water treatment plants to meet the freshwater demands in the provinces (Blersch and Du Plessis 2017). The Mossel Bay desalination plant, with a daily capacity of 15,000 $m^{3,}$ has mainly remained unutilised due to high operating costs. Continuous monitoring and evaluation of the Mossel Bay project can provide insights (opportunities and challenges) to inform the planned implementation in the WC, KZN, and EC.

13.4 Summative Discussion: Lessons Learnt, SWOT Analysis, and Preferred Pathways

South Africa is mainly arid to semi-arid; hence, irrigation is required to augment irrigated agriculture for food production. The SWOT analysis for each pathway investigates and describes the internal and external factors that impact the AU-IDAWM pathways' operationalisation. The opportunities and challenges identified in this study for each IDAWM pathway are outlined in Table 13.7.

Table 13.7 Identified opportunities and challenges

Pathway	Challenges	Opportunities
Improved water control and watershed management in rain-fed farming	Rudimentary irrigation methods (bucket system)	Water harvesting techniques employed
	Unreliable rainfall due to climate variability	Availability of treadle and motorised pumps for water abstraction from shallow wells
	Poor or no access to extension services that provide knowledge on spatially available run-off and locating run-off generating areas for effective rainwater harvesting	
	Low-level investment in agricultural development	
Farmer-led irrigation development (flid)	Laborious	Investments for revitalisation
	No titling as land largely belongs to traditional authorities	Some degree of knowledge on the benefits of CA and intercropping evidenced by engaging in the practices
	Limited knowledge capacity on implementing climate-smart agriculture (CSA)	Availability of treadle and motorised pumps for water abstraction from shallow wells
	Poor O&M programmes for canals Unwillingness by farmers to pay water fees meant for O&M programmes Vandalism of upgraded hydraulic infrastructure	A willingness by high learning institutions to invest in research around FLID Participatory irrigation management (PIM) for an increased sense of ownership
Irrigation scheme development and modernisation	Modernisation is happening in phases, and some farmers might be left out High government expenditure for small to medium schemes that are performing below expectation	Improved water conveyance and water application infrastructure for improved water use efficiency
		Participatory irrigation management (PIM) for an increased sense of ownership

(continued)

Table 13.7 (continued)

Pathway	Challenges	Opportunities
Unconventional water use for irrigation	Absence guiding framework at the national level	Availability of low-cost DEWATS systems
	Over-reliance on surface water abstraction and operational procedures	Pilot DEWATS projects a data hub for identifying challenges and bright spots
	Unutilised desalination plants	
	High establishment costs for desalination plants	
	Health and sanitation hazards for wastewater usage	

13.5 Conclusion

This study mapped the status of irrigation development and AWM practices across the four AU-IDAWM pathways in South Africa. The government of the day, through the RESIS programme, invested in smallholder irrigation development. The irrigation schemes underwent revitalisation and refurbishment of hydraulic control infrastructure. The investment also sought to drive irrigation modernisation for improved water use efficiency and food security. Legislation has been enacted to promote unconventional water processing for domestic and irrigation use. This study concluded that:

1. A significantly low population still relies on rain-fed systems. The farmers in this category implement AWM practices because it is the only means to mitigate yield penalties.
2. Although the modernisation and FLID pathways have seen significant investments, a gap exists in providing and setting up institutions for improved water governance.
3. Wastewater use has the potential to provide for irrigation in urban and peri-urban areas because of the low-cost DEWATS in high-density formal and informal settlements in South Africa. There is a need to revise the policy and frameworks for the safety monitoring of agricultural goods. Desalination is a potentially viable option, although establishment and running costs inhibit implementation.

Appendix I: Strength of Evidence Results

Table 13.8 Summarised strength of evidence grading

Author	Subject title	Subject matter	Evidence
Strength of evidence: Strong (I)			
Arksey and O'Malley (2005)	Scoping studies: towards a methodological framework	Systematic reviews	+++
AU (2020)	Framework for Irrigation Development and Agricultural Water Management in Africa	Irrigation and AWM	+++
Blersch and Du Plessis (2017)	Planning for desalination in the context of the Western Cape water supply system		+++
DAFF (2012)	Revatilisation of Irrigation Schemes: Irrigation Infrastructure	Irrigation development	+++
DEA (2018)	South Africa Environment Outlook Report	Environment	+++
Dirwai et al. (2021a, b)	Moistube irrigation technology development, adoption and future prospects: A systematic scoping review	Systematic reviews	+++
Dirwai et al. (2019)	An investigation and condition assessment of the existing water control infrastructure in selected smallholder irrigation schemes: case of Tugela Ferry irrigation scheme and Mooi River irrigation scheme, South Africa	Smallholder irrigation	+++
Dirwai et al. (2019)	Engineering and water governance interactions in smallholder irrigation schemes for improved water management.	Smallholder irrigation	+++
Dirwai et al. (2021a, b)	Water resource management: IWRM strategies for improved water management. A systematic review of case studies of East, West and Southern Africa	IWRM	+++
Fanadzo and Ncube (2018)	Challenges and opportunities for revitalising smallholder irrigation schemes in South Africa	Smallholder irrigation	+++
FAO (2008)	Plant Nutrition Management Service	Plant production	+++
FAO (2016)	Main Database—Food and Agriculture Organization of the United Nations	Statistics	+++
Jiménez (2006)	Irrigation in developing countries using wastewater	Wastewater irrigation	+++
Koutsos et al. (2019)	An efficient framework for conducting systematic literature reviews in agricultural sciences.	Systematic reviews	+++
Lazarova et al. (2001)	Role of water reuse for enhancing integrated water management in Europe and Mediterranean Countries	Smallholder irrigation	+++
Maepa et al. (2014)	Is the Revitalisation of Smallholder Irrigation Schemes (RESIS) programme in South Africa a viable option for smallholder irrigation development?	Smallholder irrigation revitalisation	+++
Methley et al. (2014)	PICO, PICOS and SPIDER: a comparison study of specificity and sensitivity in three search tools for qualitative systematic reviews.	Systematic reviews	+++

(continued)

Table 13.8 (continued)

Author	Subject title	Subject matter	Evidence
Moswetsi et al. (2017)	Cropping systems and agronomic management practices in smallholder farms in South Africa: Constraints, challenges and opportunities	Smallholder production	+++
Nyam et al. (2020)	Drivers of change in sustainable water management and agricultural development in South Africa: a participatory approach	Agricultural development	+++
Singh et al. (2009)	Performance of an anaerobic baffled reactor and hybrid constructed wetland treating high-strength wastewater in Nepal—A model for DEWATS	DEWATS	+++
Tesfamariam et al. (2018)	Decentralised wastewater treatment effluent fertigation: preliminary technical assessment	DEWATS	+++
van Averbeke et al. (2011)	Smallholder irrigation schemes in South Africa: A review of knowledge generated by the Water Research Commission	Smallholder irrigation schemes	+++
World Bank (2021)	Climate Risk Country Profile: South Africa	Climate change	+++
Zhang and Shen (2019)	Wastewater irrigation: Past, present, and future	Wastewater irrigation	+++
Strength of evidence: Moderate (II)			
Blaai-Mdolo (2009)	Revolution and Poverty Alleviation Challenges Faced By Women in Small-scale Agriculture: An Investigation Into the Siyazondla Homestead Food Production Programme, Mbhashe Local Municipality Eastern Cape	Food security	++
Cross and Buckely (2016)	SFD Promotion Initiative. Pollution Research Group	Waste water management	++
Kadlec and Wallace (2008)	Treatment wetlands	Wastewater treatment	++
Kirsten and Vink (2003)	Policy Module South Africa	Policy	++
WMO (2021)	World Meteorological Services: Water	Weather information	++
Van Averbeke and Mohamed (2006)	Smallholder irrigation schemes in South Africa: past, present, and future.	Smallholder irrigation scheme development	++
Strength of evidence: Very low (III)			
Anon (2022a)	South African Drought Map	Drought map	+
Anon (2022b)	Limpopo Revitalisation of Small Irrigation Schemes (RESIS)	Smallholder irrigation scheme revitalisation	+
FSIN (2019)	Food Security Information Network	Food Security	+

After Thomas et al. (2019)
Key: Substantiated (+++); partially substantiated (++); unsubstantiated (+)

References

Anon (2022a) Retrieved 14 May from https://www.google.co.za/search?q=south+africa+drought+map+2020&sxsrf=ALiCzsbt0RP4NWSoDbGlVpp0ldt54hYKVA:1652513604085&tbm=isch&source=iu&ictx=1&vet=1&fir=g8uTf9J2OnD5sM%252CcLXIUZJ1LT8GrM%252C_%253BIqewaqLJaetY7M%252CUuj7tRY7BeK7hM%252C_%253BGA_GHxXbYiOo6M%252CQl4XYFdYntCdtM%252C_%253BDXOvYxnthgE8_M%252CxYChhA11KrZCxM%252C_%253BiFeL0bNO7hVf3M%252C05ySTXY9fElKpM%252C_%253BPqwfVo_3THPjfM%252CdacPoyseiePZ-M%252C_%253BCZiPd5EDLX07kM%252CdacPoyseiePZ-M%252C_%253BabKWozeIhkBP2M%252Cg22MOp-34Axx7M%252C_%253BiOsbC0X6OtXtzM%252C5qs_SzPsYQCKDM%252C_%253Bd3HJFlYV60QCNM%252CthmJXCuWIgo32M%252C_%253BoPKI3-rriMpLDM%252CdacPoyseiePZ-M%252C_%253B63npfp4OZBD5GM%252CuS6zJ8z4eDW_xM%252C_&usg=AI4_-kRVHJiyvIfbKk7Nf0_KX5Xa43Weqg&sa=X&ved=2ahUKEwj-pu7yvN73AhWEZsAKHbgmCjsQ9QF6BAgbEAE&biw=1920&bih=937&dpr=1#imgrc=RhsYn4GgjeRTwM&imgdii=XHUWIK8g8AkUfM

Anon (2022b) Limpopo Revitalisation of Small Irrigation Schemes (RESIS)

Arksey H, O'Malley L (2005) Scoping studies: towards a methodological framework. Int J Soc Res Methodol 8(1):19–32

AU (2020) Framework for irrigation development and agricultural water management in Africa

Beall J (2015) Criteria for determining predatory open-access publishers. Scholarly open access

Blaai-Mdolo B (2009) The green revolution and poverty alleviation challenges faced by women in small-scale agriculture: an investigation into the Siyazondla Homestead Food Production Programme, Mbhashe Local Municipality Eastern Cape University of Fort Hare. University of Fort Hare, Alice, Eastern Cape, South Africa

Blersch C, Du Plessis J (2017) Planning for desalination in the context of the Western cape water supply system. J South Afr Inst Civ Eng 59(1):11–21

Cross X, Buckely C (2016) SFD promotion initiative

DAFF (2012) Revitalisation of irrigation schemes: irrigation infrastructure

DEA (2018) 3rd South Africa environment outlook report

Denison J, Manona S (2007) Principles, approaches and guidelines for the participatory revitalisation of smallholder irrigation schemes: a rough guide for irrigation development practitioners

Dirwai TL (2019) Engineering and water governance interactions in smallholder irrigation schemes for improved water management University of KwaZulu-Natal. Pietermaritzburg, South Africa

Dirwai T, Senzanje A, Mudhara M (2019) An investigation and condition assessment of the existing water control infrastructure in selected smallholder irrigation schemes: case of Tugela ferry irrigation scheme and Mooi river irrigation scheme, South Africa. Irrig Drain 68(4):657–668

Dirwai T et al (2021a) Moistube irrigation technology development, adoption and future prospects: a systematic scoping review. Heliyon 7(2):e06213

Dirwai TL et al (2021b) Water resource management: IWRM strategies for improved water management. A systematic review of case studies of East, West and Southern Africa. PLoS One 16(5):e0236903

Fanadzo M, Ncube B (2018) Challenges and opportunities for revitalising smallholder irrigation schemes in South Africa. Water SA 44(3):436–447

FAO (2008) Plant nutrition management service

FAO (2016) AQUASTAT Main Database – Food and Agriculture Organization of the United Nations (FAO). Food and Agriculture Organisation

FSIN (2019) Global report on food crises: joint analysis for better decisions, issue

Jiménez B (2006) Irrigation in developing countries using wastewater. Int Rev Environ Strateg 6(2):229–250

Kadlec RH, Wallace S (2008) Treatment wetlands. CRC Press

Kirsten J, Vink N (2003) Policy module South Africa. Presented under the roles of agriculture project in international conference on the 20–22 October 2003 Rome, Italy

Koutsos TM, Menexes GC, Dordas CA (2019) An efficient framework for conducting systematic literature reviews in agricultural sciences. Sci Total Environ 682:106–117
Lazarova V et al (2001) Role of water reuse for enhancing integrated water management in Europe and Mediterranean countries. Water Sci Technol 43(10):25–33
Maepa MA, Makombe G, Kanjere M (2014) Is the Revitalisation of Smallholder Irrigation Schemes (RESIS) programme in South Africa a viable option for smallholder irrigation development? Water SA 40(3):495–502
Methley AM et al (2014) PICO, PICOS and SPIDER: a comparison study of specificity and sensitivity in three search tools for qualitative systematic reviews. BMC Health Serv Res 14(1):1–10
Moswetsi G, Fanadzo M, Ncube B (2017) Cropping systems and agronomic management practices in smallholder farms in South Africa: constraints, challenges and opportunities. J Agron 16:51
Nyam Y et al (2020) Drivers of change in sustainable water management and agricultural development in South Africa: a participatory approach. Sustain Water Resour Manag 6(4):1–20
Singh S et al (2009) Performance of an anaerobic baffled reactor and hybrid constructed wetland treating high-strength wastewater in Nepal—a model for DEWATS. Ecol Eng 35(5):654–660
Tesfamariam E et al (2018) Decentralised wastewater treatment effluent fertigation: preliminary technical assessment. Water SA 44(2):250–257
Van Averbeke W, Mohamed S (2006) Smallholder irrigation schemes in South Africa: past, present, and future. 2nd symposium of the SANCID: the changing face of irrigation in South Africa, Mpumalanga, South Africa
Van Averbeke W, Denison J, Mnkeni P (2011) Smallholder irrigation schemes in South Africa: a review of knowledge generated by the Water Research Commission. Water SA 37(5):797–808
WMO (2021) World meteorological services: water
World Bank (2021) Climate risk country profile: South Africa
WWF-SA (2012) Agriculture: facts and trends
Zhang Y, Shen Y (2019) Wastewater irrigation: past, present, and future. Wiley Interdiscip Rev Water 6(3):e1234

Chapter 14
A Systematic Review of Irrigation Development and Agricultural Water Management in Mali

Tafadzwanashe Mabhaudhi, T. L. Dirwai, C. Taguta, E. K. Kanda, L. Nhamo, and O. Cofie

Abstract Rain-fed and irrigated agriculture are key to economic growth, job creation, food security, and livelihoods across Africa. Agriculture in Mali is mainly rain-fed and thus vulnerable to the country's fluctuating climate, which undermines crop production and productivity. The PRISMA protocol and SPIDER framework were used to systematically review Mali's irrigation development and agricultural water management. Mali invested in irrigated agriculture across scales to decouple agriculture from unreliable rainfall, but the potential for expansion still exists. The

T. Mabhaudhi (✉)
Centre on Climate Change and Planetary Health, London School of Hygiene and Tropical Medicine, London, UK

Centre for Transformative Agricultural and Food Systems, University of KwaZulu-Natal, Carbis Road, Scottsville, Pietermaritzburg, KwaZulu-Natal, South Africa
e-mail: Tafadzwanashe.Mabhaudhi@lshtm.ac.uk

T. L. Dirwai
International Water Management Institute, Pretoria, South Africa

Centre for Transformative Agricultural and Food Systems, University of KwaZulu-Natal, Pietermaritzburg, South Africa

C. Taguta
Centre for Transformative Agricultural and Food Systems, University of KwaZulu-Natal, Pietermaritzburg, South Africa

E. K. Kanda
Department of Civil and Structural Engineering, Masinde Muliro University of Science and Technology, Kakamega, Kenya

L. Nhamo
Centre for Transformative Agricultural and Food Systems, University of KwaZulu-Natal, Pietermaritzburg, South Africa

Water Research Commission of South Africa, Pretoria, South Africa

O. Cofie
International Water Management Institute, Accra, Ghana

T. Mabhaudhi et al. (eds.), *Enhancing Water and Food Security Through Improved Agricultural Water Productivity*,
https://doi.org/10.1007/978-981-96-1848-4_14

Malian government also developed several policies to create an enabling environment that promotes agricultural water management (AWM). Farmers employ various agricultural water management practices to control and conserve water and soil. In line with the African Union irrigation development and agricultural water management (AU-IDAWM) framework, there exists operationalization challenges in Mali. These challenges include weak implementation of policies by authorities and lack of awareness among farmers, to mention a few. Farmers lack access to inputs, technology, extension services and credit. The government lacks the financial capacity to rehabilitate irrigation schemes such as Office du Niger, wherein it politically allocated land to foreign large-scale investors without the involvement of the farmers and the management agency, and this may affect the sustainability of irrigated agriculture. Wastewater irrigation suffers from non-recognition, lack of support from all spheres, and risks to human health and the environment. Thus, the government must revamp policy implementation and utilize alternative financing models to rehabilitate irrigation infrastructure, such as private–public partnerships (PPPs). There is a need to minimize political interference in the allocation of agricultural land in the Office du Niger. Subsidies are needed to support farmers with technology and inputs for irrigation and AWM. Farmers, extension workers, and equipment suppliers must be trained to build their capacity. Wastewater irrigation contributes to food supply, income generation, and livelihoods in peri-urban and urban areas. Thus, this practice must be formalized and supported by policies, guidelines, regulations, standards, and technologies for on-site water treatment and safer irrigation practices.

Keywords Modernization · Farmer-led irrigation development · Wastewater · Rain-fed · Irrigated · Policy · Practice

14.1 Introduction

Agricultural growth is key to reducing hunger and extreme poverty (Olayide et al. 2020). The African continent is significantly challenged with achieving food security in the face of increasing food demand, competition for depleting resources, and the inability of the environment to absorb increasing natural and anthropogenic shocks (Zougmoré et al. 2021). To feed the growing population, food production needs to increase by at least 70% by 2050, and this increase in demand will be higher in sub-Saharan Africa (FAO 2021). This challenge is further worsened by land degradation, depleting soil fertility, and over-reliance on rain-fed agriculture in the face of climate change (CC) and variability (CV) (Bayala et al. 2017). Rain-fed agriculture, a mainstay for resource-poor smallholder farmers, dominates African irrigated agriculture. Rain-fed agriculture represents over 70% of the total production and over 90% of farming systems as dispersed farm locations in rural areas (World Bank 2013; Abrams 2018). This calls for exploring all forms of climate- and water-smart control and conservation practices to enhance the resilience of rain-fed and irrigated production systems. These include, for example, in situ soil water conservation and ex situ water harvesting in different storage structures to augment surface and groundwater storage and water availability for irrigation at different scales (ACPC 2013). There is an enormous potential to close the yield gap and achieve significant socioeconomic benefits through agricultural water management (AWM) measures that enhance the production intensity and resilience of agriculture, as hinted by the New Partnership for Africa's Development (NEPAD)'s Comprehensive Africa Agriculture Development Programme (CAADP) and the African Agenda 2063 (AU 2020). To sustain food production and reduce poverty for the growing population on the continent, the African Agenda 2063 proposed that water productivity from rain-fed and irrigated agriculture should increase by 60% from the 2013 estimates; harvest at least 10% of rainwater for productive use; and recycle at least 10% of wastewater for agricultural and industrial use (AUC 2014). However, these ambitious guiding documents focus mainly on the more extensive continental (Africa) scale without providing actionable, country-specific information.

Irrigated agriculture accounts for only 20% of cropland, contributes 40% of total production in sub-Saharan Africa, and can improve and sustain agricultural productivity while reducing food insecurity and importation dependency (Olayide et al. 2020). However, irrigated agriculture withdraws and consumes the majority of the freshwater. The AU reported the total area under AWM in Africa to be 18.6 million hectares (AU 2020). Still, local, environmental, climatic, management, and economic constraints have been identified as limiting AWM expansion (Higginbottom et al. 2021). AU member states committed to increasing public spending on agriculture in the 2003 Maputo and 2014 Malabo Declarations, but inadequate investments and planning in AWM continue to impair the full realization of the potential gains of different AWM practices along agricultural value chains (OECD/FAO 2021). Investments in agriculture and irrigation across Africa have deliberately favoured the construction and rehabilitation of medium- and large-scale irrigation schemes, though with questionable sustainability, more than other AWM typologies. These big irrigation schemes have generally failed to meet expectations for more than six decades due to inactivity, falling short of their promised irrigated agricultural land areas and equipped unusable lands (25%) (Bjornlund et al. 2020). Such formal irrigation schemes must be revamped, modernized, reformed, and managed to intensify irrigation water use, increase productivity, achieve a higher return on investment, secure food, reduce poverty, and stimulate economic growth and development (Froebrich et al. 2020). Similarly, alternative models of irrigated agriculture need to be explored, including small-scale and farmer-led irrigation development (FLID), which is gaining more attention from governments and donors as a sustainable practice to increase production, especially in developing regions (Harmon and Lefore 2022).

The Sahel region in West Africa has an erratic climate. It is vulnerable to CC due to its geographic location at the southern edge of the Sahara Desert and the strong dependence of its population on rain-fed crop and livestock production (Zougmoré et al. 2014; Zougmoré 2018; Andrieu et al. 2021). Mali is a low-income, food-deficit, land-locked Sahelian country characterized by desert and semi-desert conditions. The desert and semi-desert conditions account for two-thirds of the land area and are unsuitable for rain-fed agriculture (UNEP-CCC 2022). Crop production, agro-pastoralism, and fisheries are primary drivers of Mali's economic growth, social development, poverty reduction, and food security, contributing about 40% to the national GDP, employing 80% (primarily subsistence production) of which 37–44% are women, and accounting for 12% of total exports (IMF 2015; Andrieu et al. 2017; FAO 2017; Montaud 2019; Coulibaly 2021). Agriculture is the mainstay of rural livelihoods, where more than 80% of the population depends on rain-fed production of staple crops (maize, rice, millet, sorghum, and cowpea), vegetables, and fishing on small parcels of land (less than 10 ha) (CIAT et al. 2020; Nkonya et al. 2020; Coulibaly 2021; IWMI 2021). However, the agricultural productivity of cereal (millet, sorghum, maize, and rice) is relatively low in Mali. There are possibilities to boost the yields by revamping land and agricultural water management practices (Bastagli and Toulmin 2014; IMF 2015). Given the increased annual temperatures, decreased precipitation, degraded lands, and predominantly small-scale rain-fed farming in Mali, sustainable management of fragile lands and water resources is imperative for developing the sector and national economy.

Agricultural water management refers to farmer interventions that increase water availability to the root zone over and above naturally infiltrated rainfall (AU 2020). The AWM spectrum spans the management of both blue water (withdrawals) and green water (in plants), and includes infiltrated rainfall, shallow-aquifer farming (dambos, fadamas, wetlands, etc.), mulching, conservation agriculture, bunding, flood-recession farming, water harvesting, irrigation, and drainage (AU 2020). Closely related to AWM is climate-smart agriculture (CSA), which aims to co-achieve three pillars, namely: (i) sustainably increasing agricultural productivity; (ii) enhancing resilience; and (iii) reducing or removing greenhouse gas emissions, where possible, enhancing the achievement of national food security and development goals (Andrieu et al. 2017). The African Union developed an irrigation development and agricultural water management (AU-IDAWM) framework comprising four guiding pathways that include: (i) improved water control and watershed management in rain-fed farming, (ii) farmer-led irrigation development (FLID), (iii) irrigation scheme development and modernization, and (iv) unconventional water use for irrigation (AU 2020). Mainstreaming practices consistent with AU-IDAWM require critical stocktaking of ongoing and promising practices for the future, and institutional and financial enablers for AWM adoption. Thus, it has become necessary to appraise the state of irrigation development and agricultural water management in Africa.

The agriculture and irrigation sector contributes significantly to job creation, food security, and rural livelihoods across Africa and Mali. However, the status and potential of applying the IDAWM framework to improve irrigation development and agricultural water management along the four development pathways in Africa and its member states, including Mali, is unknown. It is necessary to gather country-specific evidence that supports the adoption of integrated and inclusive land and water management and governance for sustainable intensification and resilience, including watershed or river basin management supported by long-term strategies and investments. Such analysis is imperative at the country and smaller scales due to the diversity in biophysical, socio-technical, economic, and organizational conditions. Therefore, this study assessed the existing knowledge, gaps, challenges, and opportunities related to irrigation development and agricultural water management in Mali in the context of the four AU-IDAWM pathways. Specifically, and based on the AU-IDAWM framework, the state of agricultural water management development in Mali could offer insights into the experiences within West Africa and Sahelian sub-regions for policy thrusts, factors driving irrigated crop production and water resources utilization towards increasing resilience in the production systems and prevailing opportunities and threats within the agricultural water management sector. Experiences along different intervention projects and initiatives towards improving agricultural water management systems in Mali could serve as a benchmark across the sub-region.

14.2 Methods

We followed the Preferred Reporting Items for Systematic Reviews and Meta-Analyses (PRISMA) approach described by Arksey and O'Malley (2005) supported by the Sample, Phenomenon of Interest, Design, Evaluation, and Research type (SPIDER) framework (Methley et al. 2014). The approach offered a holistic methodology for scoping, gathering, screening, and reporting literature, recently by Dirwai et al. (2021a).

14.2.1 Study Area

The study area is a West African Sahelian country with an area of 1,241,238 km^2 in sub-Saharan Africa and a population of 19.1 million, of which 59–79% live in rural areas (Andrieu et al. 2017). The population growth rate is approximately 3.6% (FAO 2017). Mali consists of two central regions with different conditions of agricultural production, namely, North and South, the former being primarily a desert challenged mainly by drought, desertification, and population migration, and has some livestock production of cattle, sheep, and goats under pastoral systems (extensive production for cattle and small ruminants). The southern region receives relatively higher rainfall and has greater crop production, mainly millet, sorghum, and rice, often mixed with livestock (FAO 2017; IBRD and WorldBank 2019).

The country has four climatic zones: the Sahara, Sahel, Sudan, and Sudan-Guinea (Fig. 14.1 and Table 14.1) (FAO 2017; Coulibaly 2021).

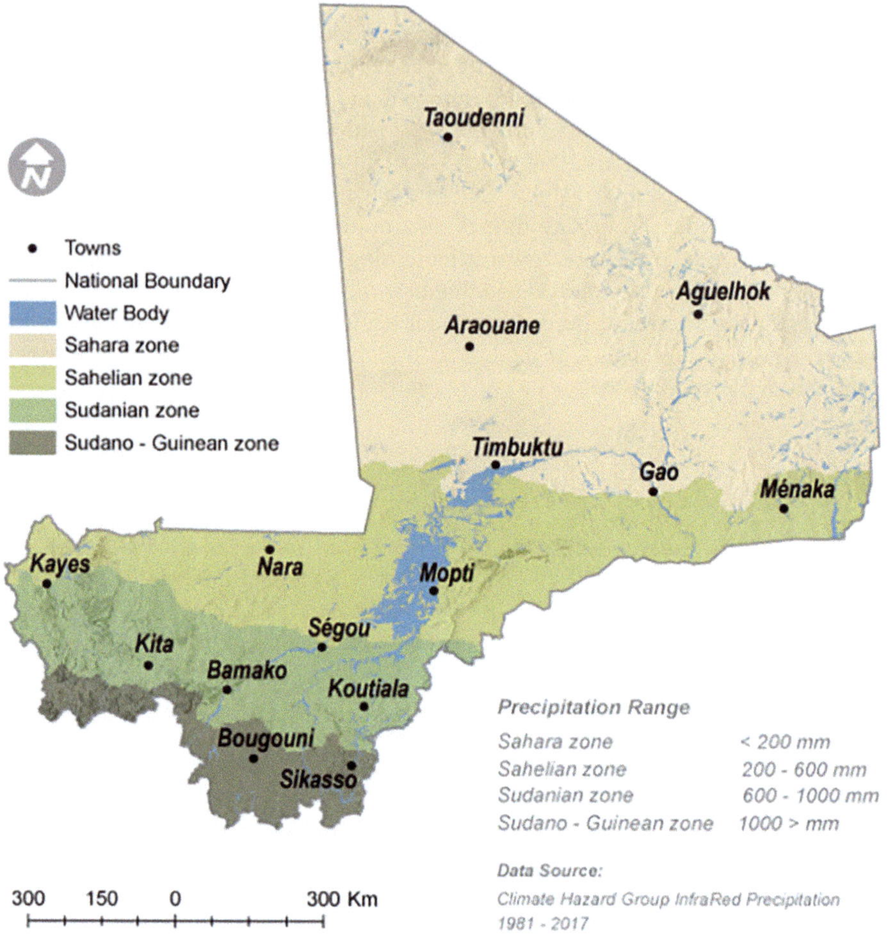

Fig. 14.1 Agro-climatic zones in Mali (IBRD and WorldBank 2019)

14.2.2 *Information Sources and Eligibility Criteria*

The eligibility criteria for this study employed the Sample, Phenomenon of Interest, Design, Evaluation, and Research type (SPIDER) framework (Table 14.2) to break down the research problem into potentially usable search words. The approach subsequently informed the inclusion–exclusion criteria for the systematic review (Table 14.3). The SPIDER criteria are preferred over other search criteria because they yield high search hits (Methley et al. 2014). Evidence search was conducted in the Scopus and Web of Science databases, which were selected for their overarching nature in terms of information archiving, as well as databases of relevant organizations such as the Food and Agriculture Organization (FAO), International Water Management Institute (IWMI), and World Bank. We did emphasize the publication

Table 14.1 The agro-climatic zones in Mali

Area	Geographical location	Percentage surface area (%)	Precipitation (mm/year)	Characteristics
Saharan	North	51	<200	Desert with caravan trade and gathering Livestock: nomadic herding of cattle, sheep, goats, camels Crops: rice, wheat
Sahelian	Central	26	200–600	Dry northern Crops: rice, wheat, sorghum, and vegetables Livestock: nomadic and transhumant raising of camels, cattle, goats, and sheep
Sudanese	South	17	600–1200	Crops: millet, sorghum, rice, wheat, maize, onions, peanuts, sweet potatoes, tomatoes, fonio, beans, potatoes, mango, citrus, peanut, cashew, shea, nere, sugar cane, sesame, and peas Livestock: poultry, cattle, sheep, goats, bees
Sudano-Guinean/ North Guinean	Far south	6	>1200	Crops: subsistence farming of sorghum, millet, fonio, corn, cowpea, fruits, tubers and vegetables (cabbages, okra, tomato, onion, beans, potatoes), with groundnut, mango, peach, cashew, cotton, and maize for food and cash Livestock: cattle, sheep, goats, bees Wild fruit gathering (shea, tamarind, nere)

Sources: Zamudio (2016), USAID (2018), IBRD and WorldBank (2019), CIAT et al. (2020), Coulibaly (2021)

date. The search strategy and search queries used the Boolean operators (OR and AND) and their combinations to form expanded search terms or queries. The search strategy was a trial-and-error process. Different search terms were matched to yield many articles for literature extraction.

14.2.3 Selection Process and Assessing the Risk of Bias

The authors employed a manual screening process to ensure the retention of many articles that facilitated a critical weight regarding information captured in the Excel database. The screening is summarized in Table 14.3. The study emphasized publication dates, in this case, 2000–2022.

Table 14.2 The adapted SPIDER structure for formulating the inclusion and exclusion criteria

	Criteria	Definitions	Example search terms
Sample	Irrigation development pathways adopted in the four economic regions of Africa	The target economic region and country were West Africa and Mali, respectively.	"West Africa*" **OR** "Mali*"
Phenomenon of Interest	AU-IDAWM pathways	The pathways of interest are: pathway 1 (Improved water control and watershed management in rain-fed farming); pathway 2 (Farmer-led irrigation development, FLID); -pathway 3 (Irrigation scheme development and modernization); pathway 4 (Unconventional water use for irrigation)	"FLID*" **OR** "Unconventional water use for irrigation *" **OR** "Irrigation scheme development and modernization*" **OR** "climate-smart agriculture" **OR** "CSA" **OR** "conservation agriculture" **OR** "rain-fed farming*"
Design	Literature extraction on different studies on AWM practices in Africa	Literature extraction	"questionnaire*" **OR** "survey*" **OR** "interview*" **OR** "focus group*" **OR** "case study*" **OR** "observations*"
Evaluation	Comparative analysis of performance and adoption rate of each AU-IDAWM pathway	Case-by-case comparison of the four AU-IDAWM pathways in the respective countries.	"attitudes*" **OR** "perception*" **OR** "imprecision*" **OR** "conflicting evidence*"
Research type	Mixed methods	Qualitative and quantitative	"qualitative*" **OR** "mixed method*"

Table 14.3 Inclusion–exclusion criteria

Inclusion	Exclusion
Articles published in English	Articles from predatory journals
Original research in peer-reviewed journals	Articles not published in English.
Conference proceedings	Full articles that could not be retrieved
MSc and PhD theses; government gazettes	Articles with insufficient and irrelevant results, discussions, and conclusions
Article profiling smallholder irrigation research globally	
Books and book chapters	

Table 14.4 Strength of evidence grading

Strength of evidence	Type of research	Study design	Study type	Evidence
Strong (I)	Applied research, adaptive/farm-level research, strategic research, systematic reviews, and meta-analyses	Experiments, field trials, systematic reviews	Experimental/ structured reviews	Substantiated (+++)
Moderate (II)	Case studies, reviews, modelling/simulations	Case studies, narrative reviews, simulations	Observational	Partially substantiated (++)
Low (III)	Opinion papers, conference papers, workshop papers	Qualitative research, opinion papers, reports of expert committees	Descriptive	Unsubstantiated, qualitative analysis and opinions (+)
Very low (IV)	N/A	N/A	N/A	N/A

14.2.4 Strength of Evidence

We applied the Koutsos et al. (2019) technique to assess the strength of evidence for each screened article. The applied grading for the strength of evidence is summarized in Table 14.4. The grading is defined as follows: (i) substantiated and documented research is given a high score, whereas (ii) an average score is assigned to partially substantiated studies and studies with conditional conclusions, (iii) opinion papers are assigned a low score, and (iv) non-evidence-based research are ungraded (Koutsos et al. 2019; Dirwai et al. 2021b).

14.3 Results

14.3.1 Literature Search

A two-step quality assessment was done, namely, (1) literature screening and (2) article grading according to the strength of evidence (Table 14.9 in Appendix). The search results from different databases yielded different results. The Scopus database had the lowest number of articles (n = 52), and the Web of Science yielded (n = 142). Upon screening, n = 79 was retained for literature synthesis. Most of the data on pathway 4 (unconventional water use) were drawn from grey literature. The PRISMA flow chart (Fig. 14.2) depicts the summary of studies used for literature synthesis, and the strength of the evidence assessment summary is presented in Table 14.9 (see Appendix).

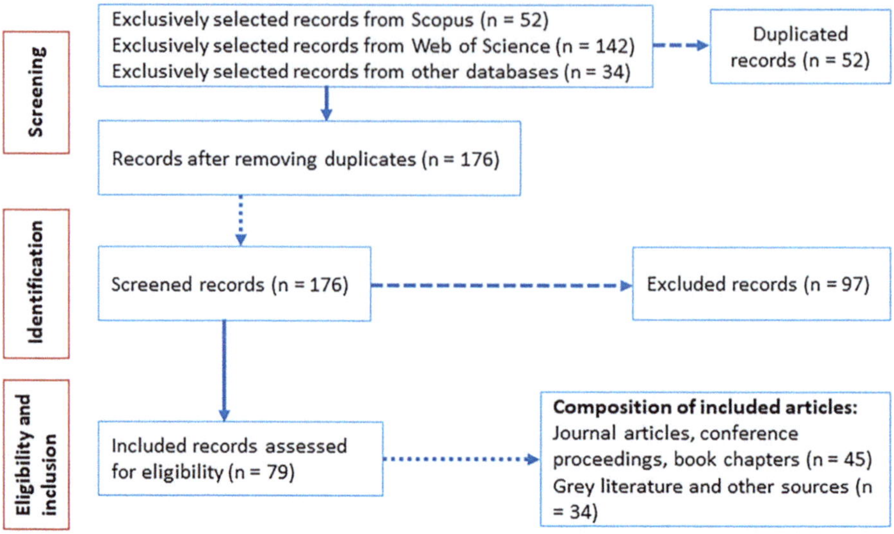

Fig. 14.2 PRISMA flow chart for literature search for Mali

14.4 Discussion

Mali has a total agricultural land of 41.6 million ha, of which the cultivated area is 7 million ha, constituting 5.6% of Mali's total land area (FAO 2022). Rain-fed cultivation of coarse grain cereals is widespread in the Niger River Basin, accounting for approximately 5.8 million hectares or 90% of Mali's arable land (IWMI 2021). The irrigation potential stands at 566 000 ha, of which 371 000 ha is irrigated as surface (54.97%), equipped lowlands (44.97%), and pressurized irrigation techniques (0.06%). An additional 250 218 ha and 3826 ha are cultivated under unequipped flooding recession and wetlands and inland valley bottoms, respectively (FAO 2022). Agricultural water management strategies and practices in the country are highly motivated by the rainfall availability across the zones for improving production, resilience and reducing the negative impacts. The implemented agricultural water management strategies and practices across all levels include soil and water conservation technologies, watershed management, small-scale irrigation farming, and large-scale formal irrigation systems. Farmers use several strategies to improve AWM and cushion themselves from the adverse effects of climate change across all four IDAWM pathways.

14.4.1 Pathway 1: Improved Water Control and Watershed Management in Rain-fed Farming

According to AU (2020), this pathway includes rain-fed smallholder plots and farms with mixed farming purposes. The standard practices include flood recession farming and shallow-aquifer use wherein farm labour is the primary labour source to mainly produce grains, legumes and tubers, sometimes intercropping using tree crops, fodder, and other shade-demanding crops. Overarching approaches within a watershed framework in this pathway include water harvesting and conservation and CSA. FLID is essential in Africa, including Mali, due to the incredible scale of the impact on area and number of farmers, combined with a high economic internal rate of returns (EIRRs) and low implementation costs compared to large-scale irrigation development (IMF 2015; AU 2020). Avenues for increasing agricultural productivity should include increasing land cultivation of semi-arid agriculture in southern Mali (IMF 2015).

To address the pressure and risks of rainfall availability in rain-fed farming, improve production and ensure sustainable management of the resources, different strategies and practices have been employed across the agro-ecological regions in Mali (Table 14.5).

Applying AWM practices (Table 14.5) individually or integratively in Mali's rain-fed crop and livestock farming accrued several benefits, including improved

Table 14.5 AWM practices in rain-fed crop production in Mali's agro-ecological regions

Region	AWM practice	Crops
Saharan	Soil moisture conservation practices	Sorghum (little)
Sahelian	Zai pits; half-moons; composting; cultural sowing techniques; crop diversification; dual purpose crops; crop rotation; agro-climatic information; farmer-managed (assisted) natural regeneration of trees; hedgerows (green fences); adapted improved varieties; system of rice intensification (SRI); cattle fattening; fodder crops; feed supplements	Rice; millet; sorghum; cowpea
Sudanian	Contour (earth and stone) bunds; zai pits; half-moons; composting; cultural sowing techniques; crop diversification; crop rotation; adapted improved seed varieties; short cycle varieties; dual purpose crops; intercropping or crop association; farmer managed (assisted) natural regeneration of trees; animal corralling; agro-climatic information; cattle fattening; fodder crops; feed supplements	Rice; Maize; Millet; Sorghum; Cowpea
Sudano-Guinean	Contour (earth and stone) bunds; cultural sowing techniques; crop diversification; adapted improved seed varieties; intercropping or crop association; crop rotation; composting; direct and early sowing; animal corralling; rational management of land; agro-climatic information; fodder crops; feed supplements	Rice; Maize

Sources: Zougmoré et al. (2014), Andrieu et al. (2017), Aggarwal et al. (2018), Ouedraogo et al. (2018), Zougmoré (2018), Zougmoré et al. (2018), Ouédraogo (2019), CIAT et al. (2020), Bayala et al. (2021), Coulibaly (2021), Traoré et al. (2021), Zougmoré et al. (2021), Bayala et al. (2012, 2016)

crop productivity, resilience, soil fertility, and reduced risk of crop loss due to drought. For example, field contouring reduced runoff by 20–50% and increased crop yields by 30% (Bee et al. 2017; UNEP-CCC 2022). Zai pits increased millet yields by 69%. The SWC structures (zai pits, half-moons, and contour bunds) are usually used in combination with biological measures (planting of grass, trees, and hedges) and application of animal manure or compost in growing crops such as sorghum, millet, and cowpeas, with reported grain yield increases of above 100% (Zougmoré 2018; Zougmoré et al. 2018). Combining zai pits or half-moons and composting in small and medium-scale millet and sorghum production improved yields, reduced soil erosion, and enhanced water retention and infiltration (CIAT et al. 2020).

Millet yield increased by 51% due to contour bunding. Contour stone bunds control runoff and improve soil water content by 59% with barriers and 84% with barriers and organic matter. Zai pits and half-moons effectively and integratively managed soil and water to combat land degradation and improve the productivity of sealed and crusted bare soils. They reduced soil loss by around 89% and accrued watershed scale benefits such as increased groundwater level and tree and vegetation cover. However, these practices are labour-intensive (Zougmoré et al. 2014; Zougmoré 2018). The Malian government and its partners developed and promoted a ridge tillage technique that conserves soil and water by drawing long-lasting ridges on the fields to reduce runoff and erosion and increase rainwater infiltration (Afokpe et al. 2022). The contour ridges were reported to increase millet and sorghum grain and straw yield in Mali, especially when combined with early maturing improved varieties (Traore et al. 2017). For example, the ridging practice increased the grain yield of pearl millet by 30–50% (Mason et al. 2015a, b).

The use of cultural sowing techniques such as direct and early millet and sorghum and crop diversification in rice production at a medium-scale increased productivity. It enhanced the rational and efficient use of water. Agro-ecologically adapted improved seed varieties and contour bunds are used in maize production to increase yields, enhance the resistance of crops to diseases, reduce greenhouse gas emissions from pesticides, reduce runoff and soil erosion, manage soil water to cope with changing weather, and maintain soil quality. For example, cultivating adapted millet varieties increased yields by 27% (Mason et al. 2015a, b). Farmers also cultivate short-season cowpea varieties, sometimes for dual purposes of food and fodder at small, medium, and large scales to increase productivity, reduce soil erosion, and enhance water retention and infiltration. The cultivation of fodder crops (e.g. legumes, bourgou (*Echinochloa stagnina*)) and the use of feed supplements in small-, medium-, and large-scale livestock production reduced pressure on resources, including water, and protecting livestock from diseases, leading to increased productivity and system resilience (CIAT et al. 2020). Cereal–legume intercropping or crop association by growing cereal and pulse crops such as sorghum–cowpea and maize–cowpea systems in the same field increased productivity, diversified food sources, stabilized yields, and potentially decreased greenhouse gas (GHG) emissions. Millet yield increases by intercropping was 23%. Combining the

African locust bean (*Parkia biglobosa*) with crops promoted farmer-managed (assisted) natural regeneration (FMNR) of trees in receding flood areas that enhanced soil organic matter and strengthened crop productivity. Crop residue mulching, on-farm compost production and use of rotating crops, micro-dosing, and field fertilization by animal corralling enrich soil organic matter. Crop residue and mulching reduces soil erosion and soil water evaporation respectively (Andrieu et al. 2021). It increased grain and stover yield in the cultivation of pearl millet (1.3–14.7%), cowpea (19.5%), and sorghum (0.7%) in Mali. Similarly, crop rotation improved yields of pearl millet (14.7–24.1%) and sorghum (23.8–32.7%), while micro-dosing boosted millet grain and stover yields by 61–133% and 35–142%, respectively (Mason et al. 2015a, b).

Mali is one of the international pioneers in implementing climate services that support agricultural water management practices (IBRD and WorldBank 2019). Off-farm intervention measures that promoted and supported rain-fed crop production in Mali include developing hydrometeorological information systems by the National Meteorological Agency (Mali Meteo) that inform seasonal agricultural activities (Montaud 2019). Since 1984, the agency has provided agro-meteorological information to assist rural communities with accurate, timely, and locally adapted weather forecasts at the beginning of the rainy season, length of the growing season, daily weather information, and other related parameters, all disseminated through national and private radios (CIAT et al. 2020). These efforts are part of Participatory Integrated Climate Services for Agriculture (PICSA) in Mali that use historical climate records, participatory decision-making tools and forecasts to assist farmers in identifying and better planning for climatically and locally contextualized livelihood options (Dayamba et al. 2018).

Regarding policy and programming, climate change and AWM-related practices have received attention on the puNDClic agenda in Mali (FAO 2017; Afokpe et al. 2022). The Agricultural Orientation Law (LOA 2006) constitutes the broad framework and long-term vision for the agricultural sector, which aims at promoting sustainable, modern, and competitive agriculture based on family farming (FAO 2017). For inclusion and equity, the LOA requires a minimum of 15% of irrigated lands to be allocated to women and youth under government irrigation land development programmes (MalaboMontpellierPanel 2018a). The relevant policy documents include the National Investment Plan for the Agricultural Sector (PNISA), the National Adaptation Programme of Action (NAPA 2007), and Nationally Determined Contributions (NDC 2015). They mention the adoption of improved crop cultivars, small-scale agricultural development, crop and soil management, agroforestry/forestry/reforestation and assisted natural regeneration for restoring degraded ecosystems, and fodder (cowpea, pigeon pea, bourgou, *Mucuna pruriens*, and Stylosanthes) production. They also highlight water management, including rainwater harvesting and storage. Mentions are also made of organic manure production, micro-dosing of urea, and dissemination of cultivation techniques and conservation methods (Bee et al. 2017; IBRD and WorldBank 2019; Andrieu et al. 2021; UNEP-CCC 2022).

Similarly, some of these practices were identified in the Technology Needs Assessments (TNA) project in Mali (Bee et al. 2017; UNEP-CCC 2022), the Climate-Smart Agriculture Investment Plan (CSAIP) for Mali (IBRD and WorldBank 2019) as well as the year-long pilot of the Climate-Smart Agriculture Prioritization Framework (CSA-PF) process in Mali which involved national and international stakeholders co-identifying climate adaptation options. These included stakeholders who are aware and cognisant of climate change and agriculture challenges and supporting institutions (Andrieu et al. 2017). The CSA-PF project also commended an institutional network and framework facilitating the CSA-PF process, which can also support AWM implementation. Nevertheless, this existing supportive policy framework needs revamped efforts for implementation.

14.4.2 Pathway 2: Farmer-Led Irrigation Development (FLID)

According to AU (2020), FLID is a process where farmers assume a driving and controlling role in improving their water use for agriculture by bringing about changes in knowledge production, technology use, investment patterns and market linkages, and the governance of land and water (Woodhouse et al. 2017). Farmers invest individually or as a group in irrigation technologies to exploit water resources within their control and maximize their economic variables to their advantage. The typical setups in this pathway include individual (private) irrigation for high-value crops wherein independent entrepreneurial risk-taker farmers assume a driving role in improving their water use for agriculture, utilizing private financing. The second type is small-scale community-managed irrigation schemes mainly developed through integrated rural development, natural resources management, community-driven development, or social fund projects. In FLID, the market-oriented farmers grow multiple high-value horticultural crops for urban, peri-urban, and, in some cases, export markets on small irrigated plots of 0.5–2 ha using mainly pumped systems (small petrol, diesel, and solar pumps) and shallow tube wells. The sources of labour are the family on smaller plots and employed/paid labour on larger farms (AU 2020). Farmer-led irrigation development (FLID) is gaining more attention from governments and donors as a sustainable practice to increase production, especially in developing regions (Olayide et al. 2020). Compared to large-scale irrigation schemes characterized by costly large dams and related infrastructure, small-scale FLIDs are relatively less expensive and more easily adaptable to different types of cropping and farming systems managed by farming communities. In Mali's rain-fed agriculture, small-scale FLID can potentially provide an immediate starting and entry point for investment in agriculture.

Crop-yield gaps in smallholder farming systems in African countries such as Mali are mainly caused by water scarcity, which has motivated the development and expansion of irrigation infrastructure for intensifying agricultural production, supporting rural economic development, and enhancing resilience to climate variability and change (Higginbottom et al. 2021). The area of irrigated croplands increased in some parts of the Sudano-Sahelian region, such as Sio in Mali (Traore et al. 2021). In Mali, off-season farming is common among one-third of the households. It includes market gardening (around 26% of the households), cultivation of flood recession crops (9%) and irrigated cereal agriculture (9%) (CIAT et al. 2020). In this FLID pathway, Malian farmers also utilize equipped lowlands irrigation and grow diverse crops, including vegetables, corn, and tubers (potato, sweet potato, yam) during the off-season for intensification. Only 6% of Mali's arable land is equipped for irrigation, and it was ranked third highest behind Morocco and South Africa in Africa for 2012–2014 (MalaboMontpellierPanel 2018a).

Mali has considerable potential to expand land under irrigation, which is estimated at 0.3 million ha for small-scale irrigation expansion, with a higher expected IRR (60%) (MalaboMontpellierPanel 2018a, b). Thus, small-scale irrigation, including FLID, has been identified as a potential pathway for boosting food security and livelihoods and adapting to climate change in dry regions such as Mali (Olayide et al. 2020). After highlighting the overall failure of large-scale irrigation schemes to deliver promised benefits in sub-Saharan Africa (Higginbottom et al. 2021) suggested FLID as the less formalized and technocratic alternative which may provide complementary, low-cost solutions for improving food production, alleviating poverty, and stimulating rural entrepreneurship and innovation. According to IWMI (2021), Small-scale FLID practices are cost-effective and scalable agricultural water management solutions that proved to improve the food security and livelihoods of smallholder farmers in sub-Saharan Africa in which Mali is located. In northern Mali, Dillon (2008) and Dillon (2011) reported that access to irrigation at a small scale increases agricultural production, food supply, household consumption (caloric and protein), livestock holdings, and the ability to share food informally with non-irrigators.

The irrigation water sources include rivers/streams, dams/water reservoirs, groundwater (wells, boreholes), tanks (water harvesting), and lakes/ponds. The modes of access to irrigation water are private, free means (communal access), payment of money to others and membership to groups. The water is lifted and controlled manually or mechanically by gravity, pumps (manual—treadle, hip, pedal, aeolian; motorized—electric, solar, fuel), hose sprinkling, Californian system, sprinkling system, and buckets/cans/gourds (with rope). However, some small-scale farmers utilize sprinkler irrigation, and drip irrigation is gradually being introduced and gaining acceptance. Manual watering using rope and containers in wells, rivers, or other surface water sources is mainly common among resource-poor farmers.

In contrast, automatic watering with pumping systems is used in small and large irrigation systems by average and wealthy farmers (Kane et al. 2018a; Kergna and Dembele 2018). The Sahelian region of Mali has seen recent development and dissemination of potential technologies such as solar-powered drip irrigation (Partey et al. 2018; Traore et al. 2021). Across the regions in Mali, farmers tend to use a variety of technology and water sources. Farmers in the Kayes region use small motor pumps installed on shallow wells, while those in Koulikoro Kati, Kolokani, and Dioïla utilize surface water (sumps) or boreholes equipped with individually or group-owned solar pumps, the California watering system or motorized pumps from wells, micro-dams, pounds, and rivers). Farmers in Segou tend to use gravity irrigation and motorized pumps, while Mopti and Dogon are characterized by large areas irrigated from reservoirs/small dams with a recent introduction of modern irrigation techniques. Farmers have different preferences between crops and irrigation systems. For example, they prefer drip irrigation for growing tomatoes and shallots and sprinklers and Californian irrigation systems for producing shallots and potatoes (Dembélé et al. 2021). Generally, farmers produce vegetables with Californian, sprinkler, and drip irrigation systems. For cereal production, farmers prefer motor pumps (Kergna and Dembele 2018).

The drip irrigation technology brings water under pressure in a system of pipelines, which is then distributed in drops in the field by a large number of gutters distributed all along rows of plants. This irrigation system grows tomatoes, onions, shallot (*Allium fistulosum*), bananas, papaya, and oranges. The Californian irrigation system consists of a network of PVC pipelines buried to decrease water losses by infiltration. In this irrigation system, water is lifted from the source (surface or underground) and conveyed to plants into furrows bound by cropped ridges. The Californian irrigation system mainly produces vegetable crops such as shallot and onion. The sprinkling irrigation system pressurizes the water in a network of conduits and distributes it by the rotary sprinklers that imitate the natural rain. This irrigation system is commonly used on commercial farms producing high-value crops such as fruit trees, coffee, sugar cane and horticultural crops, and potato (Kane et al. 2018a, b). A study by Kane et al. (2018b) and Kane et al. (2018a) on Mali's commonly used irrigation systems found that they operated inefficiently. They reported that there is room for improving the efficiency and productivity in the irrigation systems by approximately 24% without altering the input levels. They also discovered that drip and sprinkling irrigation systems were relatively more economically efficient than the Californian system. However, drip, sprinkler, and Californian irrigation systems are technically and economically more desirable than the manual gravity irrigation system when producing high-value horticultural crops such as potatoes, shallots, and tomatoes.

Sprinkler irrigation doubled the yields of potato tubers, while drip irrigation yielded more than 1.5 times farmers' practice in tomato production. The same was observed in shallot production with the Californian system, even considering the time spent and water used in irrigating shallot, potato, and tomato compared to farmers' practice. Their descending order of technical and economic performance was drip followed by sprinkling, and the third was the California irrigation system (Kane et al. 2018a, b). In terms of popularity among farmers, Nkonya et al. (2020) and Dembélé et al. (2021) found that gravity irrigation is the most common (47%) technology in large vegetable production areas, particularly in irrigated areas near villages, in villages along rivers, and near water reservoirs and lowlands. This popularity of the gravity irrigation technique was also observed in the Office of Niger zone and the irrigated perimeters of Baguineda, Segou, and Selingue. However, a study by Kane et al. (2018b) concluded that a manual irrigation system is not economically viable in producing potatoes, shallots, and tomatoes compared to the other standard irrigation methods. Among the irrigators, access to motor pumps increases the consumption of nutrient-rich food groups (fruits, vegetables, oils, spices, and cereals), significantly improving household nutrition and income (Nkonya et al. 2020).

Another notable practice of the FLID pathway in Mali is the bas-fond system (Dimithè et al. 2000). Bas-fonds are farmer-managed shallow inland valleys with high tables relying on wet season surplus on which crops (usually rice) are cultivated (USAID 2013). The bas-fonds have four production systems depending on farmers' input combinations. These bas-fond production systems include traditional, macro semi-intensive, micro semi-intensive, and intensive. In the traditional production system, the farmer has no water control, plants traditional rice varieties, and applies no chemical fertilizer or herbicide. Farmers control water and plant traditional rice varieties in the macro semi-intensive production system without applying chemical fertilizer or herbicide. In the micro semi-intensive production system, the farmer has no control over water, plants traditional rice varieties, applying some herbicide but no chemical fertilizer. Farmers control water, plant improved rice varieties in the intensive production system, and apply chemical fertilizers and herbicides (Dimithè et al. 2000). According to Dimithè et al. (2000), the four most common bas-fond production systems yield higher returns than the opportunity cost of labour. They are more profitable than the main upland crops (cotton, sorghum/ millet, and maize).

Practices for soil and water conservation in irrigated agriculture include mulching, half-moon, agroforestry, stone walls, and stone rows (Burney et al. 2013; Kane et al. 2018a; Kergna and Dembele 2018; Nkonya et al. 2020; Sidibé and Minh 2021). Other AWM practices in FLID in Mali include the development of inland valleys for rice cultivation with solar pumps, agro-climatic information, tree nursery and transplanting of receding flood areas, a system of rice intensification (SRI), micro-dosing fertilization, and development of fishponds. Far from bas-fonds, common on-farm water management involves terracing ridges to minimize field runoff, encouraging infiltration, and accessing shallow groundwater (USAID 2013). Concerning farmers' organizations in Mali, participation in farmer groups increases the propensity to adopt irrigation because this serves as an entry point for capacity building on irrigation and market participation (Nkonya et al. 2020).

In Mali, irrigation policies are mainstreamed into national sectoral policies, planning and programming (MalaboMontpellierPanel 2018a, b). These include the Rural Development Plan (1992), Accelerated Growth Strategy (1997), Poverty Reduction Strategy (1998), National Environmental Protection Policy (1998), National Strategy for Irrigation Development (SNDI 1999) and National Policy on Climate Change (2011) that mention expansion/extension of irrigation capacities, rural infrastructure, small-scale irrigation, and hydro-agriculture as sustainable agricultural production systems and environmentally sustainable farming methods for ensuring food security and nutrition, reducing imports, increasing rural incomes, and limiting emigration from rural areas (MalaboMontpellierPanel 2018a, b; Montaud 2019). The Agricultural Orientation Law (LOA 2006) constitutes the broad framework and long-term vision for the agricultural sector, which aims at promoting sustainable, modern, and competitive agriculture based on family farming (FAO 2017). For inclusion and equity, the LOA requires a minimum of 15% of irrigated lands to be allocated to women and youth under government irrigation land development programmes (MalaboMontpellierPanel 2018a). Other relevant policy documents include the National Investment Plan in the Agricultural Sector (PNISA), the National Adaptation Programme of Action (NAPA 2007) and Nationally Determined Contributions (NDC 2015). They mentioned the adoption of improved crop cultivars, small-scale agricultural development, crop and soil management, and agroforestry/forestry/reforestation and assisted natural regeneration for restoring degraded ecosystems. They also highlight water management, including irrigation schemes (lowlands, small dams, and market gardening schemes), development of pastoral hydraulics, water reservoirs, boreholes, minor irrigation, dams, ponds, cisterns, rainwater harvesting and storing, and solar irrigation. Mentions are also made of organic manure production, micro-dosing of urea, a system of rice intensification (SRI), intelligent agriculture development, and dissemination of cultivation techniques and conservation methods (Bee et al. 2017; IBRD and WorldBank 2019; Andrieu et al. 2021; UNEP-CCC 2022). Similarly, some of these practices were identified in the Technology Needs Assessments (TNA) project in Mali (Bee et al. 2017; UNEP-CCC 2022), the Climate-Smart Agriculture Investment Plan (CSAIP) for Mali (IBRD and WorldBank 2019) as well as the year-long pilot of the Climate-Smart Agriculture Prioritization Framework (CSA-PF) process in Mali which involved national and international stakeholders co-identifying climate adaptation options. These included stakeholders who are aware and cognisant of climate change and agriculture challenges and supporting institutions (Andrieu et al. 2017). The CSA-PF project also commended an institutional network and framework facilitating the CSA-PF process, which can also support AWM implementation. Although the government increased investments in the expansion of land under irrigation, to date, only a small share of the country's land potential has been tapped (MalaboMontpellierPanel 2018a, b; Montaud 2019). Nevertheless, this existing supportive policy framework needs revamped efforts for implementation.

14.4.3 Pathway 3: Irrigation Scheme Development and Modernization

According to (AU 2020), the irrigation scheme development and modernization pathway delves into the modernization of infrastructure, organizational and operational modalities, and new scheme development, especially in Africa, where most public irrigation schemes are older and need significant rehabilitation and modernization.

In Mali, the irrigation scheme development and modernization pathway potentially increases mean agricultural yields, reduces deviations, and eliminates farmers' dependence on unreliable precipitation. This is attributed to the provision of supplemental water supply during dry periods, the possibilities of off-season cultivation, and the extension of the growing season in standard years. However, the irrigation capacity in Mali is low against a background of colossal irrigation potential. Currently, most Malian irrigated lands (86%) benefit from large-scale dam-based systems (Montaud 2019). Mali stands at 6% of Mali's arable land is equipped for irrigation and was ranked third highest behind Morocco and South Africa in Africa for 2012–2014. It has considerable potential to expand land under irrigation, which is estimated at 0.19 million ha for large-scale irrigation, with an internal rate of return (IRR) of 10% (MalaboMontpellierPanel 2018a, b; Montaud 2019). There was a shift towards irrigated agriculture from 1997 to 2006, probably due to increased post-conflict irrigation investment by international nongovernmental organizations (NGOs) (Dillon 2008, 2011). According to Burney et al. (2010), access to irrigating increased household savings and informal social insurance through transfers in northern Mali. IWMI (2019) reported that Malian irrigators outperformed non-irrigators in income (22%), market participation (67–200%), employment creation (100%), as well as dietary quality and diversity. Irrigation almost doubled the daily intake by households in northern Mali between 1998 and 2006, thus improving food and nutritional security (Hanjra and Williams 2020). A significant rise in crop productivity between 1990 and 2014 is attributed to high irrigation investments and new technologies' introduction (Coulibaly 2021). However, crop yields, except rice, were stagnant between 2015 and 2020 due to the inefficiency and ineffectiveness of agricultural innovations in responding to increasing demand for food crops. This provoked the Malian government to promote irrigation, especially in rice production, through managing flood plains and small irrigation schemes (CIAT et al. 2020). In Mali, the land area under irrigated agriculture increased by 356% from 1975 to 2015 (IBRD and WorldBank 2019). However, Mali has equipped over 40% of her irrigation potential (Nkonya et al. 2020; Oke and Aduramigba-Modupe 2020).

Irrigation is mainly limited to cotton, sugar cane, and rice in the Office du Niger area and along the Niger River in central Mali (IWMI 2021). The common water sources and irrigation methods in Mali are explained in Sect. 14.4.2 Gravity irrigation is used in small-scale irrigated areas to produce cereals (rice, wheat, maize) and vegetable production, mainly in the northern zones of Mali (Gao, Tombouctou, and Mopti). The Californian irrigation system is used for vegetable crop production, especially in urban areas. Sprinkling irrigation is common in commercial farms that produce high-value crops such as fruit trees and sugar cane in the Office du Niger. Drip irrigation practice is taking off in Mali and is mainly used for fruit and vegetable cropping in the urban areas by wealthy farmers. Manual watering is commonly used in rural areas of Mali by low-resource farmers to produce mainly vegetables (Kane et al. 2018a; Kergna and Dembele 2018).

Small reservoirs were constructed in Mali in response to the Sahelian droughts of the 1970s and 1980s, and they play a critical role in supplying water for irrigation during dry spells, fishing, livestock watering, domestic use, and groundwater recharge (Ayantunde et al. 2018). Mali has irrigation schemes differ in characteristics, such as management structure and source of investments (Table 14.6). The

Table 14.6 Some characteristics of some irrigation schemes in Niger River Basin, Mali

Name of scheme	Sélingué (Guèye 2014)	Maninkoura ('l;l'l;'l'; 2014)	Bargodaga	Kamaka	Sinah	Saba1	B1	Djidian	N10
Conception			1997	1994	1997	2001	1951	1950	1954
Principal cropping system	Rice, banana, and market gardening	Rice, banana, and market gardening	Monoculture rice		Rice	Rice	Rice and vegetables	Rice	Rice and vegetables
Investment			State	State/donors /NGO	State/donors / NGO	State/ farmers	State	State	Farmers
Management	Sélingué Rural Development Office	Sélingué Rural Development Office	Farmers	Farmers	Farmers	Farmers	Farmers	Farmers	Farmers
Gross area (ha)			45	20	42	10	517	272	127
Cultivable area (ha)	1030	1094	45	16	49	35	577	298	122
Number of farmers	1943	1168	162	75	113	120	218	145	86
Average farm size (ha)			0.4	0.35	0.6	0.25	2.64	4.5	4
Overall irrigation efficiency (%)			50	45	50	40	50	40	50
Cropping intensity (%)			100	100	100	100	150	100	173
Rice Yield (kg/ ha)			6000	4300	4500	5500	4950	3000	4800
Others	Gravitational; Sélingué dam	Pumped; Sélingué dam	River Niger	River Niger; motor pumps	River Niger; group motor pumps	Group motor pumps			

Source: Acheampong (2008)

management and investment actors include the government and farmers as individuals or groups (Acheampong 2008).

AWM practices include the rice intensification (SRI) system, which utilizes compost and reduced/limited intermittent irrigation instead of flooding the rice paddy field. This practice was reported to utilize water efficiently and significantly increase rice yields and income (Styger et al. 2011; Zougmoré et al. 2018).

Mali possesses vast potential of untapped irrigation potential. For example, the Markala Dam by Office du Niger was designed to irrigate 960,000 ha but is currently irrigating around 100000 ha as limited by the hydraulic capacity of the intake structure at Markala and the capacity of the downstream canals (USAID 2013). The Segou and Sikasso regions have 145,000 and 655,000 ha of land suitable for solar irrigation, respectively, excluding areas that are unsuitable for agricultural production, including national parks, forests, permanent meadows and pastures (Table 14.7) (IWMI 2021; USAID 2021).

Mali has between 0.69 and 4.4 million ha suitable for solar-powered irrigation, representing 11–69% of the country's agricultural lands (IWMI 2019). This irrigation potential is an opportunity for irrigation scheme development and modernization, given the rapidly decreasing prices of solar photovoltaic panels and solar pumps (IWMI 2019). Solar-powered drip irrigation facilities are being promoted in the Sudano-Sahel zones due to their cost-effectiveness and significant correlation to increased household income, food security, and nutritional intake in the region. They accrue benefits in significant water savings compared to conventional irrigation practices (Andrieu et al. 2017; Partey et al. 2018; CIAT et al. 2020). Regarding the governance and institutional dimensions of irrigation scheme development and modernization, Mali has a good example and successful template of the Office du Niger irrigation scheme. The power shift from government agency monopoly to farmers through radical reforms in scheme operation, maintenance, and

Table 14.7 Suitability for farmer-led solar irrigation development in administrative regions of Mali

Region	Suitable land area (thousand hectares) for different situations of water resources				
	Groundwater <= 7 m	Groundwater <= 2 m	Surface water and small reservoirs	Surface water, small reservoirs, and groundwater <= 7 m	Surface water, small reservoirs, and groundwater <= rm
Bamako	1	1	–	1	1
Gao	50	50	7	57	57
Kayes	657	1128	19	664	1135
Kidal	–	–	3	3	3
Koulikoro	460	678	78	499	716
Mopti	495	501	134	585	590
Segou	220	220	276	463	463
SIkasso	308	1062	125	385	1125
Timbuktu	316	316	43	345	345
Total	2507	3956	685	3002	4435

Source: IWMI (2019)

management accrued benefits in increased yields and income (Aw and Diemer 2005; WorldBank 2010).

A notable AWM practice is the deliberate reforms in the operation and management of irrigation schemes. The Office du Niger (ON hereafter) refers to the irrigation scheme and the semi-autonomous management organization. It is one of the oldest (1932) and largest gravity irrigation schemes in sub-Saharan Africa (Aw and Diemer 2005; World Bank 2010). The ON irrigation scheme covers 96,000–120,000 hectares in Mali on which about 40,000 family farmers grow rice in the wet season (July–December) and vegetables in the dry season (December–May) (Hertzog et al. 2014; MalaboMontpellierPanel 2018a, b). Originally under a top-down monopoly of the government and delegated management agency ON, the government and donors started pushing for reforms and restructuring of ON in 1980 to increase the participation of citizens (farmers) in management (Aw and Diemer 2005; WorldBank 2010). Thus, water management was partly transferred from the public institution ON to farmers through joint committees and user associations (Hertzog et al. 2014). Other implemented reforms included credit for all farmers, land tenure security for farmers, full-cost recovery, ownership of harvest and post-harvest equipment by farmers, liberalization and protection of the domestic rice market, negotiable performance contracts, creating pro-farmer support services, strengthening the role of private sector, intensification, establishing co-management committees, establishing farmer organizations, assigning farmer representatives, crop diversification (horticulture), and inclusion of village associations in processes of scheme, management, and canal operation and maintenance (Darghouth 2005). These reforms towards the inclusion of farmers incentivized them to increase (double) yields, incomes, and the fee collection rate. Other beneficial outcomes included efficiency gains in business processes, increases in cultivated area, improvements in rice productivity and production, improvements in employment and labour productivity, differential changes in farm income, poverty reduction, and economic growth (Aw and Diemer 2005; Barry et al. 2009; WorldBank 2010). These reforms were cultivated on the fertile ground provided by relatively good governance following the 1991 establishment of democracy in Mali (Higginbottom et al. 2021).

Management of agricultural water along the Niger Basin and Senegal River in Mali falls under three unique categories and environments: small community-managed systems (Irrigation de proximite), larger technical irrigation systems, and government-operated pump systems for irrigation. Small community-managed systems (Irrigation de proximite) rely on run-of-the-river diversions that irrigate relatively restricted areas (e.g. extensive vegetable gardens) and require frequent maintenance to maintain diversion weirs and channels. They practice recession irrigation, extend into the floodplain as river water levels recede, or hand-watering of higher-value crops. They increasingly use small motor pumps to lift water onto higher areas, mainly by private farmers between Timbuktu and Niger, thus creating opportunities for private-sector pump repairers. Larger technical irrigation systems use improved diversion structures and more sophisticated canals and distribution systems (e.g. north of Selingue Dam). They generally include some central or local government role in operating diversion structures and maintaining larger canals.

Government-operated pump irrigation systems are standard downstream of Mopti. From a few hectares to 300 ha in farm size, they are focused on rice production for food security. In these three categories of agricultural water management along the Niger River, some opportunities exist for increasing the irrigated area though unaffordable for local communities and thus requiring additional government investment. Other typical systems involve the diversion of water during high floods of the Niger into low-lying areas and restriction of drainage back into the Niger River to allow flooding of rice, such as the Riz du Segou and Riz de Mopti systems. However, it is uneconomic to expand these diversions and lift water from the floodplain up to potentially productive higher land on either side of the main rivers (USAID 2013).

To expand the existing irrigated 96,000–120,000 ha in ON under conditions of limited public funds, the government of Mali opened access to land and water resources from the Niger River basin by providing favourable conditions of administrative and fiscal incentives to promote significant national and foreign investments in irrigated agriculture (Coulibaly 2021). By 2012, the Malian government entered into lease agreements for more than 800,000 ha (almost ten times the current cultivated area) with private investors from China and South Africa, including the Malibya project wherein Libya was freely offered 100,000 ha of land in the ON to develop and build infrastructure to cultivate rice and rear cattle. However, many of these lease agreements are still yet to be fruition since only a tiny part of the leased land has been put into actual productive use (Bastagli and Toulmin 2014). Unfortunately, this large-scale allocation of land to investors was politically centralized and bypassed the official procedure established by the ON at the regional level (Hertzog et al. 2012).

Malian irrigation policies are mainstreamed into national sectoral policies, planning, and programming (MalaboMontpellierPanel 2018a, b). These include the Rural Development Plan (1992), Accelerated Growth Strategy (1997), Poverty Reduction Strategy (1998), National Environmental Protection Policy (1998), National Strategy for Irrigation Development (SNDI 1999), and National Policy on Climate Change (2011) that mention expansion/extension of irrigation capacities, rural infrastructure, small-scale irrigation, and hydro-agriculture as sustainable agricultural production systems and environmentally sustainable farming methods for ensuring food security and nutrition, reducing imports, increasing rural incomes, and limiting emigration from rural areas (MalaboMontpellierPanel 2018a, b; Montaud 2019). The Agricultural Orientation Law (LOA 2006) constitutes the broad framework and long-term vision for the agricultural sector, which aims at promoting sustainable, modern, and competitive agriculture based on family farming (FAO 2017). For inclusion and equity, the LOA requires a minimum of 15% of irrigated lands to be allocated to women and youth under government irrigation land development programmes (MalaboMontpellierPanel 2018a). Other relevant policy documents include the National Investment Plan in the Agricultural Sector (PNISA), the National Adaptation Programme of Action (NAPA 2007), and Nationally Determined Contributions (NDC 2015). They mentioned the adoption of improved crop cultivars, small-scale agricultural development, crop and soil management, and agroforestry/forestry/reforestation and assisted natural regeneration for restoring degraded ecosystems. They also highlight water management, including irrigation schemes (lowlands, small dams, and market gardening schemes), development of pastoral hydraulics, water reservoirs, boreholes, small irrigation, dams, ponds, cisterns, rainwater harvesting and storing, and solar irrigation. Mentions are also made of organic manure production, micro-dosing of urea, a system of rice intensification (SRI), intelligent agriculture development, and dissemination of cultivation techniques and conservation methods (Bee et al. 2017; IBRD and WorldBank 2019; Andrieu et al. 2021; UNEP-CCC 2022). Similarly, some of these practices were identified in the Technology Needs Assessments (TNA) project in Mali (Bee et al. 2017; UNEP-CCC 2022), the Climate-Smart Agriculture Investment Plan (CSAIP) for Mali (IBRD and WorldBank 2019) as well as the year-long pilot of the Climate-Smart Agriculture Prioritization Framework (CSA-PF) process in Mali which involved national and international stakeholders co-identifying climate adaptation options. These included stakeholders who are aware and cognisant of climate change and agriculture challenges and supporting institutions (Andrieu et al. 2017). The CSA-PF project also commended an institutional network and framework facilitating the CSA-PF process, which can also support AWM implementation. Although the government increased investments in the expansion of land under irrigation, to date, only a small share of the country's land potential has been tapped (MalaboMontpellierPanel 2018a, b; Montaud 2019). Nevertheless, this existing supportive policy framework needs revamped efforts for implementation.

14.4.4 Pathway 4: Unconventional Water Use for Irrigation

According to AU (2020), the unconventional water used in irrigation includes reclaimed wastewater from sewage treatment plants, urban runoff, and seawater desalination. This pathway is mainly associated with risks to people's health and the environment, but its increasing importance to livelihoods in Africa cannot be ignored.

Due to scarcity of freshwater in urban and peri-urban areas of many cities of developing countries such as Bamako, Segou, Kati, Baguineda, Samanko, Sikasso, and Niono in Mali, urban and peri-urban area farmers resort to irrigating with untreated wastewater (treated, raw, or diluted) and polluted water to provide perishable food and earn livelihoods (Drechsel et al. 2008, 2010; Traoré et al. 2020). The major crops produced are vegetables that are consumed raw or cooked, including lettuce, tomato, and cucumber, with some rice production in other areas such as Sikasso, and gardening and irrigation of olive trees and non-edible crops such as fodder (Drechsel et al. 2006; Levasseur et al. 2007; Jiménez et al. 2010; Traoré et al. 2021, 2022; Aboye 2022). The major drivers for such practices include increasing urban water demand and related return flow, urban food demand and market incentives in city proximity, and lack of alternative (cheaper, similarly reliable or safer) water sources (Raschid-Sally and Jayakody 2008). The benefits of using unconventional water for irrigation include food supply, better nutrition, income generation, improved livelihoods, saving water, recycling of nutrients and saving on fertilizer to benefit farmers and the environment, land treatment, and reduced surface water pollution (Khalid 2018). For example, unconventional water use for irrigation in Bamako directly and indirectly employs more than 6000 people and produces 60–100% of the consumed leafy vegetables (Raschid-Sally and Jayakody 2008). According to Drechsel et al. (2006) and Levasseur et al. (2007), the farmers involved in year-round irrigated farming with unconventional water in Mali can earn similar to or higher than the income of a senior civil servant and twice more than rain-fed farming despite much smaller farm sizes.

However, despite these benefits, the practices are associated with risks, including health (diarrhoea, skin and worm infections), accumulation of heavy metals, catchment alteration and salt transport, microbiological contamination of water sources, and transfer of chemical and biological contaminants to crops. Heavy metals and parasites are potentially toxic elements that threaten the soil, water sources (surface and ground) and humans (Raschid-Sally and Jayakody 2008; Khalid 2018). For example, a study by Abdu et al. (2011) in Sikasso (Mali) found that irrigation with wastewater induced pollution with heavy metals, especially cadmium (Cd), chromium (Cr), and zinc (Zn), in vegetable garden soils to unsafe levels, which could result in the production of vegetables that are unsuitable for human consumption. Thus, consuming irrigated vegetables, especially uncooked raw such as lettuce, poses a health risk and hazard to consumers (Traoré et al. 2013, 2020). Similarly, irrigating fresh vegetables with non-treated and treated wastewater in Segou City (Mali) was linked to a high prevalence of enteric and waterborne diseases that was reported at a local hospital, most probably due to the consumption of vegetables contaminated with wastewater (Toure et al. 2019). A study by Toure et al. (2019) in Segou City discovered that the water in rivers and ponds was contaminated with waste to levels exceeding the recommended microbial infection standards and thus too risky to irrigate fresh vegetables for human consumption. Traoré et al. (2021) found wastewater-irrigated lettuce and tomato in Kati, Baguineda, Samanko, Sikasso, and Niono to be contaminated with parasite eggs and cysts, while cucumber was free from parasitic. All three crops in Sikasso were free from parasite eggs, most probably due to access to adequate land, better quality irrigation water, and training in good agricultural practices. The observed parasites were *Entamoeba coli* and *Trichomonas intestinalis*, *Ascaris lumbricoides*, *Giardia intestinalis*, *Balantidium coli*, *Entamoeba histolytica*, *Fasciola hepatica*, *Trichinella spiralis*, *Ancylostoma duodenale*, and *Schistosoma intercalatum* (Traoré et al. 2021).

Urban land used in wastewater irrigation is mainly owned by the state/government, leading to the insecurity of land tenure for farmers. In contrast, peri-urban land is owned by the community or chieftaincy, which leases or gives it to farmers (Raschid-Sally and Jayakody 2008). Irrigation by unconventional water is practised only in the informal smallholder irrigation sector. The major sources of unconventional water for irrigation include shallow (dugout) wells potentially contaminated with pesticides, urban waste, and other industrial contaminants (Levasseur et al. 2007). Some shallow wells are dug five metres away from wastewater streams or canals to collect shallow groundwater filtered by riverbank filtration following the hydraulic gradient (Keraita et al. 2014). Other primary sources of irrigation water include rivers and streams, which are often polluted, while the water lifting and irrigation technologies are watering cans (with rope), furrow/flooding techniques, and motor pumps with water hoses (Amoah et al. 2011). Of late, there has been a noted increase in the use of small petrol pumps and lay-flat hoses for spray irrigation as driven by mobility and savings in energy, labour, and irrigation costs (Keraita et al. 2010; Drechsel et al. 2006; Raschid-Sally and Jayakody 2008).

Despite the contribution of unconventional water use for irrigation to food and nutritional security, livelihoods and environmental and human health in Mali, there is a general lack of policy and programming related to the practices which are illegal and forbidden but unofficially tolerated and excluded in the urban political plans (Levasseur et al. 2007). This leniency emanates from limited and weak enforcement of regulations (Raschid-Sally and Jayakody 2008). In addition, the economy of Mali cannot support the treatment of untreated water in all regions and the construction of irrigation channels as recommended by WHO (2006) standards. On the other hand, the government is reluctant to stop farmers from using the water because the livelihoods of many people depend on it for income and urban vegetable supply (Traoré et al. 2021, 2022). Thus, this is an opportunity to formalize and regulate the practice for the benefit of farmers, consumers, and the environment. The farmers have been innovative enough and formed farmers' associations as a collective financial strategy to address their constraints and land conflicts. For example, the lobbying for access to public land by the Yiriwaton farmers' cooperative in Bamako gained the local government's attention, directing the municipality to consider leasing unused land to farmers (Drechsel et al. 2006).

14.4.5 Summative Discussion: Lessons Learnt, SWOT Analysis, and Preferred Pathways

Adoption of AWM practices along the four IDAWM pathways is constrained by socioeconomic, institutional, infrastructural, biophysical, and political factors (Ouedraogo et al. 2018). The significant challenges to implementing AWM practices across the four IDAWM pathways are presented in Table 14.8.

14.5 Conclusion and Recommendations

Mali's over-reliance on the fluctuating rainfalls in semi-arid and arid lands exposes it to the vagaries of dry spells, droughts, and water scarcity, leading to decreased agricultural productivity and increased farmers' vulnerability. Irrigated agriculture is increasing in terms of area and technology in Mali, including solar pumps and drip irrigation. Irrigation has been shown to improve food and nutrition security,

Table 14.8 Identified opportunities and challenges to IDAWM

AU-IDAWM pathway	Challenges/barriers	Opportunities
1: Improved water control and watershed management in rain-fed farming	Over-reliance on fluctuating rainfall Limited technical capacity Illiteracy of farmers (lack of training) Limited availability/access to credit, subsidies, inputs, and equipment Low dissemination of information Weak implementation of policies Inadequate extension services Conflicts and socio-political insecurity Degradation of natural resources Land tenure insecurity Gender inequity in the distribution of resources (especially land) Labour requirement and lack of draught power	The agro-meteorological advisory system High penetration of mobile phones Prioritization of CSA on the public agenda Counterproductive policies, e.g. export and import restrictions and tariffs Recent policy developments are inclusive of vulnerable populations (women, youth)
2: Farmer-led irrigation development (FLID)	Labour requirement Lack of access to financial resources and niche markets Land tenure insecurity (different tenure systems) Limited access to technology (e.g. irrigation, equipment), improved seeds, fertilizers Lack of access to capacity building and training Weak extension services Variability of water supply Complexity of technologies	Recent policy developments are inclusive of vulnerable populations (women, youth) Availability of technologies, evidence-based encouraging results, and enabling environment Potential in the use of digital technologies in irrigation for better management of agricultural water Government subsidies Financial institutions (banks and microfinance organizations) Farmer and support organizations Available resources: water, land, local materials

(continued)

Table 14.8 (continued)

AU-IDAWM pathway	Challenges/barriers	Opportunities
3: Irrigation scheme development and modernization	Failure of large-scale irrigation schemes High and unaffordable investment costs and unavailability of irrigation equipment Land tenure insecurity disincentives (legal provisions = modern law; traditional provisions = customary law) Lack of finance to rehabilitate and/or expand government-managed irrigated schemes Weak capacity to adapt to climate change Inadequate capacity of the in-house technocrats to cope with the dynamics of the day-to-day management of irrigation schemes Inadequate monitoring of available water resources (surface, groundwater) The historically unstable socio-political climate Unstructured peasant farmer organizations Gender inequality Lack of reliable water supply Relatively high initial investment costs Limited access to credit, technology, extension services, fertilizers, and improved seeds Waterlogging, salinization, alkalization, soil instability, alkalinization, and sodification of irrigated land Lack of communication among stakeholders	The willingness of Malian institutions to be involved in the debate over the impact of climate change The private sector (PPP) Increased public expenditure on agriculture Topography and technology Availability of resources (water, land, local materials Knowledge of traditional techniques (e.g. bas-fonds) Presence and willingness of relevant actors (farmers, research, technical, government, municipalities, farmer organizations, NGOs, design offices)
4: Unconventional water use for irrigation	Lack of socio-cultural acceptance and awareness (safety, health concerns) Lack of socio-political recognition Absence of capacity for in situ wastewater treatment Lack of supporting policy, regulatory standards, enforcement, and monitoring framework on water quality and wastewater irrigation Unfavourable markets (inputs, outputs) Land tenure insecurity Limited/lack of extension services	International standards and guidelines (e.g. WHO, FAO) for safe wastewater irrigation Safer irrigation practices and technologies Alternative farmland and water resources

Sources: Drechsel et al. (2006), Levasseur et al. (2007), Barry et al. (2009), Dittoh et al. (2010), FAO (2017), Kane et al. (2018a), Kergna and Dembele (2018), Khalid (2018), Ouedraogo et al. (2018), IBRD and WorldBank (2019), Ouédraogo (2019), Traoré et al. (2019), CIAT et al. (2020), Langyintuo (2020), Nkonya et al. (2020), Coulibaly (2021), Dembélé et al. (2021), Higginbottom et al. (2021), Sidibé and Minh (2021), USAID and SWP (2021), Zougmoré et al. (2021)

income, assets, and spill-over effects in the society and economy, such as food sharing between irrigators and non-irrigators in farming communities and increasing holdings of strategic assets such as livestock. Huge potential exists to expand irrigated areas across all scales, including small- and large-scale commercial. However, such irrigation scheme development and modernization must be accompanied by farmer-centred operation, maintenance, and management structures. A good example is the Office du Niger, which increased farmers' yields and income by reforming the scheme's management structure from agency and government monopoly to including farmers and their organizations/associations.

Farmers apply various AWM practices in rain-fed and irrigated agriculture to control water and manage watersheds. The employed practices include soil water and conservation techniques, watershed management, and agronomics. Implementation challenges include a lack of awareness and access to extension services, credit, inputs, and equipment/technology. Despite their small landholdings (2 ha or less), FLID schemes provide local food security in many low-income countries. There is a need to collaborate with the private sector to scale FLID and governments to create an enabling environment for private investment in irrigation equipment supply. Some large-scale irrigation schemes recovered their yields and productivity after adopting liberal reforms away from state control towards autonomous operation by farmers, a key characteristic of FLID schemes. This less formalized development alternative may usher in cost-effective solutions to improve food security, reduce poverty, and stimulate rural entrepreneurship and innovation in a continent dominated by a rural population dependent on agriculture.

Wastewater irrigation is increasing in Mali's urban and peri-urban areas, providing employment, income, and a diverse food supply. Average incomes in this pathway sometimes equate to or surpass those of full-time civil servants. However, wastewater irrigation is plagued by several challenges surrounding its social, political, and cultural acceptance. Associated significant constraints are linked to human health and the environment due to contamination risks by polluted water. To conclude, rain-fed and irrigated agriculture in Mali is constrained by prevailing conditions in the economy, environment, society, politics, and practices. Fortunately, opportunities for these challenges lie in the enabling environment consisting of a policy framework and the supporting plans, programming, and investments. However, several recommendations must be deliberately implemented to operationalize the IDAWM pathways in Mali.

14.5.1 Recommendations

Pathway 1: Improved Water Control and Watershed Management in Rain-Fed Farming

Mali needs to revamp efforts on the diffusion of AWM practices and lift their adoption barriers. The actions needed to boost and sustain the implementation of AWM practices include: (1) building the capacity of farmers, producers, extension agents, and agribusinesses, (2) improving the dissemination of information and technical support on AWM, and (3) improving access to credit, extension services, agricultural inputs and equipment required by AWM. Farmers need to be encouraged on the integrated use of AWM practices, such as combining structural and biological, as well as the use of mechanization for efficiency and effectiveness through saving on time and labour.

Pathway 2: Farmer-Led Irrigation Development (FLID)
The potential to leverage farmer-led irrigation in Mali is high; however, the resource-poor farmers, the main actors, are challenged in several ways. There exists a need and opportunity to promote the utilization of the bas-fond production system for rice cultivation by farmers near main rivers. On the other hand, farmers away from main rivers should implement improved on-farm technologies, including improved plant varieties, and use techniques for water harvesting and soil water retention. There is a need to build capacity by training farmers on good irrigation practices, including adopting and using technically and economically efficient drip, sprinkling, and Californian irrigation systems, especially in areas prone to droughts. This needs to be supported with the development of surface water systems such as river networks, inland valley basements, lowlands, ponds, micro-dams, and water harvesting systems to unlock the potential of small-scale irrigation. Flood-recession agricultural practices can be optimized by providing technical support and improved infrastructure. Modified public–private partnerships (micro-PPPs) can assist farmers in accessing credit for financing the establishment of efficient modern irrigation systems.

Pathway 3: Irrigation Scheme Development and Modernization
There is a need for training and building capacities of farmers, extension workers and management agencies, including raising their awareness on rational choice of irrigation equipment. Supporting farmers with irrigation technology, equipment, and infrastructure should include subsidies. For enhanced productivity, field practices and infrastructure that need to be promoted include micro-irrigation (drip irrigation, micro-jet), gabions, weirs, sprinkling, perforated pipe, pastoral hydraulics, and irrigation scheduling. The same applies to using groundwater and equipment with low maintenance costs, such as solar pumps and gravity irrigation. These need to be supported by bio-pesticides, cultivation techniques, crop rotation, live hedges, organic fertilizers, zero tillage, assisted natural regeneration, and intercropping.

New approaches and financing models are needed to tackle the high costs of rehabilitating and/or expanding government-managed irrigated schemes in Mali. Any further development and modernization of irrigation schemes, for example, in the ON area, should be paralleled by addressing governance issues to ensure agricultural productivity and sustainability. Opportunities lie in integrated approaches to water management through adopting and implementing the IWRM principles.

There is a need to promote research and monitoring for reliable data that enhances understanding of the dynamics of available water resources (ground, surface) and intra-seasonal variations in rainfall patterns at all levels (regional, district, and community). Such advanced data systems on water resources and cropland suitability analyses could effectively support agricultural decision-making and scale out existing AWM practices.

Pathway 4: Unconventional Water Use for Irrigation
Using unconventional water in irrigating urban and peri-urban agriculture can enhance food and nutrition supplies in cities, especially in arid and semi-arid countries like Mali. This pathway is an opportunity because existing practices can be improved at minimal marginal cost. The government should support farmers through sensitization and assisting farmers to treat wastewater for irrigation and use organic fertilizers for crop production. Wastewater is a potential resource for use in agriculture. However, it requires specific regulations to ensure the promotion of public health. There is a need for proper development and implementation of laws and regulations on wastewater discharge and use in the agricultural sector. Regulation and restriction of heavy metal loads to soils are required by limiting heavy metal discharges from industries into wastewater that will eventually irrigate crops. Opportunity lies in adapting existing standards and guidelines by WHO and FAO to contextualize Mali's existing conditions without compromising people's and the environment's health. Research is needed to develop low-cost, low-tech treatment methods for sanitizing wastewater for the safety of farmers and consumers of irrigated produce. Mali needs to recognize officially and anchor wastewater irrigation in its irrigation, water, agriculture, and environmental policies and urban/peri-urban plans and programmes, as well as adopt/develop locally viable practices for the safety of farmers and consumers of produce. These efforts must enhance food supply and livelihoods while reducing health and environmental risks. The government can also provide alternative farmland or safer irrigation water (groundwater) while promoting changes in the choice of crops grown, safer irrigation practices (subsurface, drip irrigation, furrow irrigation, changing height of water application and use of watering rose, cessation of irrigation before harvesting), on-farm treatment (turning reservoirs into sedimentation ponds, water filtering).

In conclusion, local conditions in Mali's different regions must inform the prioritization of these recommendations and pathways needed for expansion, development, implementation, and investments. The conditions to be assessed and considered include biophysical, economic, social, religious, political, cultural, and technological.

Appendix I: Strength of Evidence Results

Table 14.9 Summarized strength of evidence grading

Author	Subject title	Subject matter	Evidence[a]
Strength of Evidence: Strong (I)			
Abdu et al. (2011)	Vertical distribution of heavy metals in wastewater-irrigated vegetable garden soils of three West African cities	Wastewater irrigation	+++
Andrieu et al. (2017)	Prioritizing investments for climate-smart agriculture: Lessons learned from Mali	CSA	+++
Ouedraogo (2019)	Uptake of climate-smart agricultural technologies and practices: actual and potential adoption rates in the climate-smart village site of Mali	CSA	+++
Partey et al. (2018)	Developing climate-smart agriculture to face climate variability in West Africa: Challenges and lessons learnt	CSA	+++
Traore et al. (2021)	Contribution of climate-smart agriculture technologies to food self-sufficiency of smallholder households in Mali	CSA	+++
Zougmoré et al. (2014)	Climate-smart soil water and nutrient management options in semiarid West Africa: a review of evidence and analysis of stone bunds and zaï techniques	CSA	+++
Zougmoré et al. (2021)	Transforming food systems in Africa under climate change pressure: Role of climate-smart agriculture	CSA	+++
Zougmoré (2018)	Promoting climate-smart agriculture through water and nutrient interactions options in semi-arid West Africa: A review of evidence and empirical analysis	CSA	+++
CIAT et al. (2020)	Climate-smart agriculture in Mali	CSA	+++
Coulibaly (2021)	Mali—Land, climate, energy, agriculture and development. A study in the Sudano-Sahel initiative for regional development, jobs, and food security	Resources and management	+++
Dayamba et al. (2018)	Assessment of the use of participatory integrated climate services for agriculture (PICSA) approach by farmers to manage climate risk in Mali and Senegal	Climate services in agriculture	+++
Dimithe et al. (2000)	Financial profitability of Mali-Sud bas-fond rice production systems	AWM	+++
FAO (2017)	Mali—Country fact sheet on food and agriculture policy trends	Policies for food and agriculture	+++
IBRD and WorldBank (2019)	Climate-smart agriculture investment plan—Mali	CSA	+++

(continued)

Table 14.9 (continued)

Author	Subject title	Subject matter	Evidence[a]
Ouedraogo et al. (2018)	Uptake of climate-smart agriculture in West Africa: What can we learn from climate-smart villages of Ghana, Mali and Niger? Findings from a series of adoption studies on CSA technologies and practices within the climate-smart villages of Ghana, Mali and Niger	CSA	+++
Traore et al. (2017)	Optimizing yield of improved varieties of millet and sorghum under highly variable rainfall conditions using contour ridges in Cinzana, Mali	AWM	+++
Zougmoré et al. (2018)	Facing climate variability in sub-Saharan Africa: analysis of climate-smart agriculture opportunities to manage climate-related risks	CSA	+++
Amoah et al. (2011)	Low-cost options for reducing consumer health risks from farm to fork where crops are irrigated with polluted water in West Africa.	Wastewater irrigation	+++
Drechsel et al. (2006)	Informal irrigation in urban West Africa: An overview	Wastewater irrigation	+++
Drechsel et al. (2008)	Reducing health risks from wastewater use in urban and peri-urban sub-Saharan Africa: Applying the 2006 WHO guidelines.	Wastewater irrigation	+++
Traore et al. (2021)	Assessment of the parasite load of lettuce, tomato and cucumber from some large vegetable production sites in Mali	Wastewater irrigation	+++
Traore et al. (2022)	Parasitic contamination of lettuce, tomato and cucumber from vegetable farms in Mali	Wastewater irrigation	+++
Barry et al. (2009)	Better rural livelihoods through improved irrigation management: Office du Niger (Mali)	IWRM	+++
Acheampong (2008)	Analysis of agricultural water productivity of irrigation schemes in the Niger River Basin: Case studies in Mali and Niger	Water productivity	+++
Aw and Diemer (2005)	Making a large irrigation scheme work: A case study from Mali	Irrigation schemes	+++
Darghouth et al. (2005)	Modernizing public irrigation institutions: The top priority for the future of sustainable irrigation	Irrigation management	+++
Dembele et al. (2021)	Development of the vegetable seed sector in Mali and opportunities for irrigated seed production	Irrigation	+++
IWMI (2019)	Suitability for farmer-led solar irrigation development in Mali	Irrigation development	+++
Kane et al. (2018b)	Economic efficiency of water use in the small-scale irrigation systems used in vegetable production in Koulikoro and Mopti regions, Mali	Smallscale irrigation	+++

(continued)

Table 14.9 (continued)

Author	Subject title	Subject matter	Evidence[a]
Kane et al. (2018a)	Economic viability of alternative small-scale irrigation systems used in vegetables production in Koulikoro and Mopti regions, Mali	Smallscale irrigation	+++
Kergna and Dembele (2018)	Small scale irrigation in Mali: Constraints and opportunities	Smallscale irrigation	+++
Malabo Montpellier Panel (2018a)	Water-wise: Smart irrigation strategies for Africa	Irrigation	+++
Malabo Montpellier Panel (2018b)	Water-wise: Smart irrigation strategies for Africa: Mali	Irrigation	+++
Nkonya et al. (2020)	Drivers of adoption of small-scale irrigation in Mali and its impacts on nutrition across sex of irrigators	Smallscale irrigation	+++
USAID (2013)	Climate change in Mali: Key issues in water resources	Climate change and water resources	+++
Sidibé and Minh (2021)	Multi-stakeholder dialogues supporting the scaling of inclusive and sustainable agricultural water management in Mali: Kick-start meeting report	Sustainable agricultural water management	+++
World Bank (2010)	Making a large irrigation scheme work: A case study from Mali	Irrigation schemes management	+++
Strength of Evidence: Moderate (II)			
Andrieu et al. (2021)	Ex ante mapping of favourable zones for uptake of climate-smart agricultural practices: A case study in West Africa	CSA	++
Aggarwal et al. (2018)	The climate-smart village approach: framework of an integrative strategy for scaling up adaptation options in agriculture	CSA	++
Mason et al. (2015a)	Pearl millet production practices in semi-arid West Africa: a review	AWM	++
Mason et al. (2015b)	Soil and cropping system research in semi-arid West Africa as related to the potential for conservation agriculture	Conservation agriculture	++
Traoré et al. (2019)	Improving agricultural policy system performance in Mali: Stakeholder diagnostics and prescriptions	Policy	++
Keraita et al. (2010)	Farm-based measures for reducing microbiological health risks for consumers from informal wastewater-irrigated agriculture	Wastewater irrigation	++
Khalid (2018)	A review of environmental contamination and health risk assessment of wastewater use for crop irrigation with a focus on low and high-income countries	Wastewater irrigation	++

(continued)

Table 14.9 (continued)

Author	Subject title	Subject matter	Evidence[a]
Levasseur et al. (2007)	A review of urban and peri-urban vegetable production in West Africa	Wastewater irrigation	++
Raschid-Sally and Jayakogy (2008)	Drivers and characteristics of wastewater agriculture in developing countries: Results from a global assessment	Wastewater irrigation	++
Dittoh et al. (2010)	Sustainable micro-irrigation systems for poverty alleviation in the Sahel: A case for "micro" public-private partnerships?	Micro-irrigation	++
IWMI (2021)	Assessing the potential for sustainable expansion of small-scale solar irrigation in Segou and Sikasso, Mali	Irrigation expansion	++
Olayide et al. (2020)	Targeting small-scale irrigation investments using agent-based modelling: Case studies in Mali and Niger	Smallscale irrigation	++
USAID (2021)	Feed the future innovation laboratory for small-scale irrigation. ANNUAL REPORT: October 1st, 2020–September 30th, 2021	Smallscale irrigation	++
Dillon (2008)	Access to irrigation and the escape from poverty: Evidence from northern Mali	Irrigation	++
Dillon (2011)	The effect of irrigation on poverty reduction, asset accumulation, and informal insurance: Evidence from northern Mali	Irrigation	++
Higginbottom et al. (2021)	Performance of large-scale irrigation projects in sub-Saharan Africa	Irrigation performance	++
Bayala et al. (2012)	Cereal yield response to conservation agriculture practices in drylands of West Africa: A quantitative synthesis	Conservation agriculture	++
Strength of Evidence: Very low (III)			
Afokpe et al. 2022	Progress in climate change adaptation and mitigation actions in sub-Saharan Africa farming systems	Climate change adaptation and mitigation	+
Ayantunde et al. (2018)	Multiple uses of small reservoirs in crop-livestock agro-ecosystems of Volta basin: Implications for livestock management	AWM	+
Bayala et al. (2017)	Editorial for the thematic series in agriculture & food security: Climate-smart agriculture technologies in West Africa: learning from the ground AR4D experiences.	CSA	+
Bayala et al. (2021)	Multi-actors' co-implementation of climate-smart village approach in West Africa: Achievements and lessons learnt	CSA	+
Bastagli and Toulmin (2014)	Mali—Economic factors behind the crisis	Economy	+
Bayala et al. (2016)	Towards developing scalable climate-smart village models: Approach and lessons learnt from pilot research in West Africa	CSA	+

(continued)

Table 14.9 (continued)

Author	Subject title	Subject matter	Evidence[a]
Bee et al. (2017)	From needs to implementation: Stories from the technology needs assessment	Technology and AWM	+
IMF (2015)	Mali—Selected issues	Economy	+
Montaud (2019)	Agricultural drought impacts on crops sector and adaptation options in Mali: a macroeconomic computable general equilibrium analysis	AWM	+
USAID (2018)	Developing Local Extension Capacity Project. Mali: In-depth assessment of extension and advisory services	Agricultural extension and advisory services	+
Zamudion (2016)	Review of current and planned adaptation action in Mali	Climate change adaptation and mitigation	+
Aboye (2022)	Pilot review study on the importance of treated agro-industrial wastewater to reuse for irrigating agriculture	Wastewater irrigation	+
Drechsel et al. (2010)	Wastewater irrigation and health: assessing and mitigating risk in low-income countries	Wastewater irrigation	+
Keraita et al. (2014)	On-farm treatment options for wastewater, greywater and faecal sludge with special reference to West Africa	Wastewater irrigation	+
Toure et al. (2019)	Investigation of the water quality of daily used surface sources for drinking and irrigation by the population of Segou in the centre of Mali	Wastewater irrigation	+
Traore et al. (2020)	The efficiency of common washing treatments in reducing microbial levels on lettuce in Mali	Wastewater irrigation	+
Traore et al. (2013)	Evaluation of the effectiveness of bleach on microbial population of lettuce	Wastewater irrigation	+
Burney et al. (2010)	Solar-powered drip irrigation enhances food security in the Sudano-Sahel	Irrigation	+
Burney et al. (2013)	The case for distributed irrigation as a development priority in sub-Saharan Africa	Irrigation	+
Gueye (2014)	Specialisation or diversification? Divergent perspectives on rice farming in three large dam-irrigated areas in the Sahel	Agricultural production systems	+
Hertzog et al. (2012)	Ostrich-like strategies in Sahelian sands?: land and water grabbing in the Office du Niger, Mali	Land governance	+
Hertzog et al. (2014)	A role-playing game to address future water management issues in an extensive irrigated system: Experience from Mali	AWM	+
(Oke and Aduramigba-Modupe (2020)	Scaling precision agriculture in West Africa smallholder irrigation and water management systems	Small-scale irrigation	+

(continued)

Table 14.9 (continued)

Author	Subject title	Subject matter	Evidence[a]
Styger et al. (2011)	The system of rice intensification as a sustainable agricultural innovation: introducing, adapting and scaling up a system of rice intensification practices in the Timbuktu region of Mali	System of rice intensification	+
USAID and SWP (2021)	Mali water resources profile	Water resources	+

[a]Key: Substantiated (+++); partially substantiated (++); unsubstantiated (+) after Thomas et al. (2019)

References

Abdu N et al (2011) Vertical distribution of heavy metals in wastewater-irrigated vegetable garden soils of three West African cities. Nutr Cycl Agroecosyst 89(3):387–397

Aboye AD (2022) Pilot review study on the importance of treated agro-industrial wastewater to reuse for irrigating agriculture. Earth Environ Sci Res Rev 5(1):1–26

Abrams L (2018) Unlocking the potential of enhanced rainfed agriculture, Stockholm

Accelerated Growth Strategy (1997) Government of Mali, 1997. Accelerated Growth Strategy. Bamako: Government of Mali.

Acheampong EN (2008) Analysis of agricultural water productivity of irrigation schemes in the Niger River Basin: case studies in Mali and Niger. Unpublished thesis, Irrigation and Water Engineering Group, Wageningen University, Wageningen, Netherlands

ACPC (2013) Agricultural water management in the context of climate change in Africa

Afokpe PMK et al (2022) Progress in climate change adaptation and mitigation actions in sub-Saharan Africa farming systems. Cah Agric 31:4

Aggarwal PK et al (2018) The climate-smart village approach: framework of an integrative strategy for scaling up adaptation options in agriculture. Ecol Soc 23(1):14

Amoah P et al (2011) Low-cost options for reducing consumer health risks from farm to fork where crops are irrigated with polluted water in West Africa. International Water Management Institute (IWMI), Colombo

Andrieu N et al (2017) Prioritizing investments for climate-smart agriculture: lessons learned from Mali. Agric Syst 154:13–24

Andrieu N et al (2021) Ex ante mapping of favourable zones for uptake of climate-smart agricultural practices: a case study in West Africa. Environ Dev 37:100566

Arksey H, O'Malley L (2005) Scoping studies: towards a methodological framework. Int J Soc Res Methodol 8(1):19–32

AU (2020) Framework for irrigation development and agricultural water management in Africa. African Union, Addis Ababa

AUC (2014) Agenda 2063 the Africa we want. African Union Commission, Addis Ababa

Aw D, Diemer G (2005) Making a large irrigation scheme work: a case study from Mali. World Bank, Washington, DC

Ayantunde AA, Cofie O, Barron J (2018) Multiple uses of small reservoirs in crop-livestock agro-ecosystems of Volta basin: implications for livestock management. Agric Water Manag 204:81–90

Barry B, Namara R, Bahri A (2009) Better rural livelihoods through improved irrigation management: Office du Niger (Mali). In: Lenton R, Muller M (eds) Integrated water resources management in practice: better water management for development. Earthscan, London/Sterling

Bastagli F, Toulmin C (2014) Mali - economic factors behind the crisis. European Union (EU), Belgium

Bayala J et al (2012) Cereal yield response to conservation agriculture practices in drylands of West Africa: a quantitative synthesis. J Arid Environ 78:13–25

Bayala J et al (2016) Towards developing scalable climate-smart village models: approach and lessons learnt from pilot research in West Africa. ICRAF Occasional Paper No. 25. World Agroforestry Centre, Nairobi

Bayala J et al (2017) Editorial for the thematic series in agriculture & food security: climate-smart agriculture technologies in West Africa: learning from the ground AR4D experiences. Agric Food Secur 6(1):40

Bayala J et al (2021) Multi-actors' co-implementation of climate-smart village approach in West Africa: achievements and lessons learnt. Front Sustain Food Syst 5:637007

Bee S, Traerup S, Hecl V (2017) From needs to implementation: stories from the technology needs assessment. UNEP DTU Partnership, Copenhagen, Denmark

Bjornlund V, Bjornlund H, van Rooyen AF (2020) Exploring the factors causing the poor performance of most irrigation schemes in post-independence sub-Saharan Africa. Int J Water Resour Dev 36(sup1):S54–S101

Burney J et al (2010) Solar-powered drip irrigation enhances food security in the Sudano-Sahel. Proc Natl Acad Sci 107(5):1848–1853

Burney JA, Naylor RL, Postel SL (2013) The case for distributed irrigation as a development priority in sub-Saharan Africa. Proc Natl Acad Sci 110(31):12513–12517

CIAT, ICRISAT and BFS/USAID (2020) Climate-smart agriculture in Mali, Washington, D.C.

Coulibaly ON (2021) Mali - Land, climate, energy, agriculture and development. A study in the Sudano-Sahel initiative for regional development, jobs, and food security. Center for Development Research, University of Bonn, Bonn

Darghouth S (2005) Modernizing public irrigation institutions: the top priority for the future of sustainable irrigation. In: International Commission on Irrigation and Drainage Nineteenth Congress 2005. International Commission on Irrigation and Drainage (ICID), Beijing, China

Darghouth et al. (2005): Darghouth S 2005 Modernizing Public Irrigation Institutions: The Top Priority for the Future of Sustainable Irrigation. In: Proceedings of the 19th International Congress on Irrigation and Drainage, Beijing, China, 9–18 September 2005

Dayamba DS et al (2018) Assessment of the use of Participatory Integrated Climate Services for Agriculture (PICSA) approach by farmers to manage climate risk in Mali and Senegal. Clim Serv 12:27–35

Dembélé S, Tignegre JB, Diarra BG (2021) Development of the vegetable seed sector in Mali and opportunities for irrigated seed production. Shanhua, Taiwan

Dillon A (2008) Access to irrigation and the escape from poverty: evidence from Northern Mali, Washington, DC

Dillon A (2011) The effect of irrigation on poverty reduction, asset accumulation, and informal insurance: evidence from Northern Mali. World Dev 39(12):2165–2175

Dimithè G et al (2000) Financial profitability of Mali-Sud Bas-Fond rice production systems. Michigan State University, East Lansing

Dirwai TL et al (2021a) Water resource management: IWRM strategies for improved water management. A systematic review of case studies of East, West and Southern Africa. PLoS One 16(5):e0236903

Dirwai TL et al (2021b) Moistube irrigation technology development, adoption and future prospects: a systematic scoping review. Heliyon 7(2):e06213

Dittoh S, Akuriba MA, Issaka BY, Bhattarai M (2010) Sustainable micro-irrigation systems for poverty alleviation in the Sahel: a case for "Micro" public-private partnerships? Joint 3rd African Association of Agricultural Economists (AAAE) and 48th Agricultural Economists Association of South Africa (AEASA) Conference, Cape Town, South Africa

Drechsel P et al (2006) Informal irrigation in urban West Africa: an overview. International Water Management Institute (IWMI), Colombo

Drechsel P et al (2008) Reducing health risks from wastewater use in urban and peri-urban sub-Saharan Africa: applying the 2006 WHO guidelines. Water Sci Technol 57(9):1461–1466. https://doi.org/10.2166/wst.2008.245

Drechsel P et al (2010) Wastewater irrigation and health: assessing and mitigating risk in low-income countries. International Water Management Institute (IWMI) and International Development Research Centre (IDRC), UK/USA

FAO (2017) Mali - Country fact sheet on food and agriculture policy trends. Food and Agriculture Organisation of the United Nations (FAO), Rome, Italy

FAO (2021) The state of the world's land and water resources for food and agriculture – systems at breaking point. Synthesis report 2021. Food and Agriculture Organization of the United Nations (FAO), Rome, Italy

FAO (2022) Mali - FAO AQUASTAT. [Internet]. Food and Agriculture Organisation of the United Nations. Available from: https://www.fao.org/aquastat/statistics/query/results.html. Accessed: 19 June

Froebrich J et al (2020) Innovations in irrigation systems in Africa†. Irrig Drain 69(S1):3–5

Guèye B (2014) Specialisation or diversification? Divergent perspectives on rice farming in three large dam-irrigated areas in the Sahel, London

Hanjra MA, Williams TO (2020) Global change and investments in smallholder irrigation for food and nutrition security in Sub-Saharan Africa. In: Gomezy Paloma S, Riesgo L, Louhichi K (eds) The role of smallholder farms in food and nutrition security. Springer International Publishing, Cham

Harmon G, Lefore N (2022) Farmer-led irrigation as an emerging and diverging policy pathway in Sub-Saharan Africa. [Internet]. Agrilinks. Available from: https://www.agrilinks.org/post/farmer-led-irrigation-emerging-and-diverging-policy-pathway-sub-saharan-africa. Accessed 04 July

Hertzog T et al (2012) Ostrich-like strategies in Sahelian sands?: land and water grabbing in the Office du Niger, Mali. Water Altern 5(2):304–321

Hertzog T et al (2014) A role-playing game to address future water management issues in an extensive irrigated system: experience from Mali. Agric Water Manag 137:1–14

Higginbottom TP et al (2021) Performance of large-scale irrigation projects in sub-Saharan Africa. Nat Sustain 4(6):501–508

IBRD and WorldBank (2019) Climate-smart agriculture investment plan - Mali. International Bank for Reconstruction and Development/The World Bank, Washington, DC

IMF (2015) Mali - Selected issues. IMF Country Report No. 15/340, vol 15. International Monetary Fund (IMF), Washington, D.C, p 1

IWMI (2019) Suitability for farmer-led solar irrigation development in Mali. International Water Management Institute (IWMI), Colombo

IWMI (2021) Assessing the potential for sustainable expansion of small-scale solar irrigation in Segou and Sikasso, Mali. International Water Management Institute (IWMI), Colombo

Jiménez B et al (2010) Wastewater, sludge and excreta use in developing countries: an overview. In: Drechsel P, Scott CA, Raschid-Sally L, Redwood M, Bahri A (eds) Wastewater irrigation and health: assessing and mitigating risk in low-income countries. International Water Management Institute (IWMI) and International Development Research Centre (IDRC), UK/USA

Kane AM et al (2018a) Economic viability of alternative small-scale irrigation systems used in vegetables production in Koulikoro and Mopti Regions, Mali. In: Leal Filho W (ed) Handbook of climate change resilience. Springer International Publishing, Cham

Kane AM et al (2018b) Economic efficiency of water use in the small-scale irrigation systems used in vegetable production in Koulikoro and Mopti regions, Mali. Adv Agric Sci 6(4):72–84

Keraita B, Konradsen F, Drechsel P (2010) Farm-based measures for reducing microbiological health risks for consumers from informal wastewater-irrigated agriculture. In: Drechsel P, Scott CA, Raschid-Sally L, Redwood M, Bahri A (eds) Wastewater irrigation and health: assessing and mitigating risk in low-income countries. Earthscan; International Development Research Centre (IDRC); International Water Management Institute (IWMI), London/Ottawa/Colombo

Keraita B et al (2014) On-farm treatment options for wastewater, greywater and faecal sludge with special reference to West Africa. International Water Management Institute (IWMI); CGIAR Research Program on Water, Land and Ecosystems (WLE), Colombo

Kergna AO, Dembele D (2018) Small scale irrigation in Mali: constraints and opportunities. Forum for Agricultural Research in Africa (FARA), Accra

Khalid S (2018) A review of environmental contamination and health risk assessment of wastewater use for crop irrigation with a focus on low and high-income countries. Int J Environ Res Public Health 15(5):895

Koutsos TM, Menexes GC, Dordas CA (2019) An efficient framework for conducting systematic literature reviews in agricultural sciences. Sci Total Environ 682:106–117

Langyintuo A (2020) Smallholder farmers' access to inputs and finance in Africa. In: Gomezy Paloma S, Riesgo L, Louhich K (eds) The role of smallholder farms in food and nutrition security. Springer International Publishing, Cham

Levasseur V et al (2007) A review of urban and Peri-urban vegetable production in West Africa. In: Sish BEL (ed) International Horticultural Congress. ISHS [Belgique], Séoul, Corée du Sud, pp 245–252

LOA (2006): République du Mali, 2006. Loi N°06-045 Portant Loi d'Orientation Agricole. Bamako: République du Mali.

MalaboMontpellierPanel (2018a) Water-wise: smart irrigation strategies for Africa. Malabo Montpellier Panel, Dakar

MalaboMontpellierPanel (2018b) Water-wise: smart irrigation strategies for Africa: Mali. Malabo Montpellier Panel, Dakar

Mason SC, Maman N, Palé S (2015a) pearl millet production practices in semi-arid West Africa: a Review. Exp Agric 51(4):501–521

Mason SC et al (2015b) Soil and cropping system research in semi-arid West Africa as related to the potential for conservation agriculture. Int J Agric Sustain 13(2):120–134

Methley AM et al (2014) PICO, PICOS and SPIDER: a comparison study of specificity and sensitivity in three search tools for qualitative systematic reviews. BMC Health Serv Res 14(1):579

Montaud J-M (2019) Agricultural drought impacts on crops sector and adaptation options in Mali: a macroeconomic computable general equilibrium analysis. Environ Dev Econ 24(5):506–528

NAPA (2007) République du Mali, 2007. Programme d'Action National d'Adaptation aux Changements Climatiques (PANA). Bamako: Ministère de l'Environnement et de l'Assainissement.

NDC (2015): République du Mali, 2015. Contribution Prévue Déterminée au Niveau National (CPDN). Bamako: République du Mali.

Nkonya E, Kato E, Ru Y (2020) Drivers of adoption of small-scale irrigation in Mali and its impacts on nutrition across sex of irrigators. International Food Policy Research Institute (IFPRI)

National Environmental Protection Policy (1998) Government of Mali, 1998. Politique Nationale de Protection de l'Environnement [National Policy for Environmental Protection]. Bamako: Government of Mali. Available at: https://documents1.worldbank.org/curated/ar/397911468758678774/pdf/P1295160PID0Print00222201201329930020884.pdf [Accessed 7 Feb. 2025].

National Policy on Climate Change (2011) Government of Mali, 2011. Politique Nationale sur les Changements Climatiques [National Policy on Climate Change]. Bamako: Government of Mali

National Strategy for Irrigation Development (SNDI 1999) Government of Mali, 1999. Stratégie Nationale de Développement de l'Irrigation [National Irrigation Development Strategy]. Bamako: Government of Mali. Available at: https://www.afdb.org/fileadmin/uploads/afdb/Documents/Environmental-and-Social-Assessments/ADF-BD-IF-2008-181-EN-MALI-IRRIGATION-DEVELOPMENT-PROGRAMME-PHASE-I-ESIAS.PDF

OECD/FAO (2021) OECD-FAO agricultural outlook 2021–2030. OECD Publishing, Paris

Oke AO, Aduramigba-Modupe V (2020) Scaling precision agriculture in West Africa smallholder irrigation and water management systems. 1st African conference on precision agriculture 2020

Olayide OE et al (2020) Targeting small-scale irrigation investments using agent-based modelling: case studies in Mali and Niger. University of Bonn, Bonn

Ouédraogo M (2019) Uptake of climate-smart agricultural technologies and practices: actual and potential adoption rates in the climate-smart village site of Mali. Sustain For 11(17):4710

Ouedraogo M et al (2018) Uptake of climate-smart agriculture in West Africa: what can we learn from climate-smart villages of Ghana, Mali and Niger? Findings from a series of adoption studies on CSA technologies and practices within the climate-smart villages of Ghana, Mali and Niger. CCAFS Info Note. CGIAR Research Program on Climate Change, Agriculture and Food Security (CCAFS), Bamako, Mali

Partey ST et al (2018) Developing climate-smart agriculture to face climate variability in West Africa: challenges and lessons learnt. J Clean Prod 187:285–295

Poverty Reduction Strategy (1998) Government of Mali, 1998. Stratégie nationale de lutte contre la pauvreté [National Strategy to Combat Poverty]. Bamako: Government of Mali. Available at: https://www.un.org/development/desa/dspd/wpcontent/uploads/sites/22/2020/03/Dembele-paper-WORLD-WITHOUT-POVERTY_Mali.pdf [Accessed 7 Feb. 2025]

Raschid-Sally L, Jayakody P (2008) Drivers and characteristics of wastewater agriculture in developing countries: results from a global assessment (IWMI Research Report 127). International Water Management Institute (IWMI), Colombo, Sri Lanka

Rural Development Plan (1992): République du Mali, 1992. Plan de Développement Rural. Bamako: République du Mali.

Sidibé A, Minh TT (2021) Multi-stakeholder dialogues supporting the scaling of inclusive and sustainable agricultural water management in Mali: Kick-start meeting report. United States Agency for International Development (USAID)

Styger E et al (2011) The system of rice intensification as a sustainable agricultural innovation: introducing, adapting and scaling up a system of rice intensification practices in the Timbuktu region of Mali. Int J Agric Sustain 9(1):67–75

Thomas et al. (2019) An efficient framework for conducting systematic literature reviews in agricultural sciences. Science of the total environment, 682, pp.106–117.

Toure A et al (2019) Investigation of the water quality of daily used surface sources for drinking and irrigation by the population of Segou in the centre of Mali. J Water Health 17(2):338–349

Traoré S et al (2013) Evaluation of the effectiveness of bleach on microbial population of lettuce. Sci J Microbiol 2:166–173

Traore K et al (2017) Optimizing yield of improved millet and sorghum varieties under highly variable rainfall conditions using contour ridges in Cinzana, Mali. Agric Food Secur 6(1):11

Traoré A et al (2019) Improving agricultural policy system performance in Mali: stakeholder diagnostics and prescriptions. Mali Food Security Policy Research Program

Traoré S et al (2020) The efficiency of common washing treatments in reducing microbial levels on lettuce in Mali. West Afr J Appl Ecol 28:64–74

Traore B et al (2021) Contribution of climate-smart agriculture technologies to food self-sufficiency of smallholder households in Mali. Sustain For 13(14):7757

Traoré S et al (2021) Assessment of the parasite load of lettuce, tomato and cucumber from some large vegetable production sites in Mali. Afr J Agric Res 17:1324

Traore S et al (2022) Parasitic contamination of lettuce, tomato and cucumber from vegetable farms in Mali. Curr Top Agric Sci 7:94–104

UNEP-CCC (2022) Technology needs assessment, Mali Africa. [Internet]. United Nations Environment Programme Copenhagen Climate Centre. Available from: https://tech-action.unepdtu.org/country/mali/. Accessed: 01 June

USAID (2013) Climate change in Mali: key issues in water resources. United States Agency for International Development (USAID)

USAID (2018) Developing local extension capacity project. Mali: in-depth assessment of extension and advisory services. United States Agency for International Development (USAID)

USAID (2021) Feed the future innovation laboratory for small scale irrigation. Annual Report: October 1st, 2020 - September 30th, 2021. United States Agency for International Development

USAID and SWP (2021) Mali Water Resources Profile. United States Agency for International Development (USAID) and Sustainable Water Partnership (SWP)

WHO (2006) World Health Organization (2006). Guidelines for the Safe Use of Wastewater, Excreta and Greywater: Volume II – Wastewater Use in Agriculture. Geneva: WHO Press.

Woodhouse P et al (2017) African farmer-led irrigation development: re-framing agricultural policy and investment? J Peasant Stud 44(1):213–233
WorldBank (2010) Making a large irrigation scheme work: a case study from Mali. Water P-Notes; No. 43. World Bank, Washington, DC
WorldBank (2013) Unlocking Africa's agricultural potential - an action agenda for transformation. World Bank, Washington, DC
Zamudio AN (2016) Review of current and planned adaptation action in Mali. CARIAA Working Paper No. 11. International Development Research Centre and UK Aid, Ottawa, Canada and London, United Kingdom
Zougmoré R (2018) Promoting climate-smart agriculture through water and nutrient interactions options in Semi-Arid West Africa: a review of evidence and empirical analysis. In: Bationo A, Ngaradoum D, Youl S, Lompo F, Fening JO (eds) Improving the profitability, sustainability and efficiency of nutrients through site specific fertilizer recommendations in West Africa agro-ecosystems: Volume 2. Springer International Publishing, Cham
Zougmoré R, Jalloh A, Tioro A (2014) Climate-smart soil water and nutrient management options in semiarid West Africa: a review of evidence and analysis of stone bunds and zaï techniques. Agric Food Secur 3(1):16
Zougmoré RB et al (2018) Facing climate variability in sub-Saharan Africa: analysis of climate-smart agriculture opportunities to manage climate-related risks. Cah Agric 27(3):34001
Zougmoré RB, Läderach P, Campbell BM (2021) Transforming food systems in Africa under climate change pressure: role of Climate-smart agriculture. Sustain For 13(8):4305

Chapter 15
Opportunities to Improve Water Productivity in Farmer-Led Irrigation: A Case Study of Ethiopia

T. Desalegn

Abstract To construct economic growth and poverty reduction, farmer-led irrigation is increasing nowadays in many African, South Asian, and Latin American countries. Smallholder farmers' vulnerability to climate shocks due to rainfall variability, frequent droughts, and inadequate water resources, which threaten agricultural productivity and food security is known. Understanding the current and future crop water demand is key for improving agricultural productivity, and bringing food security, especially in arid and semi-arid areas where irrigation is needed to overwhelm shortage and rainfall variability. Water productivity (WP) is to produce more yield with less water use, increase income, improve livelihoods, and bring ecological benefits at less social and environmental costs per unit of water used. Some of the approaches for increasing WP include more yield with proper water use, changing the cropping pattern from low to high-value crops, decreasing costs related to social, health, and environmental aspects, and achieving more livelihood support such as more job opportunities, diversified nutritious food, and income for the same amount of water. Potential opportunities are vastly needed to achieve the productivity of irrigated agricultural systems around the globe through the full engagement of smallholder farmers, public-private sectors, government entities, and potential water resource management stakeholders. This chapter will briefly discuss opportunities for improving WP about farmer-led irrigation, focussing on smart utilization of water resources and agronomic practices to achieve higher yields using less water. Using the available water sources and low-cost water-lifting alternative technology options, implementing efficient irrigation water application methods, practicing deficit and supplemental irrigation techniques, and adopting climate-smart on-farm water management techniques and technologies are potential opportunities for improving WP in farmer-led irrigation areas.

Keywords Farmer-led irrigation · On-farm water management · Technologies · Irrigation · Water lifting · Water productivity

---

T. Desalegn (✉)
International Water Management Institute (IWMI), Addis Ababa, Ethiopia
e-mail: D.Tegegne@cgiar.org

T. Mabhaudhi et al. (eds.), *Enhancing Water and Food Security Through Improved Agricultural Water Productivity*,
https://doi.org/10.1007/978-981-96-1848-4_15

15.1 Introduction

Irrigation can be classified and defined in many ways and a general definition of irrigation is an artificial application of water to crops designed to allow production in arid regions and during the occurrence of drought's effect in semi-arid regions (FAO 1995). One of the irrigation types discussed in this chapter is farmer-led irrigation which is farmers deciding the farming plan, design, and implementation techniques (Tegegne et al. 2024). Farmer-led irrigation attains changes in investment patterns, innovations, market linkages, and sustainable natural resource use such as water and land resources (Woodhouse et al. 2017). Additionally, WP is the other important topic discussed in this chapter given it is a mass (kg), monetary ($), or energy (calorific) value of production per unit of water evapotranspiration (mm) (Kijne et al. 2003; Molden et al. 2010), and it is a measure of the amount of water required to produce a quantity of production.

The contribution of irrigation can be seen in its ability to provide food security and able to boost smallholder farmers' income (Mengistie and Kidane 2016). Consequently, irrigation is critical in overcoming rainfall shortages and improving agricultural production, especially in arid and semi-arid areas (FAO 1995). Irrigated agriculture plays an important role in meeting the basic needs of billions of people especially in developing countries, although water resources are still plenty in many countries. The studies by Hillel (1998) indicated that 85% of the available water is used for agricultural production. Farmer-led irrigation is more common in sub-Saharan African (SSA) countries than has been recognized, and in many cases, it is not well identified in official data. The data from the satellite imagery shows that the irrigated area in SSA is bigger which is two to three times more than the official figures (e.g. in Ethiopia), and the irrigated area may be as much as 4.1 million hectares—14 times the official figure (Merrey et al. 2017).

The water availability gap is one of the limiting factors in arid and semi-arid areas of the world. Many nations with low water productivity also have low incomes and high poverty rates (Jacob et al. 2009). As the studies by Cook et al. (2009) and Cai et al. (2011) indicated, increased water productivity is directly linked to improving food security and livelihoods. In this context, some analysts disagree with the term 'productivity' and want to use it only for typical production factors, such as labour, land resources, and capital (Zoebl 2006). Water use efficiency is still a useful parameter used by individual farmers or many irrigation systems. The main objectives of improving agricultural water productivity are to meet rising demands for food and food preferences, poverty reduction, and income growth, as food demands are projected to increase by 2050 by 60% to sustain an ever-growing population (FAO 2013). Higher water productivity can be achieved by applying the same amount of water and getting a higher yield and/or by reducing the amount of water that can be applied and producing the same yield and/or a higher yield. Hence, improving a crop's general intensity can also increase water productivity. Improving water productivity aims to produce more food and income, bringing better ecosystem services, and livelihoods with less water (Molden et al. 2010; Cai et al. 2011).

Water productivity can be improved by improving and knowing the different water management practices vital to addressing food insecurity and poverty alleviation (Jacob et al. 2009). In most cases, water use efficiency and productivity have the same meaning as both target increasing agricultural production while using limited resources. Therefore, water use efficiency targets production with the water input, whereas water productivity focusses on production with the water consumption in its calculation.

In African irrigation schemes, water is often distributed on a rolling basis; consequently, farmers will irrigate when they have access to water, whether needed or not (Bjornlund et al. 2017). Where roster-based irrigation works poorly, and high inequalities exist in space and time, farmers are more likely to over-irrigate when they have access as they are insecure about when the next water turn is available. Improper on-farm irrigation practices can lead to unequal water distribution, non-uniform crop growth, poor drainage, and leaching, all of which decrease the yield per unit of land area and per unit of water applied (Strelkoff et al. 1999; Cai and Rose Grant 2003). Hence, farmers struggle to translate their inputs into optimal yields because of low on-farm water productivity, leaching-induced soil degradation, over-abstraction, and downstream contamination of water resources. This, together with poor infrastructure and other challenges in irrigation scheme management, often translates into poor-performing irrigation schemes where actual irrigated land is far less than the designated command area, particularly in many African countries. With effective on-farm water management, water distribution within the schemes can be optimized for equal allocation of water resources and increase the economic viability of the entire scheme (Bjornlund et al. 2017). Therefore, water productivity produces more crops per drop of water, reducing water losses on the farmer's field in farmer-led irrigation systems by giving adequate water for the crop. This chapter emphasizes the potential opportunities for improving water productivity and/or water use efficiency of farmer-led irrigation, factors influencing water productivity, and the key areas to be considered for improvement.

15.2 Factors Affecting Water Productivity Under the Farmer-Led Irrigation

Many factors may affect the water productivity of farmer-led irrigation, such as engineering and technological factors, advancements in best crop variety selection, environmental factors, socioeconomic factors, and areas related to capacity building (Koech and Langat 2018).

Engineering and technological factors This includes improvement of water distribution networks and on-farm irrigation planning, management and development, irrigation scheduling, actual-time control and optimization, remote sensing, and sensor communication networks. This includes the cost and affordability of irrigation technologies, the price and profitability of irrigated crops, and marketing chains.

Advancements in best crop variety selection In addition to managing water for water productivity improvement, selecting high-yielding crop varieties and optimum water-user crops, disease-tolerant crop selections, and salt-tolerant crop varieties for areas where salinity is an issue must be considered.

Environmental factors access to water and understanding the environmental scenarios help to understand where to work for water productivity improvement. For example, water harvesting is one of the world's best practices, but not knowing where water harvesting can be feasible and sustainable will be challenging to bring changes.

Socioeconomic factors are important drivers of water productivity and are related to technology adoption and community decision-making processes.

Capacity building For farmers to increase their water productivity and profits, new thinking about how to do so and greater access to information about irrigation, such as how to use water efficiently, as well as input and output markets, are required. In many areas where water productivity is needed, knowledge exists but is not available at ground level. Still, institutional capacity and human capital are inadequate to support the knowledge transfer needed to accelerate the adoption of new technologies that increase yields and water productivity. In such cases, the top-down approach to technology transfer is ineffective and should be changed into a participatory approach combining traditional knowledge with scientific technology (that could be a bottom-up approach or up-to-down approach) (Lal 2007) (Fig. 15.1).

15.3 Potential Opportunities for Improving the Water Productivity of Farmer-Led Irrigation

Improving water productivity at farmer-led irrigation systems mainly depends on choosing adapted water-efficient crops, reducing unproductive water losses, applying climate-smart water management techniques and technologies, and ensuring suitable agronomic conditions for crop production (Rockstrom and Barron 2007). Some options for the available water resources to meet the growing food demands are expanding irrigation (in quantity and quality) and increasing the production per unit of water consumption. WP under farmer-led irrigation can be improved by choosing well-adapted crop types, reducing unproductive water losses, maintaining healthy, high-value growing crops through optimized water, and choosing ideal nutrient and agronomic management options (farm management (preparation to harvesting time), fertilizer type, and amount management, pesticide use, and

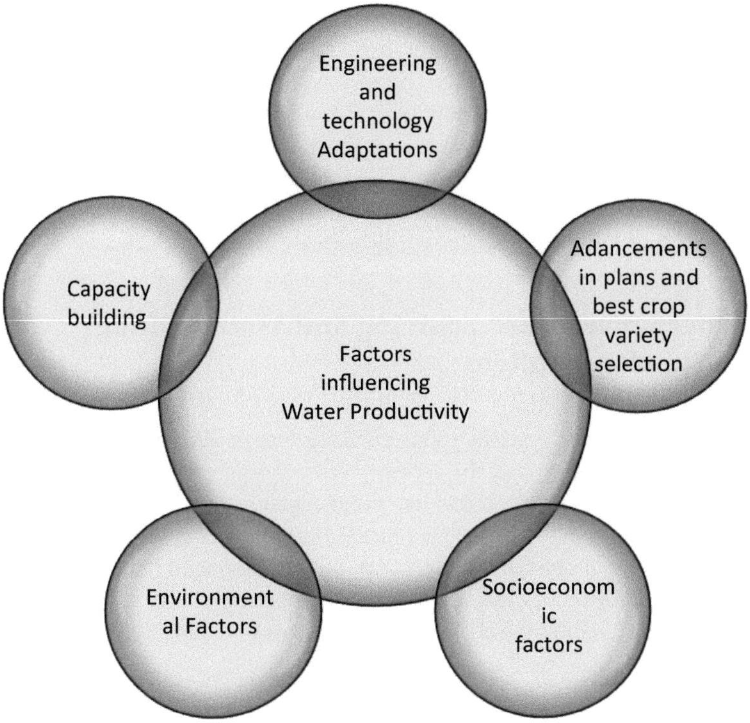

Fig. 15.1 Factors influencing water productivity

post-harvest management) (Descheemaeker et al. 2013). Improving irrigation efficiency using micro-irrigation systems can prevent seepage losses in the conveyance system and reduce evaporation during conveyance and distribution in water shortage areas. After the water reaches the farm, the losses are due to leakage and evaporation. Losses in water application to the farmers' fields depend on the type of irrigation system when the farmers apply water. One of the best practices which many scholars agree for on-farm water management is to not exceed the amount of water that can be stored on the root zones of the crop, practice using drainage/wastewater (after treatment), and control the infiltrated water which percolates below the root zones of the crop (Jacob et al. 2009).

Improper on-farm water management, especially in irrigation schemes, may result in unequal water distribution within the field, water logging, leaching of nutrients, non-uniform crop growth, wastage of water, and conflict between water users who are located in the upstream and downstream parts, all of which decrease the yield per unit of land area and per unit of water applied. Hence, for optimized water distribution within the schemes, equal allocation of water resources among users, and an increase in the economic viability of the entire scheme, on-farm water management should be improved (Bjornlund et al. 2017). Management practices that increase agricultural yields can improve crop and water productivity (the quantity and quality of production). The potential opportunities identified in this chapter for improving farmer-led irrigation water productivity include: (i) locally available water sourcing and environmentally friendly water-lifting technology options; (ii) Deficit and supplemental irrigation; (iii) Improved irrigation water application method; (iv) adoption of climate-smart on-farm water management tools/technologies; and (v) use of scientific irrigation scheduling techniques (Arora et al. 2011; Balwinder et al. 2011; Mzezewa et al. 2011; Descheemaeker et al. 2013).

15.4 Alternative Water Sourcing and Water-Lifting Technology Options

For farmer-led irrigation systems to succeed, the main determining factors are the water sources, availability, and the type of water lifting or pumping system. Water sources include rivers/lakes, rainwater, ponds/springs, and shallow groundwater. The water obtained from the different sources that are locally available can be managed with the different systems called community-managed small-scale irrigation infrastructures. Additionally, depending on the location of the water sources, low-cost water-lifting technologies are also important for water productivity improvements.

15.4.1 Community-Managed Small-Scale Irrigation Infrastructures

Thousands of small-scale reservoirs serve as a water source, but their productivity is lower than expected and often not well-managed. Small-scale communal irrigation systems, Formal irrigation schemes, and water harvesting infrastructures are some of the community-managed small-scale irrigation infrastructure types:

Small-scale communal irrigation systems are grouped into two: small communal reservoirs constructed for multiple uses (small-scale irrigation, livestock watering, domestic and small business uses, fisheries, etc), which may or may not have the infrastructure for delivering water to fields.

Formal irrigation schemes that include a reservoir or offtake from small rivers significantly impact poverty reduction (Tesfaye et al. 2008; Gebregziabher et al. 2009). In many sub-Saharan African countries, the government continues supporting this, as indicated in Fig. 15.2.

Water harvesting: It is the collective name for a range of technologies that aim to concentrate the water flows from different sources and storage, either as in situ technologies (in principle using the soil as storage of infiltrated rainfall) or in artificial storage structures such as dams, ponds (geomembrane plastics/concrete/earther), and tanks which help for supplemental irrigation or full irrigation

Fig. 15.2 Formal community irrigation scheme, Koga irrigation scheme, Ethiopia. (Photo credit: Desalegn Tegegne)

and these man-made structures are used for high-value crops. More than 50% of the lost water can be recovered using water harvesting techniques. Water harvesting encompasses diverse methods of collecting and managing floodwaters and surface runoff for later use in the season when there is no rainfall. Water harvesting focusses on efficient water resource use and helps reduce crop failure by 7–20% (Jacob et al. 2009; Oweis and Hachum 2006). The advantage of ponds is not only recharging groundwater for water harvesting but also it is crucial to minimize runoff erosion in the catchment (Tadesse et al. 2008). Additionally, it greatly improves crop and water productivity (water harvesting technologies can bridge up to 40% of the yield gap attributable to water deficit). (Mekuria and Tegegne 2022). Figure 15.3 indicates how it is possible to harvest water from the roof.

15.4.2 Low-Cost Water-Lifting Pumps

Recently, low-cost water-lifting pumps that can lift water from the ground, or the surface are becoming common in many countries. These pumps are manual or pedal (Trincheria et al. 2017), and the manual pumping systems (rope and washer pump (Fig. 15.4b), treadle pump (Fig. 15.4d), Kick Start Money Maker pumps, and Brazilian pump) show high potential in rural communities and offer lowest cost solution for poor farmers having little capital, but with access to shallow groundwater, and surface water sources (rivers/springs or ponds). These pumps also showed positive performance on food security, poverty reduction, and crop revenue in Zimbabwe, Kenya, Ethiopia, and Ghana. Diesel or petrol-powered pumps (Fig. 15.4c) help to irrigate and get benefit from irrigation (most common in Nigeria, Mali, Mauritania, Niger, and Ethiopia) (Schmitter et al. 2016; Kamwamba-Mtethiwa et al. 2016). Solar pumps (Fig. 15.4a) are non-polluting and can be used at any scale, their costs are declining rapidly, and their availability is increasing. Experiments in Africa combining solar pumps with drip irrigation for raising high-value crops indicated that it could improve household nutrition and food security for

Fig. 15.3 Rooftop water harvesting using water tanks in Oromia region, Ethiopia. (Photo Credit: Desalegn Tegegne)

households (mainly women) (e.g. Bangladesh, India, and Ethiopia) (Burney et al. 2010; Burney and Naylor 2012; Alaofe et al. 2016; Schmitter et al. 2016). Nowadays, these pumps are marketed to individual farmers. Marketing chains gradually build up in many rural areas, linking farmers to importers (or rarely, local manufacturers), wholesale and retail outlets, and repair services.

15.5 Deficit and Supplemental Irrigation Techniques

Supplemental irrigation may be defined as 'adding small amounts of water to fundamentally rainfed crops when rainfall fails to provide sufficient moisture for normal plant growth to improve and stabilize yields' (Nangia and Oweis 2016). Supplemental irrigation is applied when the water supply is insufficient to meet full crop water demand (not able to get water during its growth). Supplemental irrigation aims to add a limited amount during critical crop development stages, such as the initial, development, middle, and late stages of the crop, and this is common and helpful in semi-arid and dry sub-humid cropping systems with climate change occurrences such as rainfall variability and others (Barron 2004; Pandey et al. 2001). *Deficit irrigation* is the application of less water than what is required by the plant, and it is also a strategy that is often used when water is limited. For deficit irrigation to be successful, farmers need to know soil moisture stresses at the root zones of the crop and how much water must be added at each of the crop development stages

Fig. 15.4 the various water-lifting technologies such as a solar pump (**a**), rope and washer pump (**b**), diesel pump (**c**), and treadle pump (**d**). (Photo credit: Desalegn Tegegne)

(initial, development, middle, and late stages) to avoid over application and/or water shortage around the root zone.

15.6 Improved Irrigation Water Application Method

Drip Irrigation

Irrigation water use efficiency can also be improved using efficient water application technologies like drip irrigation. With donor support, many NGOs and private firms have developed low-cost drip kits, demonstrated them in the field, and attempted to promote wider uptake (Merrey et al. 2017). Drip irrigation is one of the best application-improving techniques. It is delivered in small amounts via small holes in the drip line installed in pipes or tapes, either above the ground or underground. The applied irrigation is directly targeted to the root zone where it is needed rather than lost due to soil evaporation and/or deep drainage. Hence, drip irrigation has considerable advantages over furrow or sprinkler irrigation regarding water application efficiency as it incurs less water losses (Fig. 15.5). In developing countries like Ethiopia, farmers can afford drip or pipe irrigation; however, water storage tankers are lacking (Jacob et al. 2009). Getting such high-quality drip irrigation

Fig. 15.5 Drip irrigation system at Allada, Benin. (Photo credit: Desalegn Tegegne)

Fig. 15.6 Sprinkler irrigation

systems available along with pumps (motor, solar, treadle) and other equipment (water storage tankers) would increase the likelihood of synergies (Schmitter et al. 2016). This system needs less labour as once the installation is done, it can irrigate by opening the tape valves and has high water use efficiency.

Sprinkler Irrigation

Sprinkler irrigation is one of the pressurized irrigation systems that operate under low pressure and often involves some form of pumping; the system includes solid sets, centre-pivots, and travelling irrigators, and water is delivered in the form of sprays using overhead sprinklers (Fig. 15.6). Sprinkler irrigation is less labour intensive and has significantly higher WUE though its investment cost is higher.

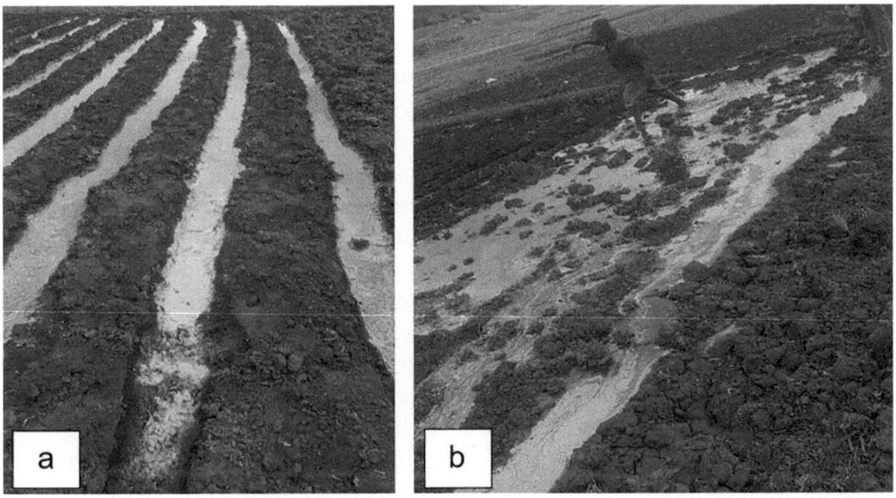

Fig. 15.7 Surface irrigation methods in the form of furrow (**a**) and flooding (**b**) under the Africa RISING project, Mush Irrigation Scheme, Ethiopia. (Photo credit: Desalegn Tegegne)

Furrow Irrigation (Surface Irrigation)
The surface irrigation system includes furrow (Fig. 15.7), flooding, and border irrigation, in which the gravitational force conveys water over the field surface. The furrow system is the most common method for irrigating row crops worldwide (Koech and Langat 2018). However, the system is highly labour intensive and has low water use efficiency compared with drip and sprinkler systems. For water productivity improvement, the water flow in inlets of the furrow must be measured to avoid adding too much water or less water amount.

15.7 Scientific-Based Irrigation Scheduling Practices

Irrigation scheduling involves knowledge of plant water use patterns, which are affected by weather, growth stage, and canopy wetness. To improve the water use efficiency, losses along the conveyance and distribution channels must be minimized, and the timing and the amount of water applied (irrigation scheduling) must be optimized. One of the best water productivity options is measuring the amount of water supplied to irrigators. This brings water savings, which may be used to irrigate more land with the saved water (expand the irrigated area) and close the water productivity gaps.

The primary aim of irrigation scheduling is to enable the farmers to minimize crop water stress and maximize yields; schedule the water rotation among the various fields; save fertilizer by holding surface runoff and deep percolation losses (leaching) to a minimum; to increase yield and crop quality; to minimize water logging problems by reducing the drainage requirements; to assist by controlling the root zone salinity problems through controlled leaching; to minimize wasteful losses of water (percolation beyond what is necessary for salt leaching evaporation and runoff); and to increase transpiration by the crop (Borner 2015). Therefore, scheduling is fundamental to crop and water productivity determination, which are performance indicators that describe the relationship between water applied and agricultural product outputs. Ideally, the water given per irrigation application is called the irrigation depth at the beginning of the growing season, which is small and given frequently. This is due to their shallow root depth and the low evapotranspiration of the young plants. The different cropping stages have different water demands, for example, at the initial stage, water demand at one time is less and frequency is high. Irrigation scheduling using scientific methods has made it more accessible, and it is possible to schedule the irrigation water supply according to the crops' water demand.

Methods to determine the irrigation schedule include plant observation and simple calculation methods. The plant physical observation method is one of the methods normally used by farmers' experiences in the field to estimate when irrigation must be carried out. The changes can often only be detected by looking at the crop rather than the individual plants. The method is based on observing changes in plant characteristics, such as changes in crop colour, twisting leaves, and then plant wilting. Deep knowledge of local circumstances is required to successfully use the plant observation method. During the early stages, when the plants are small, the crop water is less than during the mid-season stage.

The irrigation depth can be estimated using the following equation

$$dA = Qt$$

Q is the flow rate in cubic feet per second (CFS); t is the time of irrigation (hours); d is the depth of water applied (inches), and A is the area irrigated (acres) (Edward 2000). Irrigation scheduling can be done using (a) remote sensing and user-friendly on-farm water management tools.

15.8 Remote Sensing

Research for development projects investments and the establishment of different on-farm water management technologies and practices have positively influenced water productivity. Determining irrigation water demand and scheduling using weather data and soil-plant characteristics are possible. Although this works at the ground level, it is time-consuming, expensive, location-specific, and unsuitable for large areas. Recently, remote sensing has been one of the best options for estimating crop water demand and determining irrigation scheduling using thermal and multispectral imageries collected using unmanned aerial vehicles (UAV) or drones with the capacity to save time and resources. It is important due to its systematic measurements across space and time, ability to cover more area, and capability to be integrated into models and with Geographic Information Systems (GIS) (Koech and Langat 2018; Zwart et al. 2006; Molden et al. 2007).

15.9 On-Farm Irrigation Water Management Technologies

On-farm water management improvements can also be achieved with existing user-friendly on-farm water management technologies by improving the performance of the water delivery and/or application system. Technological advancement includes the internet, several computer-based irrigation scheduling systems, sensors, and mechanical and electronic tools. These tools have been developed to help farmers in their decision-making process. Providing useful on-the-ground information to guide irrigation applications has interested many parts of the world. In-field sensors and tools, prediction models (i.e. climate and crop models) are often used to determine crop water demand in the field. Some tools (sensors and tubes) help monitor the soil moisture status and indicate whether the crop must be irrigated. Some tools are soil moisture probes, Chameleon sensors, and wetting front detectors (WFD) (Schmitter et al. 2016). These tools help producers to improve their irrigation efficiency, productivity, and product quality and reduce costs. Tools such as the wetting front detector (WFD) (Fig. 15.8) and Chameleon sensors (Fig. 15.9) (Stirzaker 2003; Stirzaker et al. 2017) have proven successful in improving on-farm water management for irrigators by reducing the level of information complexity for the farmer. So far, they have been mainly tested at the farm level, and they are low-cost technologies for farmers to increase irrigation efficiency and effectiveness.

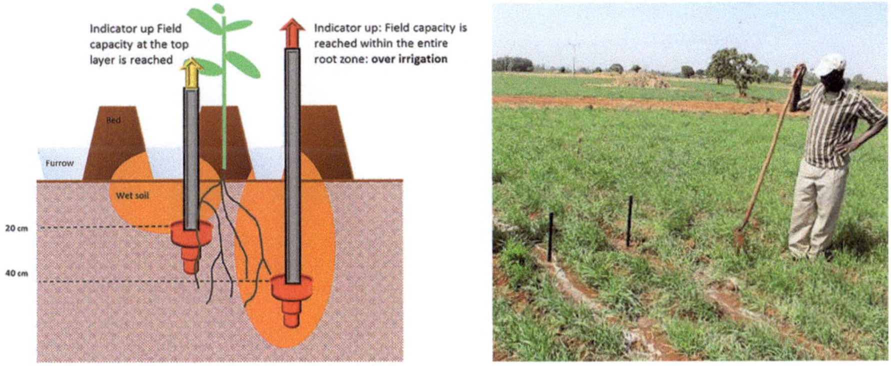

Fig. 15.8 Wetting front detector

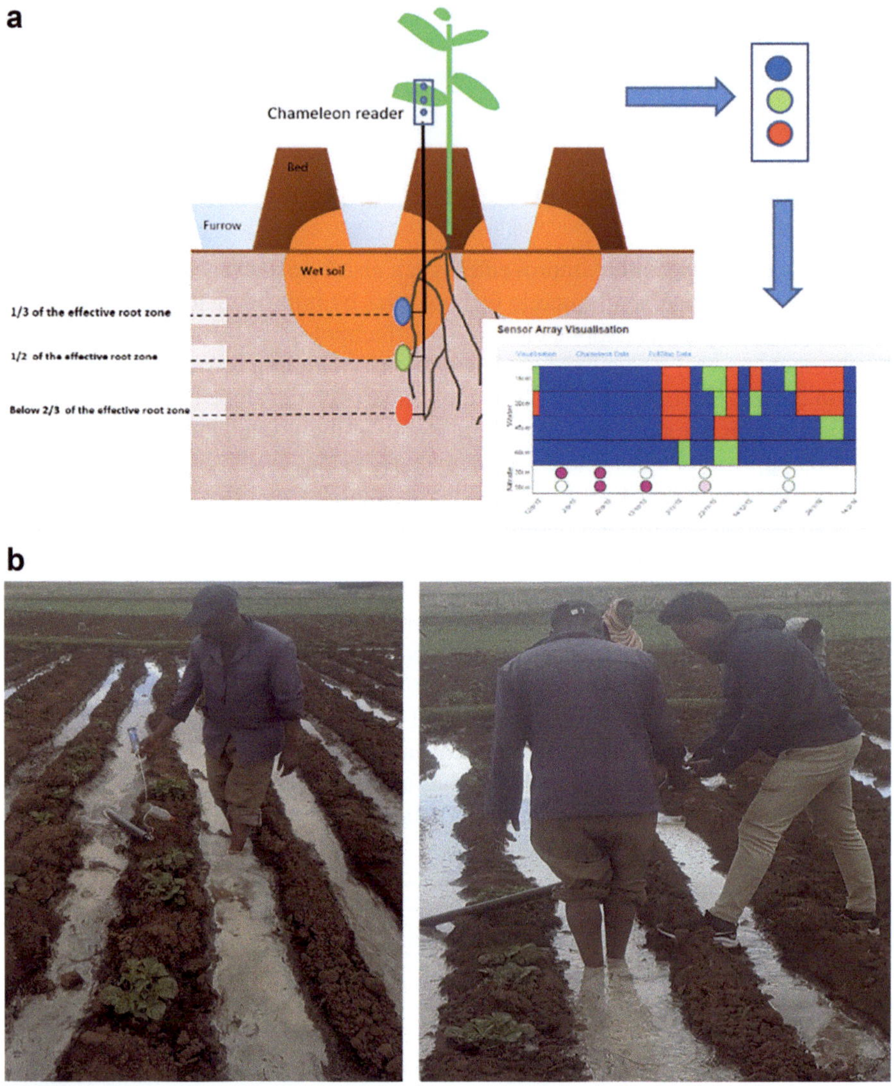

Fig. 15.9 Irrigation scheduling using Chameleon Sensor (field experiment for the AR project, Bakello Irrigation scheme, Ethiopia). (Photo credit, Desalegn Tegegne)

Wetting Front Detectors (WFD)
WFD is a mechanical tool depending on the soil characteristics, and irrigation techniques. The quantity is installed in pairs at a specific depth below the soil surface. The tool helps farmers to stop their irrigation when field capacity is attained at field capacity (FC) but also gives little information on when to irrigate. When the soil reaches fc and gravitationally moves, water is stored within the storage below the funnel (part of the tool). The study by Schmitter et al. (2017) indicated that the amount of water collected in the storage will activate the floaters up. A yellow and a red indicator are integrated in each of the pair of detectors. The shallow detector (yellow indicator) is installed around half of the effective root zone whereas the deep detector (red indicator) is generally installed around 2/3rd of the root zone. Installation depth depends on the irrigation system (surface or pressurized system) and the soil condition.

Chameleon Sensor (CS)
Chameleon sensors run out of gypsum blocks that measure the endurance which is converted to tension which indicates the 'easiness' for the crop to uptake water from the soil. The sensors are installed in the soil at specific soil depths depending on the root zone development of the soil and in connection with a WiFi reader that has three led lights (each connected light corresponds to one specific depth). Each led light has three colour options: blue, green, and red that tells the farmer whether the soil is wet, moist, or dry, respectively. The turn from blue to green occurs between 20 and 25 kPa, whereas from green to red >40–45 kPa which supports farmers to decide on their irrigation systems. Hence, the sensor does provide a simple guide to the farmer, informing them whether there is irrigation need (Schmitter et al. 2016).

On-farm irrigation scheduling technologies like the WFD and Chameleon sensors could potentially be used to set up local irrigation information platforms and reduce the dependency on individual investments in irrigation scheduling tools whilst simultaneously improving on-farm water management and crop and water productivity at the water user group level (Fig. 15.10).

15.10 Areas to Be Considered for Water Productivity Improvement

To bring effective water productivity under farmer-led irrigation, some of the key things to be considered are capacity strengthening, enhancing access to financial resources, Adequacy of supportive infrastructures: transport (roads, railways), electricity, communications, and storage and processing facilities for agricultural

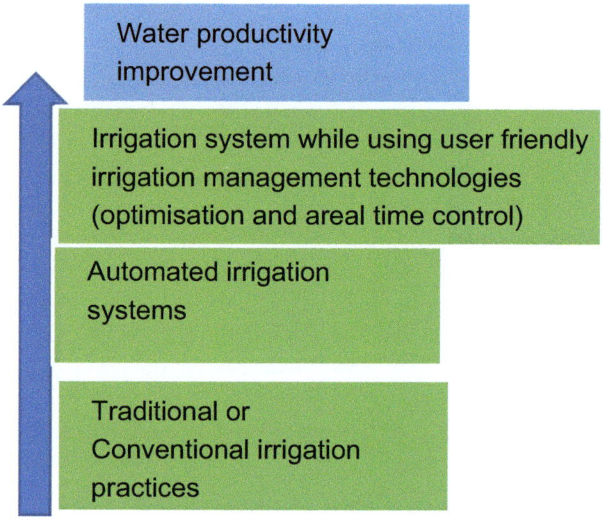

Fig. 15.10 Water use efficiency using water-improving technologies

products; and policy reform which is one of the most critical things for the achievement (Merrey et al. 2017).

Closing the yield gap and improving crop and water productivity is not just transferring better technologies and practices to farmers but also putting in place the institutional structures (markets, finance, and risk management mechanisms) that farmers need to adopt new technologies/tools. National and local policies can be supportive or provide barriers and disincentives for farmers' adoption of better practices. It is in the government's interest to support and stimulate the adoption of available technologies because they will likely affect crop selection, livelihood choices, and the productivity of land and water resources. Most governments in sub-Saharan Africa face financial constraints, making it difficult for governments to invest in supporting the adoption of better water and land management technologies or adaptations to climate change. However, governments, donors, research, implementers, and development experts agree that agriculture is critical for reducing poverty and promoting economic development. Farmers can produce high-value crops such as vegetables and fruits using water-smart technologies/water optimization tools. Microcredit and private commercial investments can help people use water by arranging financial availability and making the technologies available in the local markets (World Bank 2007).

15.10.1 Benefits of Improved Water Productivity

One of the advantages of increasing farmer-led irrigation water productivity is improving incomes in rural areas through an employment effect. Water productivity is critical in areas with high poverty/food insecurity and less crop and water productivity. This is more common in SSA and parts of South Asia and Latin America, in areas where there is water shortage and at the same time when there is high competition to get water, in areas where water resources development is lacking, and in areas where there are water driven ecosystem degradations (Molden et al. 2007).

15.10.2 Non-water Management Interventions

Practices not directly related to water management impact water productivity because of interactive effects derived from pest and disease control improvements, soil fertility, and access to better crop selection or markets (Molden et al. 2010). Water management, combined with improved farm management practices, including better crop selection, the right time fertilizer application, good tillage practices, availability of agrochemical inputs management, and appropriate cultural practices, helps to improve productivity (UNEP 2009; Jacob et al. 2009). Where water is scarce compared with land and other resources involved in the production, improving water productivity (water productivity for the crop, livestock, and aquaculture production) is critical. The water productivity of smallholder farmers in sub-Saharan Africa varies as per the income classes of farmers.

15.11 Conclusion

The purpose of this chapter was to review the potential opportunities that help to improve the water productivity of farmer-led irrigation. Higher water productivity can be achieved by applying the same amount of water and getting a higher yield and/or by reducing the amount of water that can be applied and producing the same yield and/or a higher yield. Some factors affecting crop and water productivity at the farmer-led irrigation level are engineering and technological factors, seed variety selection, environmental factors, socioeconomic factors, and lack of capacity building. Management practices that increase agricultural yields potentially improve crop and water productivity (quantity and production quality). Identification of water sourcing and environmentally friendly water-lifting technologies; deficit and supplemental irrigation techniques implementation support improve crop and water productivity of the farmer-led irrigation systems. In addition, improved irrigation water application methods/practices, adoption of climate-smart on-farm water management tools/technologies, and scientific-based irrigation scheduling techniques are the opportunities that also contribute to improved productivity. Food insecurity and poverty alleviation can be addressed by improving water productivity. All relevant stakeholders, farmers, and government entities should participate in any of the aspects that help improve crop and water productivity.

References

Alaofe H et al (2016) Solar-powered drip irrigation impacts on crop production diversity and dietary diversity in Northern Benin. Food Nutr Bull 37(2):164–175. https://doi.org/10.1177/0379572116639710

Arora VK et al (2011) Irrigation, tillage and mulching effects on soybean yield and water productivity about soil texture. Agric Water Manag 98:563–568

Balwinder S et al (2011) Growth, yield and water productivity of zero till wheat as affected by rice straw mulch and irrigation schedule. Field Crop Res 121:209–225

Barron J (2004) Dry spell mitigation to upgrade semi-arid rain fed agriculture: water harvesting and soil nutrient management for smallholder maize cultivation in Machakos, Kenya. Ph.D. thesis, Natural Resource Management, Stockholm University

Bjornlund H, van Rooyen A, Stirzaker R (2017) Profitability and productivity barriers and opportunities in small-scale irrigation schemes. Int J Water Resour Dev 33:690–704

Borner L (2015) Irrigation scheduling. Colorado University Extension Program, Fort Collins, Colorado

Burney JA, Naylor RL (2012) Smallholder irrigation as a poverty alleviation tool in Sub-Saharan Africa. World Dev 40(1):110–123. https://doi.org/10.1016/j.worlddev.2011.05.007

Burney JL et al (2010) Solar-powered drip irrigation enhances food security in the Sudano–Sahel. PNAS 107(5):1848–1853. www.pnas.org/cgi/doi/10.1073/pnas.0909678107

Cai X, Rose Grant MW (2003) World water productivity: current situation and future options. In: Kijne W, R.B.a.D.M. (eds) Water productivity in agriculture. CAB International, Sri Lanka, p 368

Cai X et al (2011) Producing more food with less water in a changing world: assessment of water productivity in 10 major river basins. Water Int 36:42–62. https://doi.org/10.1080/02508060.2011.542403

Cook et al (2009) Water, food and livelihoods in river basins. Water Int 34:13–29

Descheemaeker K et al (2013) Increasing water productivity in agriculture, Wageningen University, Wageningen Netherlands.

Edward R, Martin C (2000) Determining the Amount of Irrigation Water Applied to a Field Cooperative Extension, College of Agriculture and Life Sciences, University of Arizona.

FAO (1995) Irrigation in Africa in figures: water reports, vol No. 7. FAO, Rome

FAO (2013) Climate-smart agriculture: sourcebook. Food and Agriculture Organization of the United Nations, Rome

Gebregziabher G, Namara RE, Holden S (2009) Poverty reduction with irrigation investment: an empirical case study from Tigray, Ethiopia. Agric Water Manag 96(12):1837–1843. https://doi.org/10.1016/j.agwat.2009.08.004

Hillel (1998) Environmental soil physics. Academic Press, Elsevier, USA, 771 pp. http://akvopedia.org/wiki/Rope_pump

Jacob K et al (2009) Opportunities to increase water productivity in agriculture with special reference to Africa and South Asia

Kamwamba-Mtethiwa JR et al (2016) Treadle pump irrigation in Malawi: adoption, gender, and benefits. Irrig Drain 61:583–595. https://doi.org/10.1002/ird.1665

Kijne JW, Barker R, Molden D (eds) (2003) Water productivity in agriculture: limits and opportunities for improvement. Comprehensive assessment of water management in agriculture series 1. CAB International, Wallingford, UK in Association with International Water Management Institute (IWMI), Colombo

Koech R, Langat P (2018) Improving irrigation water use efficiency: a review of advances, challenges, and opportunities in the Australian context. Water 10(12):1771

Lal R (2007) Bringing the green revolution to resource-poor farmers in Sub-Saharan Africa. CSA News 52(6):14–15. American Society of Agronomy, Madison, Wisc

Mekuria W, Tegegne D (2022) Water harvesting. https://www.sciencedirect.com/science/article/pii/B9780128229743000422?via%3Dihub
Mengistie D, Kidane D (2016) Assessment of the impact of small-scale irrigation on household livelihood improvement at Gubalafto District, North Wollo, Ethiopia. MDPI J, Agric 6:27
Merrey D, McCornick P, Nance M (2017) Pathways to increasing farmer-led investments in sustainable agricultural water management in sub-Saharan Africa
Molden D et al (2007) Pathways for increasing agricultural water productivity. In: Water for food water for life-a comprehensive assessment of water management in agriculture. Earthscan, UK
Molden D et al (2010) Improving agricultural water productivity: between optimism and caution. Agric Water Manag 97:528–535. https://doi.org/10.1016/j.agwat.2009.03.023
Mzezewa J, Gwata ET, van Rensburg LD (2011) Yield and seasonal water productivity of sunflower as affected by tillage and cropping systems under dryland conditions in the Limpopo Province of South Africa. Agric Water Manag 98:1641–1648
Nangia V, Oweis T (2016) Supplemental irrigation: a promising climate-resilience practice for sustainable dryland agriculture. In: Farooq M, Siddique K (eds) Innovations in dryland agriculture. Springer, Cham. https://doi.org/10.1007/978-3-319-47928-6_20
Oweis T, Hachum A (2006) Water harvesting and supplemental irrigation for improved water productivity of dry farming systems in West Asia and North Africa. Agric Water Manag 80(1–3):57–73
Pandey RK, Maranville JW, Adamo A (2001) Tropical wheat response to irrigation and nitrogen in a Sahelian environment. I. Grain yield, yield components and water use efficiency. Eur J Agron 15(2):93–105
Rockstrom J, Barron J (2007) Water productivity in rainfed systems: an overview of challenges and analysis of opportunities in water scarcity prone savannahs. Irrig Sci 25:299–311
Schmitter P et al (2016) Evaluation of suitable water lifting and on-farm water management technologies for the irrigation of vegetables and fodder in Lemo district, Ethiopia. Africa RISING working paper
Schmitter P, et al (2017) Improving on-farm water management by introducing wetting-front detector tools to smallholder farms in Ethiopia. LIVES Working Paper 28. Nairobi, Kenya: International Livestock Research Institute (ILRI).
Stirzaker R (2003) When to turn the water off scheduling micro-irrigation with a wetting front detector. Irrig Sci 22:177–185
Stirzaker R, Mbakwe I, Mziray NR (2017) A soil water and solute learning system for small-scale irrigators in Africa. Int J Water Resour Dev 33:788–803
Strelkoff TS et al (1999) Surface irrigation evaluation models: application to level basins in Egypt. Trans ASAE 42(4):1027–1037
Tadesse N, Berhane A, Bheemalingeswara K (2008) Initiatives opportunities and challenges in shallow groundwater utilization: a case study from Debrekidane Watershed, Hawzien Woreda, Tigray Region, Northern Ethiopia. Agric Eng Int: CIGR E-J X. Manuscript LW 08 008
Tegegne D, Damtie MT, Shumye A (2024) Assessment of irrigation performance trends using selected comparative indicators over consecutive periods: a case study of the Koga Irrigation Scheme in the Amhara region, Ethiopia. Irrig Drain:1–11. Available from: https://doi.org/10.1002/ird.3021
Tesfaye A et al (2008) The impact of small-scale irrigation on household food security: the case of Filtino and Godino irrigation schemes in Ethiopia. Irrig Drain Syst 22(2):145–158
Trincheria et al (2017) Best practices on the use of rainwater for off-season small-scale irrigation: fostering the replication and scaling-up of rainwater harvesting irrigation management in arid and semi-arid areas of sub-Saharan Africa. AFRHINET Project, Hamburg University of Applied Sciences, Hamburg
UNEP (2009) Rainwater harvesting: a lifeline for human well-being. United Nations Environment Programme (UNEP), Nairobi/Stockholm Environment Institute (SEI), Stockholm

Woodhouse P et al (2017) African farmer-led irrigation development: re-framing agricultural policy and investment? J Peasant Stud 44:213–233
World Bank (2007) Agriculture for development. World development report 2008. World Bank, Washington, DC
Zoebl D (2006) Is water productivity a useful concept in agricultural water management? Agric Water Manag 84(3):265–273
Zwart SJ et al (2006) SEBAL for detecting spatial variation of water productivity for wheat in the Yaqui Valley, Mexico. Paper presented at the International Conference on Earth Observation for vegetation monitoring and water management, Naples, Italy, 10–11 Nov 2005

Chapter 16
Irrigation Development and Agricultural Water Management in Rwanda: A Systematic Review

Tafadzwanashe Mabhaudhi, T. L. Dirwai, C. Taguta, E. K. Kanda, L. Nhamo, and O. Cofie

Abstract Most African countries rely on food imports and cannot feed their populations. The most vulnerable region to chronic food insecurity is sub-Saharan Africa (SSA) where agriculture is mainly rainfed and therefore threatened by climate change and variability. Irrigation is one of the main solutions for stabilizing yields and reinforcing food security, yet it is underdeveloped in most parts of Africa. However, irrigation consumes the largest amount of water than the other sectors; thus, exploring and implementing ways of producing more yield per unit volume of

T. Mabhaudhi (✉)
Centre on Climate Change and Planetary Health, London School of Hygiene and Tropical Medicine, London, UK

Centre for Transformative Agricultural and Food Systems, University of KwaZulu-Natal, Carbis Road, Scottsville, Pietermaritzburg, KwaZulu-Natal, South Africa
e-mail: Tafadzwanashe.Mabhaudhi@lshtm.ac.uk

T. L. Dirwai
International Water Management Institute, Pretoria, South Africa

Centre for Transformative Agricultural and Food Systems, University of KwaZulu-Natal, Pietermaritzburg, South Africa

C. Taguta
Centre for Transformative Agricultural and Food Systems, University of KwaZulu-Natal, Pietermaritzburg, South Africa

E. K. Kanda
Department of Civil and Structural Engineering, Masinde Muliro University of Science and Technology, Kakamega, Kenya

L. Nhamo
Centre for Transformative Agricultural and Food Systems, University of KwaZulu-Natal, Pietermaritzburg, South Africa

Water Research Commission of South Africa, Pretoria, South Africa

O. Cofie
International Water Management Institute, Accra, Ghana

T. Mabhaudhi et al. (eds.), *Enhancing Water and Food Security Through Improved Agricultural Water Productivity*,
https://doi.org/10.1007/978-981-96-1848-4_16

water is necessary. To counter food insecurity and improve agricultural water management, the African Union (AU) developed a framework for irrigation development and agricultural water management (IDAWM) to be adopted in all the member states in the continent. This framework is premised on four development pathways, namely, improved water control and watershed management in rainfed farming, farmer-led irrigation development (FLID), irrigation scheme development and modernization and the use of unconventional water for irrigation. Therefore, this review sought to assess the status, challenges, and opportunities of IDAWM in Rwanda. The systematic review adopted the PRISMA-P (Preferred Reporting Items for Systematic Review and Meta-Analysis Protocols). The results indicated that Rwanda has adopted various strategies such as terraces, contour bunds, and water harvesting to address soil erosion and improve water storage. Irrigation is practised in three ways: marshland, hillside, and small-scale irrigation technologies, which are faced with several challenges, such as land use policy and inadequate participation, which hinder progress in FLID. Inadequate private sector involvement hinders investment in the modernization of irrigation schemes in Rwanda. Inadequate sewerage and wastewater treatment infrastructure limits wastewater reuse in irrigation. The bright spots are anchored in sound and progressive agricultural policy, abundant water resources, favourable climatic and ecological conditions and a ready regional market.

Keywords Climate-smart agriculture · Community-managed irrigation · Conservation agriculture · Food insecurity · Land tenure system

16.1 Introduction

Agriculture is the backbone of most African economies, yet food insecurity is one of the main challenges afflicting several African countries. Therefore, the focus should be improving agricultural production to boost food security and improve livelihoods. Africa remains dependent on food imports to satisfy the needs of their people, with only a third of African countries having enough agriculture export revenue to pay for food import bills (Darko 2020). Sub-Saharan Africa (SSA) is the most hit region, where agriculture is mostly rainfed and thus faces uncertainties attributed to climate change and variability. Irrigation has been suggested as the major option to boost levels of agricultural productivity in SSA (Xie et al. 2014). Irrigation reduces the impacts of climate change-induced shocks in agriculture by stabilizing yields. However, irrigation consumes large quantities of water and thus needs to improve agricultural water productivity by adopting sustainable irrigation technologies and agricultural water management practices. It is against this background that the New Partnership for Africa Development (NEPAD)'s Comprehensive African Agriculture Development Programme (CAADP) Pillar 1 on extending the area under sustainable land and water management (NEPAD 2003) declared the need for scaling up investments in irrigation development to provide farmers with opportunities to sustainably increase output and contribute to the reliability of food supplies in the continent (Mutiro and Lautze 2015).

The performance of irrigation schemes in SSA is riddled with challenges, and public irrigation schemes have failed in many countries. Irrigation managers are often confronted with the twin challenge of either vertical expansion (intensifying irrigation management on existing areas using fewer water resources, i.e. more crop per drop) or horizontal expansion, i.e. increasing the area under conventional and intensified irrigation management (Bastiaanssen and Perry 2009). The latter option may not be feasible for a country like Rwanda which is densely populated and thus land is limited. Improving crop water productivity has gained traction due to the ever-increasing pressure on finite water resources from population growth, urbanization, and industrialization.

The failure of public-led irrigation institutions to accelerate irrigation development led to the clamour for farmer-led irrigation development (FLID) because public-led irrigation institutions rarely can accelerate irrigation development (World Bank 2021). Indeed, in Rwanda, several established irrigation schemes have been characterized by low performance with low water use efficiencies and low production levels, leading to poor productivity and low incomes for farmers (RAB 2020). In this context, FLID is a process where farmers assume a driving role in improving their water use for agriculture through changes in knowledge production, technology use, investment patterns and market linkages, and the governance of land and water (Woodhouse et al. 2017). FLID systems take many forms and sizes, from small individual farmer systems to much larger community-managed systems (Nkoka et al. 2014).

Agriculture is an essential pillar of Rwanda's economy as it contributes 30% of GDP and employs 70% of the population while generating about 50% of the country's export revenues (USAID 2019). However, the agricultural production system in Rwanda, just like in most African countries, is rainfed subsistence farming, which is limited to two cropping seasons occurring during the 7-month rainy season (Ngango and Hong 2021) of September to January and February to June (Weatherspoon 2021). This mode of agricultural production is threatened by climate change and variability. Although the annual rainfall is relatively well distributed, rainfall events are erratic, especially in the Eastern part of the country (Verdoodt and Van Ranst 2003). Climate projections for 2050 indicate that Rwanda will experience a rise of between 1.4 °C and 2.3° in average annual temperatures, more frequent and intense heavy rains, and potentially increased duration of dry spells of up to 7 days (USAID 2019) and thus a threat to rainfed agriculture. Besides climate change, agricultural production is also limited by the country's topography, where the western part is a mountainous landscape while the eastern part is mainly lowlands. Therefore, these two conditions make the land susceptible to landslides, erosion, and flooding (Lydie 2022).

Rwanda has an irrigation potential of about 600,000 ha based on its available water resources (MINAGRI 2010), but only about 3% of the area is developed with irrigation schemes (Ngango and Hong 2021). In addition, most of the irrigation schemes in the country are operating below the planned capacity, and thus, irrigation is not contributing much to the overall agricultural productivity (Hakuzimana and Masasi 2020). Therefore, the country has to devise mechanisms for modernizing its irrigation sector through investments in infrastructure and creating a favourable enabling environment.

Holistic development of the agriculture sector requires interventions around effective agricultural water management and irrigation development. The aim of water management includes practices that (a) increase the available water at the root zone, (b) improve soil water holding capacity, (c) promote water harvesting, (d) address irrigation design, operation and management, and (e) improve soil drainage (AU 2020). These practices will help the farmers adapt and build resilience to climate change.

Adopting and promoting FLID requires interventions that enable easier entry for new irrigation farmers, accelerated progress for those already irrigating, and the inclusion of a wider circle of people who benefit (World Bank 2021). The success of FLID depends on the resource potential, farmer benefits and the enabling environment, i.e. legal and policy framework, technology availability and know-how, financial accessibility and marketing arrangements (Fig. 16.1).

The African Union (AU), in response to the challenges facing the agricultural sector in the continent, developed the irrigation development and agricultural water management (IDAWM) framework that aims at supporting regional and national strategies and project implementation to achieve continental targets by promoting country-level initiatives in agricultural water management (AWM) revolving around the following four pathways: (1) improved water control and watershed management in rainfed farming, (2) farmer-led irrigation development (FLID), (3) Irrigation scheme development and modernization, and (4) unconventional water use for irrigation (AU 2020).

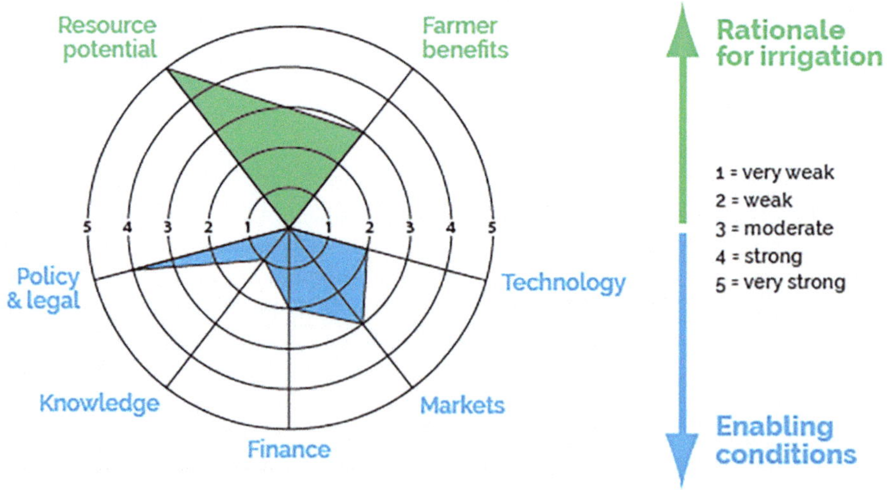

Fig. 16.1 Drivers and enablers of FLID systems (World Bank 2021)

Pathway 1 is a response to climate variability and change that hinders the attainment of optimal yields in rainfed agriculture predominantly practised in Africa. It encompasses water harvesting and control (WHC) technologies and institutional reform initiatives to achieve better coordinated local water resource use in surface and groundwater situations. This pathway is bolstered by key intervention indicators of AU's Agenda 2063, including increased water productivity from rainfed agriculture and irrigation by 60% from the 2013 baseline and harvesting at least 10% of rainwater for productive use. Pathway 2 focuses on FLID, which encompasses a group of farmers who assume a driving role (with no or minimal external support) in improving their agricultural water use. Pathway 3 is premised on the fact that most public irrigation schemes are older with significant infrastructure rehabilitation needs. In this regard, modernization of irrigation infrastructure, organizational and operational modalities, and new scheme development are needed. Pathway 4 helps reduce pressure on finite freshwater resources by reusing wastewater for irrigation (AU 2020).

African countries are at different levels economically and technologically, which has implications for the status of IDAWM. Therefore, it is essential to assess country-specific conditions and baseline information required to address the above pathways. This review assesses the status, challenges, and opportunities of IDAWM in the Republic of Rwanda. This will be an essential entry point for devising strategies and intervention measures required to sustainably ensure food security in the country using abundant but fragile water resources.

16.2 Methodology

16.2.1 Study Area

Rwanda is a landlocked country in the South West of the Lake Victoria Basin and belongs to the Upper Nile River States. The country shares its borders with the Democratic Republic of Congo in the west, Uganda in the north, Tanzania in the east, and Burundi in the south (Fig. 16.2). Rwanda has a total area of 26,338 km^2 and is divided into the Congo basin, representing 17% of the area, and the Nile basin representing 83%. Its relief comprises a succession of relatively large hills and valleys. More than 40% of the country is between 1500 and 1800 m above sea level (NBI 2012).

Rwanda has four main land types: cultivated lands, marshlands, forests, and wetlands. Cultivated land represents 1.12 million ha, around 46% of the country, distributed between 870,000 ha for annual crops and 250,000 ha for permanent crops (Ngabitsinze et al. 2011). The country is classified into 12 agricultural zones where various crops are grown (Table 16.1) with abundant rainfall, and only three regions of Bugesera, Eastern Plateau and Eastern Savannah have average rainfall of less than 1000 mm.

Fig. 16.2 Map of Rwanda. (Source: https://www.nationsonline.org/oneworld/map/rwanda_map2.htm)

16.2.2 Research Approach and Eligibility Criteria

The study adopted the Preferred Reporting Items for Systematic Review and Meta-Analysis Protocols (PRISMA-P) (Moher et al. 2015) in conducting the review.

The eligibility criteria followed the SPIDER (Sample, Phenomenon of Interest, Design, Evaluation and Research type) framework (Table 16.2), which was used to formulate the inclusion–exclusion criteria for the systematic review (Table 16.3). The SPIDER criteria are preferred over other search criteria because they yield high search hits (Methley et al. 2014). Evidence search was conducted in the following databases: Web of Science (WoS), Scopus, and Google Scholar and databases of organizations such as the World Bank and Food and Agriculture Organizations (FAO). The search strategy utilized the Boolean operators (OR and AND), emphasizing publication date. The search strategy was a trial-and-error process. Different search terms were matched to yield many articles for literature extraction.

Table 16.1 Agricultural zones in Rwanda (Verdoodt and Van Ranst 2003)

S. No.	Zone name	Altitude (m)	Rainfall (mm)	Soil	Crops
1	Imbo	970–1400	1050–1600	Alluvial	Banana, cassava, sweet potato, groundnut, common bean, cotton, rice, sugar cane
2	Impara	1400–1900	1300–2000	Very fine red, < basalt	Banana, cassava, sweet potato, groundnut, common bean, sorghum, maize, coffee, tea, quinquina
3	Kivu Lake borders	1460–1900	1150–1300	Shallow clay loam	Banana, cassava, sweet potato, groundnut, common bean, sorghum, maize, coffee
4	Birunga	1600–2500	1300–1600	Volcanic	Banana, cassava, sweet potato, potato, common bean, pea, sorghum, maize, pyrethrum, tobacco
5	Congo-Nile watershed divide	1900–2500	1300–2000	Humiferous, acid	Potato, pea, maize, wheat, tea
6	Buberuka Highlands	1900–2300	1100–1300	Lateritic	Banana, sweet potato, potato, common bean, pea, sorghum, maize, wheat
7	Central Plateau	1500–1900	1100–1300	Humiferous	Banana, sweet potato, soybean, common bean, sorghum, maize, yam, coffee
8	Granitic Ridge	1400–1700	1050–1200	Coarse, gravely	Banana, cassava, sweet potato, groundnut, common bean, sorghum, maize, yam, coffee
9	Mayaga	1350–1500	1000–1200	Clayey, < schists	Banana, cassava, sweet potato, groundnut, common bean, sorghum, maize, coffee
10	Bugesera	1300–1500	850–1000	Strongly weathered	Banana, cassava, sweet potato, groundnut, common bean, sorghum, maize
11	Eastern Plateau	1400–1800	900–1000	Lateritic	Banana, cassava, sweet potato, groundnut, common bean, sorghum, maize, coffee
12	Eastern Savannah	1250–1600	800–900	Strongly weathered	Cassava, sweet potato, groundnut, common bean, sorghum, maize

16.2.3 Literature Screening and Assessing the Risk of Bias

We employed a manual screening process. Literature screening was conducted in three stages: firstly, the title; secondly, by abstract and keywords; and finally, the full article. The study emphasized publication dates where articles from before 2000 were excluded. Articles were included and excluded, as summarized in Table 16.3.

Table 16.2 Inclusion and exclusion criteria based on the SPIDER framework

Item	Criteria	Definitions	Example Search terms
Sample	Irrigation development pathways adopted in Africa	The target economic region was Eastern Africa. The target country was Rwanda	"Eastern Africa*" **OR** "Rwanda*"
Phenomenon of Interest	AU-IDAWM pathways	The pathways of interest are; pathway (1) Improved water control and watershed management in rainfed farming, pathway (2) Farmer-led Irrigation Development (FLID); pathway (3) Irrigation scheme development and modernization; pathway (4) Unconventional water use for irrigation	"FLID*" **OR** "Unconventional water use for irrigation*" **OR** "wastewater*" **AND** "Irrigation*" **OR** "Irrigation scheme development*" **AND** "modernization*" **OR** "rainfed farming*" **AND** "water control measures*" **OR** "climate-smart agriculture*" **OR** "conservation agriculture*"
Design	Literature extraction on different studies on AWM practices in Africa.	Literature extraction	questionnaire **OR** "survey*" **OR** "interview*" **OR** "focus group*" **OR** "case study*" **OR** "observations*"
Evaluation	Analysis of performance analysis of practices adopted in each AU-IDAWM pathway	A case-by-case analysis of the 4 AU-IDAWM pathways	"attitudes*" **OR** "perception*" **OR** "imprecision*" **OR** "conflicting evidence*"
Research type	Mixed methods	Qualitative and quantitative	"qualitative" **OR** "mixed method"

Table 16.3 Inclusion–exclusion criteria

Item	Included	Excluded
Article type	Reviews, journal articles, communications, reports, grey literature and books	Articles with no full texts Journal articles not peer-reviewed Articles with insufficient or inconclusive results or irrelevant discussions
Geographical	Studies conducted Republic of Rwanda	
Publication year		Sources published before 2000
Language		Articles not published in English

Table 16.4 Strength of evidence grading

Strength	Type of research	Study type	Evidence
Strong (I)	Applied research, adaptive/ farm-level research, strategic research, systematic reviews, and meta-analyses	Experimental/ structured reviews	Substantiated (+++)
Moderate (II)	Case studies, reviews, modelling/ simulations	Observational	Partially substantiated (++)
Low (III)	Opinion papers, conference papers, workshop papers	Descriptive	Unsubstantiated, qualitative analysis and opinions (+)

16.2.4 Strength of Evidence

The strength of evidence of each article was assessed based on the criteria applied by Koutsos et al. (2019), where substantiated and documented research is given a high score. In contrast, an average score is assigned to partially substantiated studies with conditional conclusions (Table 16.4). Opinion papers are assigned a low score, and non-evidence-based research is ungraded (Koutsos et al. 2019; Dirwai 2021) (Appendix 16.1).

16.3 Results and Discussion

16.3.1 Literature Sources

The literature search produced 396 articles, with 60 included in qualitative analysis and synthesis (Fig. 16.3), with the strength of evidence assessment in Appendix 16.1. The distribution of the articles was as follows: journal articles (from Web of Science, Scopus, and Google Scholar) were the majority (60%), grey literature (33.3%), books and book chapters (5%), and conference proceedings (1.7%). The last search was completed on 30 June 2022.

16.3.2 Pathway 1: Improved Water Control and Watershed Management in Rainfed Farming

Rainfed agriculture is primarily practised in Rwanda; therefore, various water control and watershed management techniques must be adopted to enhance food security in the face of climate uncertainty. This is even more necessary for Rwanda, which is mountainous and susceptible to erosion and landslides. Besides erosion and landslides, climate change-induced shocks threaten agricultural production;

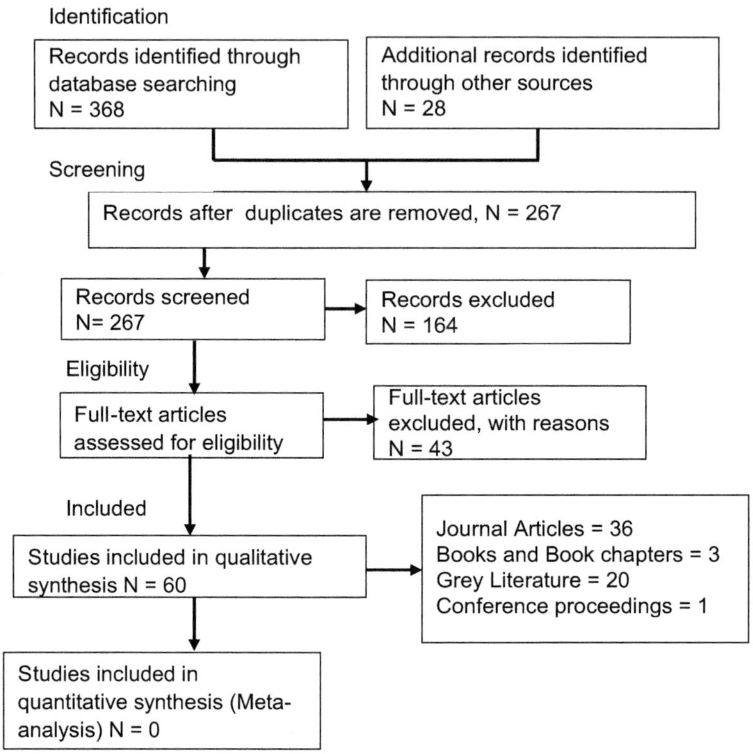

Fig. 16.3 Systematic review flow chart based on PRISMA methodology

therefore, adopting climate-smart agriculture (CSA) practices is necessary. The government of Rwanda is encouraging the use of a wide range of cost-effective erosion control structures and measures, such as check dams, soil/water detention trenches, cut-off drains, waterways, tree belts, contour belts, grass strips, contour bunds, planting of fodder grasses on bunds/ridges, use of permanent and perennial vegetation on contours, and conservation agriculture (CA) practices such as agroforestry including intercropping (GoR 2018). Conservation agriculture is a sub-set of CSA which revolves around three fundamental principles: (1) minimizing soil tillage, (2) covering the soil with organic mulches or cover crops, and (3) diversification of crops in time (crop rotation) and space (crop association mainly in the forms of intercropping or mixed cropping) (Sasaki and Muvunyi 2021).

Rwanda has intensified soil conservation activities through terracing where an area of 131,056.7 ha under radical terraces and 972,055 ha under progressive terraces had been constructed by 2020 (MINAGRI 2021). Progressing terraces are recommended for slopes up to 12% (Pande et al. 2014). These projects are implemented under the National Adaptation Programme of Action (NAPA), which Rwanda submitted in 2006, and it revolves around six thematic areas, including activities around soil conservation (World Bank and CIAT 2015). The construction of bench terraces is highly technical due to their nature (being constructed on slopes of steeper slopes with a gradient of 25–55%). Therefore, farmers need training before they embark upon the terracing process to ensure their technical efficiency and sustainability (Bizoza 2014); otherwise, it can fail. This is supported by a study by Bugenimana et al. (2019), where farmers constructed bench terraces on land slopes lower or higher than standards, thus increasing risks of landslide and erosion with no sustainable benefit for soil erosion control and crop production.

Minimum tillage, crop residue retention and crop rotations have been proposed against poor agricultural productivity and soil degradation where it has been established to have significant benefits; it reduces soil loss by 35.5 to 14.5 t/ha/year, has 50–70% greater infiltration and increases organic carbon by 42% (Kabirigi et al. 2015). This is important in Rwanda, where estimates put soil losses at 20–150 (t/ha)/year on slopes of about 15–50%, which are standard across the country (Prasad et al. 2016).

The extent of adoption of CA is low due to some factors. For example, Kabirigi et al. (2015), noted that decreased yield observed in high rainfall areas, increased labour requirements when herbicides are not used and lack of mulch due to poor productivity and priority given to using them for other purposes constrained CA adoption. In Rwanda's Eastern and Southern provinces, a shortage of mulching materials limited the adoption of CA (Sasaki and Muvunyi 2021).

Rainwater harvesting (RWH) is an important mitigation measure against drought and flooding. Water harvesting in Rwanda is one of the packages in the Crop Intensification Programme (CIP) under the Land Husbandry, Water Harvesting, and Hillside Irrigation (LWH) project, which was jointly implemented by the World Bank and the Rwandan Government (Clay and Zimmerer 2020). This programme reported significant progress in improving agricultural productivity and the livelihoods of rural farmers. Farmers in the Bugesera district outlined the benefits of water harvesting as reduced runoff and soil erosion, capturing rainwater which could otherwise go to waste, and, when covered, could provide less contaminated and safer (Ndayamaje 2008). In some catchments in Rwanda, runoff during the long rainy season when stored is sufficient for supplemental irrigation during a short rainy period, which implies that integrated watershed management and rainwater harvesting are the promising options to stabilize water deficit and improve crop productivity in drought-prone agroecological zones (Uwizeyimana et al. 2019). Farmers in the Bugesera district also employed water harvesting to mitigate against distorted changes in precipitation, which affected their rice yield (Rwanyiziri and Rugema 2013). Land husbandry activities under the LWH project include soil bunds, terraces, cut-off drains, waterways, afforestation, and reforestation (World Bank and CIAT 2015).

Adoption of rainwater harvesting for agriculture requires investments in constructing and maintaining RWH infrastructure, such as ponds and installing liners, which are not affordable to smallholder farmers. Therefore, the cost of RWH systems may be prohibitive to most farmers. Zingiro et al. (2014) found that families endowed with assets other than land, including family income, adopted RWH than those without assets. Benimana et al. (2015) established that farmers in the Bugesera district could not afford the construction of water storage and other retention facilities, such as ponds. However, the area has high runoff potential. The cost problem could be solved by a government subsidy programme under the LWH and CIP and thus help adopt RWH in Rwanda (World Bank and CIAT 2015). Inadequate skills and knowledge were cited as one of the challenges facing RWH adoption in Nyaruguru district, Southern Province (Bizoza and Umutoni 2012).

In the Gicumbi district, awareness, training, extension and financial services accessibility, social incentives, and government programme support were motivators of CA adoption (Murindangabo et al. 2021). The government of Rwanda, through the implementation of the agriculture policy of 2018, strives to support efforts to increase the capacity of on-farm water harvesting, storage, and use (GoR 2018).

Agroforestry is integral to Rwanda's climate change adaptation strategy (GoR 2018). In addition to controlling soil erosion, planting agroforestry species like Calliandra sp., *Calliandra calothyrsus*, and Pennisetum give the farmers fodder and fuel wood (Pande et al. 2014). Agroforestry practices have been reinforced by the creation of smart green villages in some parts of the country, such as Gicumbi District, Northern Province, where agroforestry has been integrated with other soil conservation measures such as terracing (Msaki et al. 2015). However, agroforestry faces challenges such as a lack of extension services and coordination of agroforestry interventions, hindering its adoption (Stainback et al. 2012). Other challenges hindering agroforestry adoption include land scarcity, lack of capital, credit inaccessibility and inadequate skills in agroforestry (Maniraho 2016; Rwaburindi et al. 2019). Therefore, providing technical training, awareness of agroforestry benefits (Sasaki and Muvunyi 2021), and access to financial and input credits (Maniraho 2016) could drive adoption.

The land use consolidation (LUC) programme, as much as it has its benefits of improving crop productivity, may be limiting other CSA activities, such as crop diversification in crop rotation or intercropping. LUC advocates planting one crop per zone (monoculture), which the government determined to be beneficial in that area (Ntihinyurwa et al. 2019). Intercropping coffee with other crops, such as common beans and soybeans, is primarily used in Rwanda (Harelimana et al. 2018). Indeed, smallholder farmers in some parts of Rwanda felt that monocropping, as promoted by the government through LUC and CIP, is risky as it may mean the loss of crops in case of rainfall unavailability (Sasaki and Muvunyi 2021). The introduction of monocropping under LUC discouraged the adoption of CA in parts of northern Rwanda where other farming systems are used (Nahayo et al. 2016). This is because, though voluntary, LUC is tied with CIP, which includes other incentives such as subsidies, and thus it is implicitly state-sanctioned.

Adoption of CSA requires a participatory decision-making and implementation approach, especially with the LUC programme in Rwanda. In this regard, Clay and Zimmerer (2020) proposed that adaptive governance enabling smallholder land use decision-making and support for smallholder food producers' existing agroecological intensification strategies will enable CSA to have a meaningful impact.

Strength, weaknesses, opportunities, and threats (SWOT) analysis for CSA adoption in Rwanda were analysed and reported by Prasad et al. (2016) and illustrated in Table 16.5. The SWOT can be grouped into agro-climatic, infrastructural, institutional, and socio-economic.

16.3.3 Pathway 2: Farmer-Led Irrigation Development (FLID)

Areas under FLID in Rwanda are mainly concentrated in the Eastern and Southern Provinces, which are prone to droughts or erratic rainfall where small-, medium-, and large-scale farming (Table 16.6) are practised with the major crops cultivated consisting of maize, rice, soybeans, and horticulture products (such as fruits and vegetables) (Nzeyimana 2021). Rwanda adopted a farmer-led small-scale irrigation development programme referred to as small-scale irrigation technology (SSIT) in 2014 which targeted the use of sprinkler kits, portable diesel/petrol water pumps and/or treadle pumps, delivery pipes, and dam sheet technology to irrigate relatively small plots/farms ranging from 0.5 to 10 ha (RAB 2020). The area under SSIT is 20,787.5 ha as of 2020 (MINAGRI 2021). Adoption of SSIT is favoured by a high level of education, large land size, membership in-farmer cooperatives/organizations, presence of extension services, access to credit, water accessibility and reliability, and awareness of rainwater harvesting techniques (Ngango and Hong 2021).

Farmers' access to irrigation is also facilitated by private service providers who supply and install irrigation kits in rural areas, while financial institutions work closely with the Ministry of Agriculture and Animal Resources of Rwanda (MINAGRI) to assist poverty-stricken farmers who cannot afford 50% of the investment cost by offering them a loan or lease of SSIT kits (Ngango and Hong 2021).

Table 16.5 SWOT analysis for CSA in Rwanda (Prasad et al. 2016)

Strengths	*Weaknesses*
1. Existence of agro-climatic data 2. Presence of agricultural seasons 3. Availability of markets 4. Rural infrastructure network 5. Strong government policies 6. Strong institutions 7. Active population and skilled labour 8. Adaptive farmers and culture	1. Soil erosion 2. Poor soil fertility 3. Lack of access to water 4. Limited information sharing/access 5. CSA difficult to mechanize 6. Inadequate coordination 7. Small landholdings 8. Lack of nutritious food
Opportunities	*Threats*
1. Harvesting water/irrigation 2. Linkage of farmers to markets 3. Restoring measures for soil fertility 4. Accessibility of information possible 5. CSA funds and Guarantee funds 6. Harmonization of policy/practice on CSA 7. Collaboration—building partnerships	1. Erratic rainfall, weather, pests & diseases 2. Price fluctuations 3. High levels of soil acidity 4. Limited ICT tools/awareness for CSA 5. Money Inflation 6. Changes in policies 7. Land scarcity and high population growth

Table 16.6 Typologies of FLID in Rwanda (MINAGRI 2010; Nzeyimana 2021)

Type	Description	Techniques
Small scale	Individual smallholder farmers, commercial farmers, and community enterprises (cooperatives) Land size between 0.10 and 10 ha. Benefit from 50% government subsidy for irrigation equipment	Motorized pumps, treadle pumps, and solar-powered pumps, drip, sprinkler systems hand-watering kits, and surface systems (flood irrigation)
Medium scale	Marshland and hillside irrigation schemes Land size of 10–100 ha Developed with financial support from the government, development partners, or private commercial farmers at medium The schemes are later transferred to cooperatives and irrigation water users' associations (IWUAs) after signing a transfer agreement for the operation and maintenance of infrastructure IWUA is responsible for the collection of water use fees	Flood fed by gravity irrigation canals in marshlands and drip, sprinkler, or pivot systems in hillside
Large scale	Marshland and hillside irrigation schemes at medium Land size of more than 100 ha Developed with financial support from the government, development partners, or private commercial farmers Government transfers the management and operation to IWUA and producer associations	Flood irrigation fed by gravity irrigation canals in marshlands and drip, sprinkler, or pivot systems in hillside

Hillside irrigation is practised at a small scale in Gashora (12 ha) for cassava and 50 ha of coffee farms in Ngugu near Lake Rwampanga in Kirehe district, both under sprinkler irrigation, and pressurized irrigation systems of 600 ha at Nasho, 1750 ha in Muvumba, and 400 ha in Kagitumba areas (NBI 2012). Hillside irrigation enables farmers to grow high-value horticultural crops, thus improving their income. A study by Byiringo et al. (2020) in four hillside irrigation schemes of Nyanza, Karongi 12, Karongi 13, and Rwamagana found that the adoption of hillside irrigation is low (30%) despite its high productivity and thus jeopardizing the sustainability of the irrigation schemes. Market constraints were cited as the primary reason for the low adoption since, while irrigation water is accessible to farmers in these schemes, horticulture is associated with increased use of complementary inputs, including labour, fertilizer, and seeds and thus, market failures induce a wedge between shadow prices and market prices of the inputs (Jones et al. 2022).

The challenges facing FLID in Rwanda include infrastructural, land availability, and institutional factors, as indicated in Table 16.7.

Table 16.7 Challenges and opportunities of FLID in Rwanda

Challenges	Opportunities
Effect of climate change on water resources	Government-led CSA practices
Land scarcity and degradation through erosion, loss of soil fertility and over-exploitation	Land use consolidation under the crop intensification programme provides an avenue for economies of scale
Inadequate extension services	Good legal framework; irrigation policy which promotes sustainable technologies
Imperfect agricultural commodity markets and value chains	Population growth in the East Africa region creates a pool of ready market
Lack of access to credit	Establishment of Business Development Fund (BDF), which provides guarantees for loans to farmers
Poor irrigation infrastructure	Opportunity for private sector involvement
Inadequate mechanization	
Poor coordination among irrigation institutions	
The land tenure system inhibits privates sector investment	
Weak IWUA and farmer cooperatives/ organizations	

MINAGRI (2010, 2018), Nabahungu and Visser (2011), Theogene et al. (2019), Miklyaev et al. (2021), and Nzeyimana (2021)

Decision-making needs to be participatory, especially under the CIP in Rwanda, since it inhibits the sovereignty over land use, decreasing livelihood flexibility and constricting access to resources (Clay and Zimmerer 2020). The top-down approach in CIP may severely threaten FLID since farmers cannot choose cropping systems. Management of irrigation schemes through a participatory process is essential in improving efficiency and minimizing water conflicts. This was proved in the Cocurirwa cooperative, Rwamagana rice project, where participatory irrigation management through the involvement of progressive farmers in water distribution and collection of irrigation fees led to increased irrigated areas due to the availability of sufficient water and reduced conflicts (Narayanan 2014).

16.3.4 Pathway 3: Irrigation Scheme Development and Modernization

Rwanda's irrigation potential is 589,713 ha (MINAGRI 2010), which was revised to 501,509 ha (RAB 2020) when the following water resources are considered, as represented in Table 16.8.

Despite its huge potential, Rwanda has only developed 50,000 ha (10% of its potential) under irrigation as of 2019, which comprises 148 marshland, hillside, and small-scale irrigation schemes (RAB 2020). Rwanda's irrigation schemes are

Table 16.8 Irrigation potential in Rwanda

Type of water source	Irrigation potential (ha)	
	(MINAGRI 2010)	(RAB 2020)
Runoff for small reservoirs	125,627	52,000
Runoff for dams	27,907	52,100
Direct river and floodwater	79,847	135,880
Lake water resources	100,107	102,364
Groundwater resources	36,432	36,000
Marshlands	219,793	123,164
Total	589,713	501,509

categorized into small-scale and large-scale. The large-scale irrigation schemes are built behind small dams in narrow river valleys, and the seven most notable are Kanyonyomba, d'Agasasa, Migina, Bugarama, Kibaya, Base, and Murago (Bastiaanssen and Perry 2009). Rwanda's irrigation schemes are mainly used for rice production (NBI 2016). It was established that schemes with dams cannot be commercially feasible based on irrigation alone and that multipurpose use and some degree of public subsidy are required (RAB 2020).

Irrigation development in Rwanda is typically not demand-driven but is being carried out through government-led initiatives and donor support to achieve food security (Nzeyimana 2021). This is an unsustainable approach due to increased financial, technical, and managerial constraints (RAB 2020); thus, the government should sustain irrigation development by attracting private sector investments. This is recognized in the revised agricultural policy developed in line with Vision 2050. As per the policy, the government of Rwanda plans to promote an enabling environment by moving away from state actors to enablers by promoting private sector-led irrigation scheme management models and maintenance fee collection establishment (GoR 2018).

The institutional framework for the irrigation sector in Rwanda is shown in Table 16.9. The institutional framework for irrigation and water development has been strengthened significantly through the Rwanda Irrigation Master Plan 2010, the Strategic Plan for Agricultural Transformation (PSTA), the National Agricultural Policy 2004, the Poverty Reduction Strategy and Vision 2020 as the primary guiding document to development of water resources and irrigation (NBI 2012).

Irrigation scheme development in Rwanda is hampered by several factors, which include the high cost of irrigation development, poor organization in schemes, inadequate technical capacities in both private and public sectors, inadequate functioning of market systems, inadequate transport facilities, lack of sufficient storage facilities, inadequate access to financial and extension services and poor coordination of institutions in the irrigation sector (RAB 2020). The prospect lies in the revised agriculture policy, which advocates increasing the area under irrigation, promoting private sector-led models of irrigation scheme management, and attracting private sector and external finance for irrigation development (GoR 2018). The government of Rwanda promised to allocate 2% of public funds for irrigation

Table 16.9 Institutional framework for the irrigation sector in Rwanda (Bastiaanssen and Perry 2009)

Institution/agency	Function
Ministry of Agriculture and Animal Resources	Planning and regulation Sector coordination Supervise local and foreign contractors constructing/rehabilitating irrigation schemes through its implementing agencies, i.e. Rwanda Agriculture and Animal Resources Development Board (RAB) and the National Agricultural Export Board (NAEB)
Farmers/farmer organization	Operation and maintenance
Ministry of Trade and Industry, Ministry of in charge of cooperatives, local government and specialized agencies	Organizational, trade marketing and environmental regulations

development as part of the CAADP commitment (Lydie 2022). Rwanda National Rice Policy (NRP) also promotes small-scale irrigation infrastructure development in selected marshlands while preventing environmental degradation (Ruganzu et al. 2015). Therefore, prospects in this pathway are bright.

16.3.5 Pathway 4: Unconventional Water Use for Irrigation

The unconventional water use for irrigation is premised on population growth, urbanization, and industrialization, increasing inter-sectoral competition for scarce water resources. Therefore, using treated wastewater for irrigation is a possible alternative for reducing freshwater use for agricultural production. Irrigation using wastewater is possible in urban and peri-urban areas where wastewater transport and treatment infrastructure is adequate.

There is no evidence of wastewater reuse for irrigation in Rwanda. This could be attributed to the lack of a wastewater reuse policy. The abundance of water resources could also contribute to the non-adoption of unconventional water use for irrigation. However, when population and industrialization pressure increase on finite freshwater resources, unconventional water use for irrigation will be necessary. At this point, the country needs to invest in wastewater transport and treatment infrastructure that is currently underdeveloped. For example, there is no centralized public sewerage system in Kigali, while the few semi-centralized and decentralized sewerage and wastewater treatment systems do not function appropriately as initially designed (Kazora and Mourad 2018). Occasionally, untreated wastewater is discharged directly into stormwater drains (Mbateye et al. 2010). A study by Manirakiza et al. (2022) established that industrial wastewater discharged to the receiving water bodies contained hazardous pollutants above the national standards and the World Health Organization's allowable limit for reuse in irrigation. The absence of proper wastewater treatment facilities will hinder wastewater reuse in agriculture.

16.4 Conclusion

Rwanda has made progress in implementing the IDAWM framework under the four pathways. Under pathway 1, significant progress has been made by promoting soil and water control measures such as terraces, contour bunds, and rainwater harvesting, as well as climate-smart practices such as agroforestry. This has been made possible through state-sponsored projects and programmes. The major challenges under pathway 1 are challenges such as high input costs, inadequate training, and insufficient extension support for CSA and CA practices. The progress in pathway 2 is being implemented through various approaches, including government subsidy. Inadequate infrastructure, weak farmer organizations, land tenure system, inadequate extension services, and market inaccessibility hamper this pathway. Pathway 3 is hampered by the high cost of irrigation infrastructure, poor sectoral coordination, and inadequate technical capacity. There is no literature on unconventional water use in Rwanda due to the abundance of water.

Nevertheless, this should be an option as pressure on freshwater resources increases due to population growth, urbanization, and industrialization. However, this should be preceded by investment in wastewater treatment infrastructure. Bright prospects exist regarding legal and institutional strengthening, favourable agro-climatic conditions, and local and regional markets.

Appendix 16.1: Strength of Evidence Assessment

Author, Year	Title	Subject area	Strength
AU (2020)	Framework for Irrigation Development and Agricultural Water Management in Africa	Irrigation and AWM	++
Bastiaanssen and Perry (2009)	Agricultural Water Use and Water Productivity in the Large-Scale Irrigation (LSI) Schemes of the Nile Basin	Irrigation	++
Benimana et al. (2015)	Rainwater harvesting potential for crop production in the Bugesera district of Rwanda	CSA	+++
Bizoza (2014)	Institutions and the adoption of technologies: Bench terraces in Rwanda	CA	+++
Bizoza and Umutoni (2012)	Socio-economic impacts of rain water harvesting technologies in Rwanda: A case study of Nyaruguru District, Southern Province.	CSA	+++
Bugenimana et al. (2019)	Assessment of technical conformity of bench terraces for soil erosion control in Rwanda	CA	+++
Byiringo et al. (2020)	Impacts, maintenance and sustainability of irrigation in Rwanda	Irrigation	+++
Clay and Zimmerer (2020)	Who is resilient in Africa's green revolution? Sustainable Intensification and climate-smart agriculture in Rwanda	CSA	+++

(continued)

Author, Year	Title	Subject area	Strength
Darko (2020)	Irrigated agriculture for food self-sufficiency in the sub-Saharan African region	Irrigation	+++
Dirwai (2021)	Moistube irrigation technology development, adoption and future prospects: A systematic scoping review	Systematic review	+++
GoR (2018)	National Agriculture Policy	Irrigation	++
Hakuzimana and Masasi (2020)	Performance evaluation of irrigation schemes in Rugeramigozi marshland, Rwanda	Irrigation	+++
Harelimana et al. (2018)	Coffee production systems: Evaluation of intercropping system in coffee plantations in Rwanda	Irrigation	+++
Jones et al. (2022)	Factor Market Failures and the Adoption of Irrigation in Rwanda	Irrigation	+++
Kabirigi et al. (2015)	Applicability of conservation agriculture for climate change adaptation in Rwanda's situation	CA	+++
Kazora and Mourad (2018)	Assessing the sustainability of decentralized wastewater treatment systems in Rwanda	Wastewater treatment	+++
Koutsos et al. (2019)	An efficient framework for conducting systematic literature reviews in agricultural sciences.	Systematic review	
Lydie (2022)	Droughts and flooding implications in agriculture sector in Rwanda: Consequences of global warming	Climate change/ rainfed agriculture	+++
Maniraho (2016)	Assessment of the Role of Trees on Farmland in Soil Conservation and Household Welfare in Rwanda	CA	+++
Manirakiza et al. (2022)	Physico-chemical analysis of industrial wastewater pollution from Kigali Special Economic Zone (KSEZ) and the potential impacts in the downstream regions of Kigali City in Rwanda	Wastewater treatment	+++
Mbateye et al. (2010)	Assessment of wastewater management practices in Kigali City, Rwanda	Wastewater treatment	+++
Methley et al. (2014)	PICO, PICOS and SPIDER: a comparison study of specificity and sensitivity in three search tools for qualitative systematic reviews	Systematic review	+++
Miklyaev et al. (2021)	Sustainability of agricultural crop policies in Rwanda: An integrated cost–benefit analysis	Agricultural policy	+++
MINAGRI (2010)	Rwanda Irrigation Master Plan	Irrigation	+++
MINAGRI (2018)	Strategic Plan for Agriculture Transformation	Agricultural policy	++
MINAGRI (2021)	Annual Report 2020–2021	Agricultural report	+++
Moher et al. (2015)	Preferred reporting items for systematic review and meta-analysis protocols (PRISMA-P) 2015 statement	Systematic review	+++

(continued)

Author, Year	Title	Subject area	Strength
Msaki et al. (2015)	State of knowledge on CSA in Africa, case studies from Rwanda, Tanzania and Zambia	CSA	++
Murindangabo et al. (2021)	Adoption of conservation agriculture in Rwanda: A case study of the Gicumbi District Region	CA	+++
Nabahungu and Visser (2011)	Contribution of wetland agriculture to farmers' livelihood in Rwanda	Irrigation	+++
Nahayo et al. (2016)	Factors influencing the adoption of soil conservation techniques in Northern Rwanda	CA	+++
Narayanan (2014)	Impact of participatory irrigation management–case study: Cocurirwa Cooperative, Rwamagana Rice Project, Rwanda	Irrigation	+++
NBI (2012)	Assessment of the irrigation potential in Burundi, Eastern DRC, Kenya, Rwanda, Southern Sudan, Tanzania and Uganda	Irrigation	++
NBI (2016)	The Nile Basin Water Resources Atlas	Irrigation	+
Ndayamaje (2008)	Management of rainwater harvesting lined ponds in dry areas of Rwanda: A case study of Ntarama lined ponds owners	CSA	++
NEPAD (2003)	Comprehensive Africa Agriculture Development Programme	Agriculture policy	++
Ngabitsinze et al. (2011)	Planning and costing adaptation of perennial crop farming systems to climate change: Coffee and banana in Rwanda	Climate change	++
Ngango and Hong (2021)	Adoption of small-scale irrigation technologies and its impact on land productivity: Evidence from Rwanda	Irrigation	+++
Nkoka et al. (2014)	Organisational modalities of farmer-led irrigation development in Tsangano District, Mozambique	Irrigation	+++
Ntihinyurwa et al. (2019)	The positive impacts of farmland fragmentation in Rwanda	Land use	+++
Nzeyimana (2021)	Assessment of farmer-led irrigation development in Rwanda	Irrigation	+++
Pande et al. (2014)	Coping with climate change through water harvesting techniques for sustainable agriculture in Rwanda	CSA	+++
Prasad et al. (2016)	Climate-smart agriculture and sustainable intensification: assessment and priority setting for Rwanda	CSA	++
RAB (2020)	Rwanda Irrigation Master Plan	Irrigation	+++
Ruganzu et al. (2015)	Salinity reducing food security and financial returns from rice production in Rwanda.	CSA	+++
Rwaburindi et al. (2019)	Drivers of smallholder farmers' decision to adopt agroforestry in Rulindo district, Rwanda	CSA	+++
Rwanyiziri and Rugema (2013)	Climate change effects on food security in Rwanda: Case study of wetland rice production in Bugesera District	CSA	+++

(continued)

Author, Year	Title	Subject area	Strength
Sasaki and Muvunyi (2021)	Constraints to adoption and scaling-up of conservation agriculture in Rwanda smallholder farmers' perspectives	CA	+
Stainback et al. (2012)	Smallholder agroforestry in Rwanda: A SWOT-AHP Analysis	CSA	+++
Theogene et al. (2019)	Effect of institutional factors on the adoption of small scale irrigation system in Rwanda	Irrigation	+++
USAID (2019)	Climate risk in Rwanda: Country risk profile	Climate change	++
Uwizeyimana et al. (2019)	Modelling surface runoff using the soil conservation service-curve number method in Rwanda's drought-prone agroecological zone.	CSA	+++
Verdoodt and Van Ranst (2003)	Land evaluation for agricultural production in the tropics. A large-scale land suitability classification for Rwanda	Agriculture	+++
Weatherspoon (2021)	Rwanda's Commercialization of Smallholder Agriculture: Implications for Rural Food Production and Household Food Choices	Agriculture	+++
World Bank (2021)	The farmer-led irrigation development guide: A what, why and how-to for intervention design	Irrigation	++
World Bank and CIAT (2015)	Climate-smart agriculture in Rwanda. CSA country profiles for Africa, Asia, Latin America and the Caribbean Series	CSA	+++
Xie et al. (2014)	Estimating the potential for expanding smallholder irrigation in Sub-Saharan Africa	Irrigation	+++
Zingiro et al. (2014)	Assessment of adoption and impact of rainwater harvesting technologies on rural farm household income: the case of rainwater harvesting ponds in Rwanda	CSA	+++

CA conservation agriculture, *CSA* climate-smart agriculture, *AWM* agricultural water management
+++ strong, ++ moderate, + low

References

AU (2020) Framework for irrigation development and agricultural water management in Africa. African Union, Addis Ababa

Bastiaanssen W, Perry C (2009) Agricultural water use and water productivity in the large scale irrigation (LSI) schemes of the Nile Basin. Nile Basin Initiative Entebbe, Uganda

Benimana J et al (2015) Rainwater harvesting potential for crop production in the Bugesera district of Rwanda. Afr J Agric Res 10(19):2020–2031

Bizoza AR (2014) Institutions and the adoption of technologies: bench terraces in Rwanda. In: Vanlauwe B, Asten PV, Blomme G (eds) Challenges and opportunities for agricultural intensification of the humid highland systems of Sub-Saharan Africa. Springer International Publishing, Geneva, pp 335–354

Bizoza AR, Umutoni G (2012) Socio-economic impacts of rain water harvesting technologies in Rwanda: a case study of Nyaruguru District, Southern Province. Rwanda J 26:103–115

Bugenimana ED et al (2019) Assessment of technical conformity of bench terraces for soil erosion control in Rwanda. Afr J Agric Res 14(2):69–77

Byiringo E et al (2020) Impacts, maintenance and sustainability of irrigation in Rwanda, 3ie impact evaluation report 112. International Initiative for Impact Evaluation (3ie), New Delhi

Clay N, Zimmerer KS (2020) Who is resilient in Africa's green revolution? Sustainable intensification and climate-smart agriculture in Rwanda. Land Use Policy 97:1–12

Darko RO (2020) Irrigated agriculture for food self-sufficiency in the sub-Saharan African region. Int J Agric Biol Eng 13(3):1–12

Dirwai T (2021) Moistube irrigation technology development, adoption and future prospects: a systematic scoping review. Heliyon 7(2):e06213

GoR (2018) National agriculture policy. Republic of Rwanda Kigali

Hakuzimana J, Masasi B (2020) Performance evaluation of irrigation schemes in Rugeramigozi marshland, Rwanda. Water Conserv Manag 4(1):15–19

Harelimana A et al (2018) Coffee production systems: evaluation of intercropping system in coffee plantations in Rwanda. J Agric Sci 10(9):17

Jones M et al (2022) Factor market failures and the adoption of irrigation in Rwanda. Am Econ Rev 112(7):2316–2352

Kabirigi M et al (2015) Applicability of conservation agriculture for climate change adaptation in Rwanda's situation. J Soil Sci Environ Manag 6(9):241–248

Kazora AS, Mourad KA (2018) Assessing the sustainability of decentralized wastewater treatment systems in Rwanda. Sustainability 10(12):1–15

Koutsos TM, Menexes GC, Dordas CA (2019) An efficient framework for conducting systematic literature reviews in agricultural sciences. Sci Total Environ 682:106–117

Lydie M (2022) Droughts and floodings implications in agriculture sector in Rwanda: consequences of global warming. In: Harris SA (ed) The nature, causes, effects and mitigation of climate change on the environment. IntechOpen, London, pp 1–13

Maniraho L (2016) Assessment of the role of trees on farmland in soil conservation and household welfare in Rwanda. Am Sci Res J Eng Technol Sci 15(1):302–314

Manirakiza B et al (2022) Physico-chemical analysis of industrial wastewater pollution from Kigali Special Economic Zone (KSEZ) and the potential impacts in the downstream regions of Kigali City in Rwanda. Sustain Water Resour Manag 8(4):1–10

Mbateye UFA, Wali U, Banadda N (2010) Assessment of wastewater management practices in Kigali City, Rwanda. Open Environ Biol Monit J 3(1):21–28

Methley AM et al (2014) PICO, PICOS and SPIDER: a comparison study of specificity and sensitivity in three search tools for qualitative systematic reviews. BMC Health Serv Res 14(1):1–10

Miklyaev M, Jenkins G, Shobowale D (2021) Sustainability of agricultural crop policies in Rwanda: an integrated cost–benefit analysis. Sustainability 13:1–21

MINAGRI (2010) Rwanda irrigation master plan. Ministry of Agriculture and Animal Resources, The Government of Rwanda, Kigali

MINAGRI (2018) Strategic plan for agriculture transformation 2018–24. Ministry of Agriculture and Animal Resources, Kigali

MINAGRI (2021) Annual report 2020–2021. Ministry of Agriculture and Animal Resources, Kigali

Moher D et al (2015) Preferred reporting items for systematic review and meta-analysis protocols (PRISMA-P) 2015 statement. Syst Rev 4(1):1–9

Msaki MM, Tambi E, Bangali S, FARA (2015) State of knowledge on CSA in Africa, case studies from Rwanda, Tanzania and Zambia. Forum for Agricultural Research in Africa, Accra

Murindangabo YT, Kopecký M, Konvalina P (2021) Adoption of conservation agriculture in Rwanda: a case study of Gicumbi district region. Agronomy 11(9):1–13

Mutiro J, Lautze J (2015) Irrigation in Southern Africa: success or failure? Irrig Drain 64(2):180–192

Nabahungu N, Visser S (2011) Contribution of wetland agriculture to farmers' livelihood in Rwanda. Ecol Econ 71:4–12

Nahayo A, Pan G, Joseph S (2016) Factors influencing the adoption of soil conservation techniques in Northern Rwanda. J Plant Nutr Soil Sci 179(3):367–375

Narayanan K (2014) Impact of participatory irrigation management–case study: Cocurirwa Cooperative, Rwamagana Rice Project, Rwanda. Adv Plants Agric Res 1(3):72–78

NBI (2012) Assessment of the irrigation potential in Burundi, Eastern DRC, Kenya, Rwanda, Southern Sudan, Tanzania and Uganda: final report, appendix Rwanda. Nile Basin Initiative, Entebbe

NBI (2016) The nile basin water resources atlas. Nile Basin Initiative, Entebbe

Ndayamaje JC (2008) Management of rainwater harvesting lined ponds in dry areas of Rwanda: a case study of Ntarama lined ponds owners. In: El-Beltagy A, Saxena MC (eds) Proceedings of the ninth international conference on development of drylands: sustainable development in drylands – meeting the challenge of global climate change. The International Dryland Development Commission, Alexandria, pp 437–441

NEPAD (2003) Comprehensive Africa agriculture development programme. New Partnership for Africa's Development, Midrand

Ngabitsinze JC et al (2011) Planning and costing adaptation of perennial crop farming systems to climate change: coffee and banana in Rwanda. International Institute for Environment and Development, London

Ngango J, Hong S (2021) Adoption of small-scale irrigation technologies and its impact on land productivity: evidence from Rwanda. J Integr Agric 20(8):2302–2312

Nkoka F, Veldwisch G, Bolding J (2014) Organisational modalities of farmer-led irrigation development in Tsangano District, Mozambique. Water Altern 7(2):414–433

Ntihinyurwa PD et al (2019) The positive impacts of farm land fragmentation in Rwanda. Land Use Policy 81:565–581

Nzeyimana I (2021) Assessment of farmer-led irrigation development in Rwanda. World Bank, Washington, DC

Pande SK, Rayar AJ, Hakizimana P (2014) Coping with climate change through water harvesting techniques for sustainable agriculture in Rwanda. In: Behnassi M, Muteng'e MS, Ramachandran G, Shelat KN (eds) Vulnerability of agriculture, water and fisheries to climate change: toward sustainable adaptation strategies. Springer, London, pp 217–239

Prasad P et al (2016) Climate-smart agriculture and sustainable intensification: assessment and priority setting for Rwanda: final report. USAID, Washington, DC

RAB (2020) Rwanda irrigation master plan. Rwanda Agriculture and Animal Resources Development Board, Kigali

Ruganzu V, Shirimpumu A, Mutegi JK (2015) Salinity reducing food security and financial returns from rice production in Rwanda. J Agric Sustain 7(1):114–127

Rwaburindi JC et al (2019) Drivers of smallholder farmers' decision to adopt agroforestry in Rulindo district, Rwanda. Int J Soc Sci Inform Technol 4(11):1–12

Rwanyiziri G, Rugema J (2013) Climate change effects on food security in Rwanda: case study of wetland rice production in Bugesera district. Rwanda J 1(1):35–51

Sasaki K, Muvunyi S (2021) Constraints to adoption and scaling-up of conservation agriculture in Rwanda smallholder farmers' perspectives. ASC-TUFS Work Pap 1:119–145

Stainback GA et al (2012) Smallholder agroforestry in Rwanda: a SWOT-AHP analysis. Small-Scale For 11(3):285–300

Theogene U et al (2019) Effect of institutional factors on the adoption of small scale irrigation system in Rwanda. Int J Sci Res 8(10):1320–1322

USAID (2019) Climate risk in Rwanda: country risk profile. United States Agency for International Development, Washington, DC

Uwizeyimana D et al (2019) Modelling surface runoff using the soil conservation service-curve number method in a drought prone agro-ecological zone in Rwanda. Int Soil Water Conserv Res 7(1):9–17

Verdoodt A, Van Ranst E (2003) Land evaluation for agricultural production in the tropics. A large-scale land suitability classification for Rwanda. Laboratory of Soil Science, Ghent University, Gent

Weatherspoon DD (2021) Rwanda's commercialization of smallholder agriculture: implications for rural food production and household food choices. J Agric Food Ind Organ 19(1):51–62
Woodhouse P et al (2017) African farmer-led irrigation development: re-framing agricultural policy and investment? J Peasant Stud 44(1):213–233
World Bank (2021) The Farmer-led Irrigation Development Guide: a what, why and how-to for intervention design. The World Bank, Washington, DC
World Bank, CIAT (2015) Climate-smart agriculture in Rwanda. CSA country profiles for Africa, Asia, Latin America and the Caribbean series. The World Bank Group, Washington, DC
Xie H, You L, Wielgosz B, Ringler C (2014) Estimating the potential for expanding smallholder irrigation in Sub-Saharan Africa. Agric Water Manag 131:183–193
Zingiro A, Okello JJ, Guthiga PM (2014) Assessment of adoption and impact of rainwater harvesting technologies on rural farm household income: the case of rainwater harvesting ponds in Rwanda. Environ Dev Sustain 16:1281–1298

Chapter 17
Summary: Crop Water Productivity: A Catalyst for Food and Water Security

L. Nhamo, S. Mpandeli, Vimbayi G. P. Chimonyo, and Tafadzwanashe Mabhaudhi

Abstract Agriculture accounts for over 70% of freshwater withdrawals globally, yet water and food insecurity remain prevalent. The challenge of water insecurity is compounded by climate change, rapid urbanisation, and the need to increase the agricultural area to produce more crops to meet the growing needs of an increasing population. Improving crop water productivity at the farm level is key to resource security and climate change adaptation and resilience as it enhances the production

L. Nhamo (✉)
Water Research Commission of South Africa (WRC), Lynnwood Manor, Pretoria, South Africa

Centre for Transformative Agricultural and Food Systems (CTAFS), School of Agricultural, Earth and Environmental Sciences, University of KwaZulu-Natal, Scottsville, Pietermaritzburg, South Africa
e-mail: luxonn@wrc.org.za

S. Mpandeli
Department of Environmental, Water and Earth Sciences, Tshwane University of Technology (TUT), Pretoria, South Africa

Faculty of Science, Engineering and Agriculture, University of Venda, Thohoyandou, South Africa

V. G. P. Chimonyo
Centre for Transformative Agricultural and Food Systems (CTAFS), School of Agricultural, Earth and Environmental Sciences, University of KwaZulu-Natal, Scottsville, Pietermaritzburg, South Africa

International Centre for Maize and Wheat Improvement (CIMMYT), Harare, Zimbabwe

T. Mabhaudhi
Centre on Climate Change and Planetary Health, London School of Hygiene and Tropical Medicine, London, UK

Centre for Transformative Agricultural and Food Systems, University of KwaZulu-Natal, Carbis Road, Scottsville, Pietermaritzburg, KwaZulu-Natal, South Africa

T. Mabhaudhi et al. (eds.), *Enhancing Water and Food Security Through Improved Agricultural Water Productivity*,
https://doi.org/10.1007/978-981-96-1848-4_17

of more food with less water. This summary chapter provides a brief overview of each chapter in this book, highlighting the pathways to improved crop-water productivity in both rainfed and irrigated cropping systems. Specifically, the major highlight is that water productivity can be enhanced by practicing well-adapted climate-smart crop types that have the potential to reduce unproductive water losses and, at the same time, maintain healthy and suitably adapted crops that optimise water, nutrient, and agronomic management.

Keywords Water scarcity · Resource use efficiency · Irrigation expansion · Water management · Climate change adaptation

17.1 Background

Improvements in water productivity can produce more food with less water, a critical factor in achieving sustainability in the water and food sectors (Nhamo et al. 2016; Zheng et al. 2018). In turn, this improves the income and livelihoods of vulnerable communities with limited resources and promotes the continued provision of ecosystem services (Zheng et al. 2018). The need to improve water productivity is because irrigation consumes more than 70% of the available freshwater resources globally, posing huge water and food insecurity challenges in an increasingly drying globe (FAO 2022). Therefore, improving water productivity is key to enhancing the resilience building and adaptation initiatives and in achieving sustainable development goals (SDGs) that include Goals 1 (no poverty), 2 (zero hunger, 3 (good health and wellbeing), 6 (clean water and sanitation), 12 (responsible consumption and production) and 13 (climate action) (FAO and IHE-Delft 2019) (Fig. 17.1). There is considerable debate on the possibilities of enhancing crop-water productivity amidst increasing demand for food from a growing population, climate change and environmental degradation. However, recent evidence has demonstrated the potential to improve the water productivity of crops, livestock, and fisheries from the field level to the catchment scale (Giller et al. 2021; Nhamo et al. 2016).

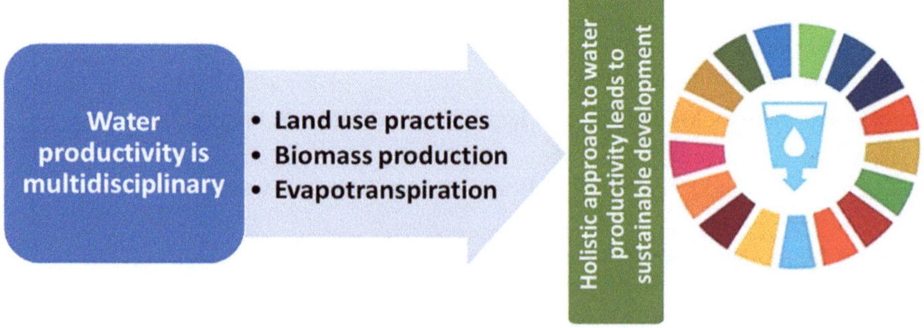

Fig. 17.1 Holistic application and implementation of the water productivity process leads to sustainable development

Understanding the concept of water productivity requires knowledge of the scale of analysis and their interactions (Molden et al. 2010). Understanding the interconnectedness between application scales, how to improve water productivity, and the various actors and disciplines involved (Fig. 17.1). Therefore, it is important to note the interdisciplinary nature of the water productivity term. Thus, its application and implementation require integration and cross-sector analysis. This indicates that water productivity does not fall into a single discipline, subject area, or group of specialists but rather embraces the work of various stakeholders, including farmers, natural resource management specialists, physical scientists, and sociologists.

Based on this background, this book illuminates the significance of agricultural water productivity as a powerful practice that addresses water and food security challenges, focusing on the Global South. The book showcases evidence-based research, case studies, and innovative approaches in crop-water productivity, providing an important tool for knowledge dissemination that guides policymakers, researchers, practitioners, and other stakeholders in making informed decisions on implementing effective strategies for water productivity. The diverse range of topics covered is crafted to enhance agricultural water productivity from a multidisciplinary perspective, considering technological advancements, policy frameworks, socio-economic dynamics, and local contexts. The covered topics underscore the need for integrated water management practices, innovative climate-smart technologies, knowledge exchange platforms, and inclusive governance to foster sustainable agricultural systems. Furthermore, the insights from these topics resonate strongly with several Sustainable Development Goals (SDGs), as already alluded to. The innovations significantly promote water-efficient agricultural practices, resource optimisation, and ecosystem protection.

The book provides a one-stop-shop that provides pathways, practices, and theories on knowledge that covers water productivity and showcases how this knowledge contributes to addressing grand global challenges related to water and food insecurity and climate change while addressing several related SDGs in the Global South. The chapters provide evidence-based guidelines on agricultural water management in the Global South, informing regional and sectoral policies in Africa on sustainable development solutions and climate change resilience and adaptation through better management of land and water resources. The knowledge generated and the provided smart tools, innovations, and applications on water productivity are broken down into (i) case studies, (ii) critical analyses, and (iii) meta-analyses.

17.2 Significance of Water Productivity in Achieving Sustainability

This book emphasises water productivity's role in improving water management for sustainable agriculture, food security and healthy ecosystem functioning. This is critical for achieving sustainable development in an era of increasing water scarcity, degradation and depletion from overuse and climate change. The challenges are exacerbated by the increasing demand for water and food resources from a growing global population, which has witnessed the over-exploitation and overuse of resources (Boretti and Rosa 2019). Under these circumstances, improving food production with less water is essential since agriculture consumes most freshwater resources. Freeing more water resources from the agriculture sector will go a long way in availing the resource to other equally important sectors. The available options to continue meeting the food demands include (a) expanding irrigated and rainfed areas, (b) increasing production per unit of water, (c) promoting trade in food commodities, and (d) behavioural changes in diets (Giller et al. 2021). As agricultural land expansion is no longer viable as it alters hydrological and environmental conditions (Cai et al. 2017; Viana et al. 2022), there is an urgent need to improve agricultural productivity on existing lands using the same amount of water. Studies have shown that increased water productivity improves food security and livelihoods, saves fresh water and makes it available for other uses (Nhamo et al. 2016, 2020; Zheng et al. 2018). Therefore, improved water productivity is a critical element that promotes the management of water and ecosystems for sustainable agriculture and food security.

The topics covered in this book provide the balance between increasing agricultural production and reducing environmental impacts to preserve water resources and simultaneously satisfy the growing demands for food and fibre. The approach is particularly essential for regions faced with water scarcity challenges. Water productivity is important for integrating food production and water utilisation in the sector (Molden et al. 2010). The concept is a performance indicator for various purposes, including quantifying improved implementation of good practices, identifying high-water productivity locations, and determination of potential areas of improvement through yield gap analysis, among other benefits (Poudel et al. 2021).

17.3 Chapter Synthesis

The knowledge, innovations, and applications of water productivity in this book are divided into (a) case studies, (b) critical analyses, and (c) meta-analyses. As such, the topics explore various subjects of productivity, case studies, and perspectives to address the Global South's grand water and food insecurity challenges. The following is a summary of the chapters:

- Chapter 1 reviews existing water use and nutritional water productivity knowledge and sets the foundation for subsequent chapters. Chapter 2 is a systematic review of Rwanda's irrigation development and agricultural water management, offering valuable insights into the country's achievements, challenges, and future directions. Chapter 3 focuses on management strategies and improvements in agricultural water productivity in India and the Ganges Basin, providing valuable insights into context-specific approaches.
- Chapters 4, 5, and 6 delve into the applications of remote sensing, endophytes, and plot-scale estimation of crop water productivity, respectively. These chapters

showcase innovative techniques and technologies to enhance agriculture's water efficiency and productivity.

- Chapter 7 explores the concept of sustainable intensification in mixed farming systems in West Africa, highlighting the need for context-specific approaches to address challenges and promote sustainable water management practices. Chapter 8 sheds light on the research gap surrounding the crop–livestock–soil nutrient–water nexus in West Africa, emphasising the importance of integrated approaches in enhancing agricultural water productivity.
- Chapter 9 presents modelling tools to estimate water use for underutilised cereals in South Africa, providing valuable insights for optimising water management practices. Chapter 10 examines the spatial and temporal variations of water productivity in South Asia, emphasising the need for tailored strategies in diverse agroecological contexts.
- Chapter 11 explores the potential of nanotechnological applications to improve agricultural water productivity, showcasing the promising advancements in this field. Chapter 12 addresses the scaling of Resource Recovery and Reuse (RRR) innovations in low- and middle-income countries for climate change adaptation and mitigation, underscoring the importance of scaling up innovative solutions for widespread impact.
- Chapters 13 and 14 present systematic reviews of irrigation development and agricultural water management in South Africa and Mali. These chapters provide comprehensive overviews of the challenges, best practices, and future directions for improving water productivity in these contexts.
- Lastly, Chapter 15 presents a case study of farmer-led irrigation in Ethiopia, highlighting opportunities for enhancing water productivity through participatory approaches and knowledge sharing.

17.4 Conclusion and Recommendations

Successful implementation of the concept relies on applying the synergistic efforts of farmers and water-resource managers at different scales. The main challenges encountered in applying the water productivity concept include (a) the unavailability of high-quality spatial data and (b) the need for more expertise to interpret the water productivity spatial to derive the right information and support agricultural water management decision-making. The following factors are recommended:

(a) The concept of water productivity is more relevant at a local scale than the global scale, as changes in productivity are more evident at the local scale than at the global scale. On a global scale, information is mainly important for information. Yet, at the local scale, it can provide critical information on improving productivity and could enhance climate change resilience and adaptations.

(b) Equally important is understanding the synergies and interactions between various disciplines and spatial scales. Its application requires the intervention of various stakeholders.
(c) At the same time, it is also important to identify and examine the trade-offs between different water uses. It is, therefore, important to consider the integral use of transformative approaches that allow an analysis of the whole system, other than a single sector. Important transformative approaches for water productivity include the circular economy, water–energy–food (WEF) nexus, scenario planning, and one health.

The concept of water productivity, therefore, contributes to the water scarcity challenges. Applying the concept in a holistic and across sectors provides the stimulus to water-resource management problems. The key is to merge the knowledge in various disciplines and abandon the sector-based approach.

References

Boretti A, Rosa L (2019) Reassessing the projections of the world water development report. NPJ Clean Water 2:1–6
Cai X, Magidi J, Nhamo L, van Koppen B (2017) Mapping irrigated areas in the Limpopo Province, South Africa. International Water Management Institute (IWMI), Colombo, p 42
FAO (2022) FAO Aquastat. FAO, Rome
FAO, IHE-Delft (2019) WaPOR quality assessment. Technical report on the data quality of the WaPOR FAO database version 1.0. Food and Agriculture Organisation of the United Nations (FAO), and IHE Delft Institute for Water Education (IHE DELFT), Rome, Amsterdam, p 134
Giller KE, Delaune T, Silva JV, Descheemaeker K, van de Ven G, Schut AG, van Wijk M, Hammond J, Hochman Z, Taulya G (2021) The future of farming: who will produce our food? Food Secur 13:1073–1099
Molden D, Oweis T, Steduto P, Bindraban P, Hanjra MA, Kijne J (2010) Improving agricultural water productivity: between optimism and caution. Agric Water Manag 97:528–535
Nhamo L, Mabhaudhi T, Magombeyi M (2016) Improving water sustainability and food security through increased crop water productivity in Malawi. Water 8:411
Nhamo L, Magidi J, Nyamugama A, Clulow AD, Sibanda M, Chimonyo VGP, Mabhaudhi T (2020) Prospects of improving agricultural and water productivity through unmanned aerial vehicles. Agriculture 10:256
Poudel U, Stephen H, Ahmad S (2021) Evaluating irrigation performance and water productivity using EEFlux ET and NDVI. Sustain For 13:7967
Viana CM, Freire D, Abrantes P, Rocha J, Pereira P (2022) Agricultural land systems importance for supporting food security and sustainable development goals: a systematic review. Sci Total Environ 806:150718
Zheng H, Bian Q, Yin Y, Ying H, Yang Q, Cui Z (2018) Closing water productivity gaps to achieve food and water security for a global maise supply. Sci Rep 8:14762

MIX
Papier aus verantwortungsvollen Quellen
Paper from responsible sources
FSC® C105338

FSC
www.fsc.org

If you have any concerns about our products,
you can contact us on
ProductSafety@springernature.com

In case Publisher is established outside the EU,
the EU authorized representative is:
Springer Nature Customer Service Center GmbH
Europaplatz 3, 69115 Heidelberg, Germany

Printed by Libri Plureos GmbH
in Hamburg, Germany